A Climate Modelling Primer

RESEARCH AND DEVELOPMENTS IN CLIMATE AND CLIMATOLOGY

Series Editors

A. Henderson-Sellers and M. M. Verstraete

A Climate Modelling Primer
A. Henderson-Sellers and K. McGuffie

Climate Change: Developing Southern Hemisphere Perspectives
T. W. Giambelluca and A. Henderson-Sellers

A Climate Modelling Primer 2nd Edition
K. McGuffie and A. Henderson-Sellers

A Climate Modelling Primer

SECOND EDITION

K. McGuffie

University of Technology, Sydney, Australia

and

A. Henderson-Sellers

Royal Melbourne Institute of Technology, Australia

JOHN WILEY & SONS

Chichester · New York · Weinheim · Brisbane · Singapore · Toronto

National 01243 779777
International (+44) 1243 779777
e-mail (for orders and customer service enquiries): cs-books@wiley.co.uk
Visit our Home Page on http://www.wiley.co.uk
or http://www.wiley.com

First edition published 1987 by John Wiley & Sons Ltd.

Other Wiley Editorial Offices

John Wiley & Sons, Inc., 605 Third Avenue,
New York, NY 10158-0012, USA

VCH Verlagsgesellschaft mbH, Pappelallee 3,
D-69469 Weinheim, Germany

Jacaranda Wiley Ltd, 33 Park Road, Milton,
Queensland 4064, Australia

John Wiley & Sons (Asia) Pte Ltd, 2 Clementi Loop #02-01,
Jin Xing Distripark, Singapore 0512

John Wiley & Sons (Canada) Ltd, 22 Worcester Road,
Rexdale, Ontario M9W 1L1, Canada

Library of Congress Cataloging-in-Publication Data

McGuffie, K.
A climate modelling primer / K. McGuffie and A. Henderson-Sellers.
—2nd ed.
p. cm. — (Research and developments in climate and climatology)
Henderson-Sellers, A. appears first on the earlier edition.
Includes bibliographic references and index.
ISBN 0-471-95558-2 (acid-free paper)
1. Climatology—Mathematical models. I. Henderson-Sellers, A.
II. Title. III. Series.
QC981.M482 1996
551.6'01'1—dc20 96-12815
CIP

British Library Cataloguing in Publication Data

A catalogue record for this book is available from the British Library

ISBN 0-471-95558-2

Typeset in 10/12 Times from the author's disks by MCS Ltd, Salisbury, Wiltshire
Printed and bound in Great Britain by Bookcraft (Bath) Ltd
This book is printed on acid-free paper responsibily manufactured from sustainable forestation, for which at least two trees are planted for each one used for paper production.

TO

James E. Lovelock

who has shown us all the vitality of simple models

Contents

Preface

This little child his littel book lernynge
As he sat in the schole at his prymer
—Chaucer, Prioress Tales, 1386

According to the *Oxford English Dictionary*, a primer serves as the first means of instruction or 'a prayer-book or devotional manual for the use of the laity'. This edition of *A Climate Modelling Primer* follows closely the format of the first edition but contains significant updates where they were required. Our motivation for the original work was the lack of a single book which provided a good introduction for those unfamiliar with the field. Although a number of excellent 'climate modelling' books have appeared since the 'Primer' was first published in 1987, the need remains for a book for those who are not meteorologists by training. The book assumes basic high school mathematics, but, in all cases, the book can be read without following the mathematical developments. Throughout the book, we have tried to underline the importance of simple models of the climate system. With these, it is possible to gain an understanding of the relative importance of different forcing effects. These simple models are also invaluable in testing and extending the concepts upon which more complex models are based.

At its beginning, the science of climate modelling was dominated by atmospheric physicists, and no one without a sound training in fluid dynamics, radiative transfer or numerical analysis could hope or expect to make a contribution. After thirty or so years, the climate modelling community is seeking out oceanographers, ecologists, geographers, remote sensers and glaciologists to provide expertise appropriate to the rapidly expanding domain of the models. The requirement for policy advice has meant that economists, planners, sociologists, demographers and even politicians need to know about climate models. This second group needs to understand the credibility of the different model types, and how to apply (and when not to apply) the output from these models. It is for all these people that this book is intended. We have included a list of reading at the end of each chapter (expanded and updated from that in the first edition). These reading lists are intended as a jumping off point into the climate modelling literature, providing more detailed discussion of the material in the particular chapter. In addition, there is a more complete bibliography at the back of the book. Although, in recent years, a great deal of attention has been focused on the application of climate models to studying the sensitivity of climate to enhanced levels of greenhouse gases, we have elected to avoid a detailed discussion of these simulations. The interested reader should refer to the exhaustive treatment of the science of 'enhanced greenhouse' modelling given in the Intergovernmental Panel on Climate Change reports.

With the development of computer technology, it is possible for anyone with a PC to run a range of climate models. We have included a range of different climate models on a CD with this book. The simplest of these models are the BASIC programs associated with the first edition, ranging through a radiative–convective model to the most complex, which is a fully functional three-dimensional GCM suitable for UNIX™ workstations (not for the faint-hearted). A fuller description and a list of the requirements to run the climate models is given in Appendix B.

We are only too keenly aware of the simplifications that we have made in our explanations. Once more, we beg the indulgence of climate modellers who see that sometimes our explanations and analogies are not completely rigorous. This book was not really intended for you.

Climate modelling can be great fun. Tackled fully, it is a broad and demanding science, for which you will need to learn new techniques and approaches. We recommend it as a way of learning about the geophysical environment and the human activities that affect it, as a pastime and as a career.

KENDAL McGUFFIE AND ANN HENDERSON-SELLERS
Boulder, Colorado, 1995

Acknowledgements

It would be unfair to exclude those who made contributions to the first edition, so thank you all again: Dave Carson, Bob Dickinson, Graham Cogley, Graham Thomas, Mo Wilson, Peter Briggs, Mary Benbow and the 'class of 1986' in Liverpool (wherever they are now). Barry Saltzman, Esmael Malek and Richard deDear provided comments and minor corrections to the first edition. Starley Thompson, Dave Pollard, Aslam Khalil, Robert MacKay. Huqiang Zhang and Carter Emmart contributed material and commented on this new edition. Bryant McAvaney, Jan Polcher, Neil Holbrook, Ulrich Cubasch and Jerry Meehl provided comments on the final draft of the book. Many thanks to Tim Scheitlin at NCAR for his meticulous review of the CD, and to Lara Ferraro for her assistance with the creation of the CD-R disk. As always, we would be pleased to hear of any further corrections (via e-mail to 100233.1554@compuserve.com). We completed much of the work for this edition while at the National Center for Atmospheric Research, as the guests of the Scientific Computing Division (McGuffie) and the Climate and Global Dynamics Division (Henderson-Sellers). Thanks to Don Middleton and Starley Thompson for hosting us. Finally thanks to Brian for continuing positive support, help and advice.

CHAPTER 1

Climate

The climate is a beautiful system, exceedingly rich in interconnections and complexities.

A. H. Oort (public communication, 1986)

1.1 THE COMPONENTS OF CLIMATE

The term 'climate' has a very wide variety of meanings. To a geologist or geomorphologist the 'climate' is an external agent which forces many of the phenomena of interest. For an archaeologist, the 'climate' of an earlier time might have been a crucial influence upon the people being studied, or might have been of little socio-economic significance but still so strong an environmental feature that it has left a 'signature' which can be interpreted. An agriculturalist probably sees the 'climate' as the background 'norm' upon which year-to-year and day-to-day weather is imposed, while the average person may speak of moving to a location with a 'better climate'. To many of us, 'climate' often first suggests temperature, although rainfall and humidity may also come to mind. When we think of climatic change it is often in the time frame of glacial periods. More recently, however, there has been considerable public concern over the possible shorter-term impact upon the climate of increasing atmospheric carbon dioxide and other trace greenhouse gases.

The climate is both a forcing agent and a feature liable to be disturbed. It can fluctuate on relatively short time-scales, producing, for example, the Sahel and Ethiopian droughts, and over much longer times giving rise to glacial epochs. The climate is perceived in terms of the features of the entire climate system which most readily or most usefully characterize the phenomenon of interest. All of these characteristics of the climate are depicted in Figure 1.1. The three axes themselves are fundamental, but the intervals are arbitrary and many more could be included.

A single satisfactory definition of climate is probably unobtainable because the climate system encompasses so many variables and so many time- and space-scales. One definition might be 'all of the statistics of a climatic state determined over an agreed time interval (seasons, decades or longer), computed for the globe or possibly for a selected region'. This definition is broad, but it

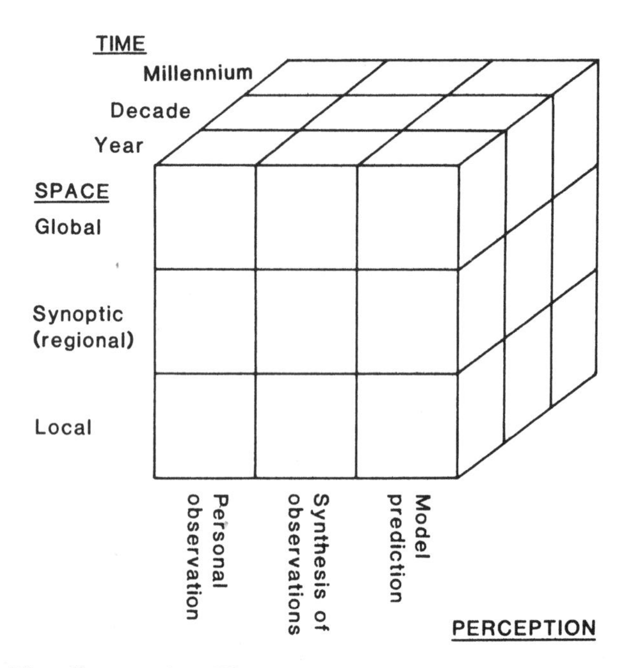

Figure 1.1 The climate cube. Climate can be viewed as existing in at least three domains: time, space and human perception. The divisions of these domains depicted here are arbitrary, a great many more could exist. Historically, individual disciplines have been concerned with single 'cells'. Recently the extent of the climate system and the importance of interactions between domains have been recognized

does serve to emphasize that higher-order statistics, such as variance (variability), can often be more useful in characterizing a climatic state than just the mean (average). This definition also permits further description of a climatic change as the difference between two climatic states, and a climatic anomaly as the difference between a climatic state and the mean state. The variations of the climate system arise from interactions between different parts of the climate system and from external forcings. Although the greatest variations are due to changes in the phase state of water (i.e. frozen, liquid or vapour), the constituents of the atmosphere and ocean and the characteristics of the continental surface can also change, giving rise to a need for consideration of atmospheric chemistry, ocean biogeochemistry and land-surface exchanges.

Introduction and outline of the book

In this book we have set out to introduce and describe the way in which the climate is modelled. The climate models we will discuss are those developed using physically based formulations of the processes which make up the climate system. We are concerned with explaining the approaches and methods employed by climate modellers, and shall not focus directly on meteorology, socio-economic impacts of climatic changes or palaeoclimatic reconstruction, although all of these disciplines and many others will be drawn upon in our descriptions.

In this first chapter we identify the components of the climate system and the nature of their interactions, as well as describing briefly some of the motivations of climate modellers. Chapter 2 contains a history of climate modelling and provides an introduction to all the types of models to be discussed in subsequent chapters. The other chapters are concerned with different model types, their development and applications. Throughout we have taken climate models to be predictive descriptions of regional- to global-scale phenomena; hence empirically based 'models' such as crop prediction equations and water resource management codes have not been included. The reason for this limitation is not that such models are uninteresting, but rather that they have grown from well-identified fields and thus background literature can be readily obtained. Climate modelling in the sense in which we use the term, on the other hand, has developed from a wide variety of backgrounds in a somewhat haphazard manner and consequently there is little background to which the uninitiated can refer.

In one sense the book develops the background material required for an understanding of the most complex type of climate model, the fully coupled general circulation climate model, by illustrating principles in other, simpler, model types. Thus it is necessary to introduce the concept of energy balance, especially planetary radiation balance, before one-dimensional energy balance models (Chapter 3) can be understood. In Chapter 4, computationally efficient models which consider only a few of the important processes in the atmosphere are examined. One-dimensional radiative–convective models, which concentrate on the radiative processes in an atmospheric column and the vertical motion (convection) which they can induce, are used to gain deeper understanding of the nature of feedbacks and forcings within the climate system. In cases where the meridional (across latitudes) circulation is important, two-dimensional statistical dynamical models can be employed. We also explore a range of other low-complexity models, including intentionally simplified versions of three-dimensional models and 'box' models of the ocean.

By Chapter 5 the reader is, hopefully, well prepared to understand the way in which radiative forcing, vertical motion and horizontal advection are included in three-dimensional models of the atmosphere and ocean, and how these two circulating systems can be coupled. In Chapter 6 we explore how the results from climate models are evaluated and how these results can be integrated with impact assessment with a view to use by policy advisers.

Appendix A contains a glossary of terms which may be new to readers unfamiliar with climatology/meteorology. As we have used this glossary for definitions rather than disturb the main thread of the text, reference to it is recommended. The accompanying CD (described in Appendix B) contains source codes for a range of model types. These will allow readers to make their own climate simulations, ranging from global glaciations to tropical cyclones. A set of simulations from a global climate model also permits analysis of the results of a tropical deforestation experiment. A number of images and MPEG movies illustrate some of the techniques used to analyse and display the results from a range of climate models.

Throughout the book an effort will be made to underline the importance of simpler models in understanding the complex interactions between various components of the climate system. Complex models are only one, particularly sophisticated, method of studying climate. They are not necessarily the best tool, and simple models are often used in conjunction with, or sometimes even to the exclusion of, more complex and apparently more complete models. The literature contains many fascinating examples of very simple models being used to demonstrate failures in much more complex systems.

Last, but by no means least, any introduction to climate modelling must stress the crucial role played by computers. Without the recent growth in computational power and the reduction in computing costs most of the developments in climate modelling which have taken place over the last three decades could not have happened. We have intentionally emphasized computing tools over mathematical skills in the description of the simplest type of climate model, the energy balance model (EBM), in Chapter 3. In this chapter, the steps required to construct a simple EBM are described, the BASIC code for a simple EBM is listed, and time is taken to develop the questions that might prompt further use and thought. The CD (Appendix B) also includes instructions on how to build an EBM in a spreadsheet.

It is estimated that a full atmospheric general circulation model (AGCM) takes about 25–30 person-years to code, and the code requires continual updating as new ideas are implemented and as advances in computer science are accommodated. Most modellers who currently perform experiments with the most complex of models modify only particular components of the models. The complexity of these models means that only through a sharing of effort can progress be made. As the models have become increasingly complex, increased application of the principles of software engineering have made it easier to upgrade and exchange parts of the models. The CD which accompanies this book also includes computer code for more complex models, including a GCM, as well as the results of some climate model experiments.

The climate system

The climate system was defined, in a document produced by the Global Atmospheric Research Programme (GARP) of the World Meteorological Organization in 1975, as being composed of the atmosphere, hydrosphere, cryosphere, land surface and biosphere. In 1992 the United Nations Framework Convention on Climate Change (FCCC) defined the climate system as 'the totality of the atmosphere, hydrosphere, biosphere and geosphere and their interactions'. These definitions are similar, but the emphasis on interactions, both in the definition and in the literature, has grown in the 20 years since 1975. Figure 1.2 shows a schematic representation of the climate system components which climate modellers must consider. It complements Figure 1.1 by emphasizing components and processes rather than the space- and time-scales. The order of the components of the climate enumerated in 1975 is also a rough indicator of the historical order in which these elements were considered and, to some extent, the (increasing) magnitude of their time-scales. The first modelled component was the atmosphere, which, because of its low density and ease of movement, is the most 'nervous' of the climatic subsystems. Precipitation has been included, but many aspects of clouds (such as cloud liquid water and the effects of different cloud droplet sizes) are difficult to incorporate successfully, and linking the major part of the hydrosphere, the oceans, into climate models is only now becoming common. This is partly because the critical space- and time-scales of the ocean and atmosphere subsystems differ, but also because the coupling between the subsystems is strongly latitude-dependent. In the tropics the systems are closely coupled, especially through temperature (Figure 1.3). In mid-latitudes the coupling is weak, predominantly via momentum transfer, but in high latitudes there is a tighter coupling, primarily through salinity, which is closely involved in the formation of oceanic deep bottom water. Biochemical processes controlling the exchange of carbon dioxide between atmosphere and ocean also vary as a function of geographical location.

The cryosphere (frozen water) has been incorporated into climate models to some extent following the description of simple EBMs in which the high albedo of the ice and snow dominated the radiative exchanges. However, the insulating effect of the cryosphere is now known to be at least as important as its albedo effect. Sea-ice, especially, decouples the ocean from the overlying atmosphere, causing considerable changes in both subsystems. Snow has a similar, but smaller, effect on land.

Scientists concerned with land-surface processes had described the climate as both an agent and a feature of change for over a century before this generation of climate modellers began serious consideration of their theories. The importance of the biosphere was underlined by the realization, following the first studies of climate impact, that the carbon dioxide component of the atmosphere is critically dependent upon the biota. These studies are generally directed towards the state of the continental surface and the growth and well-being of plants and animals, but the importance of the marine biota is also recognized.

The stratospheric 'ozone hole', first identified over Antarctica in 1986, was the catalyst for incorporating atmospheric chemistry into climate models.

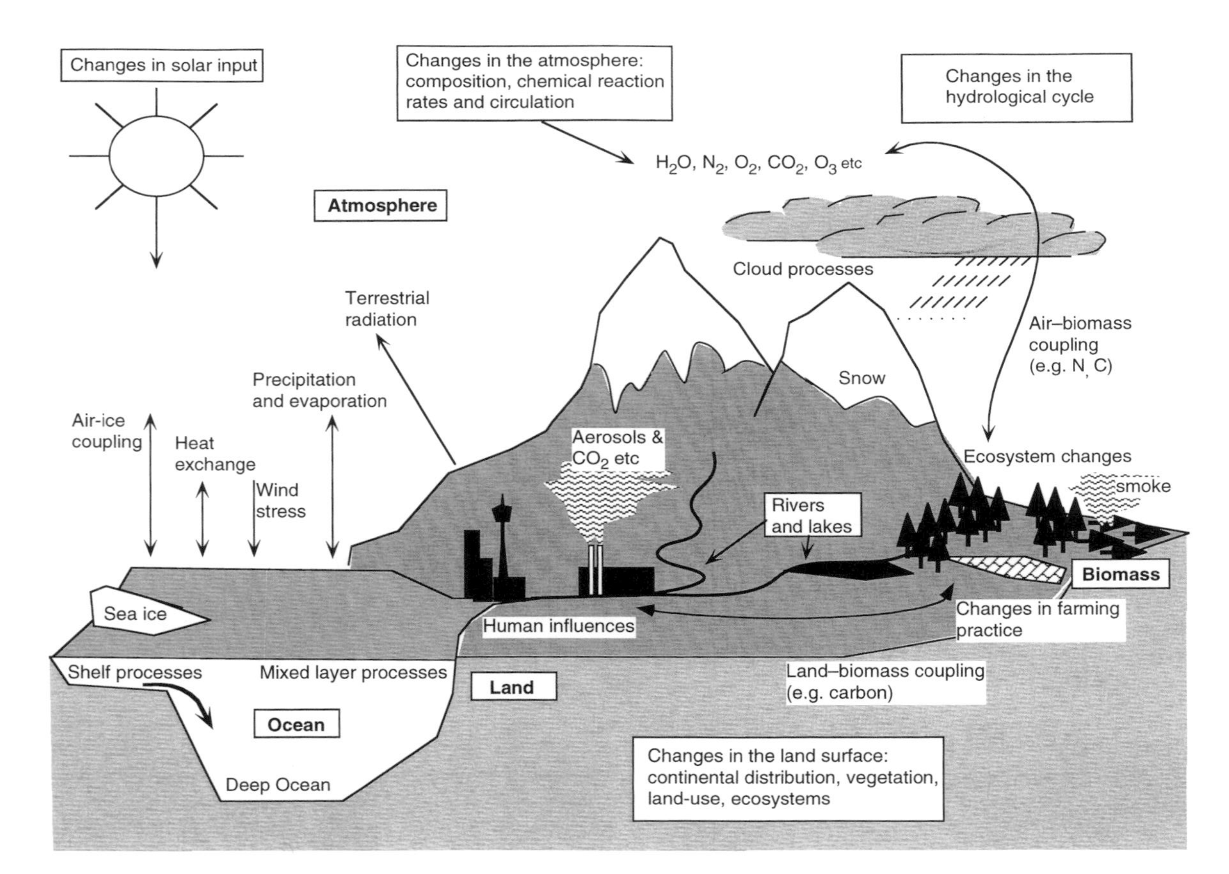

Figure 1.2 Schematic illustration of the components and interactions in the climate system (modified from Houghton *et al.*, 1996)

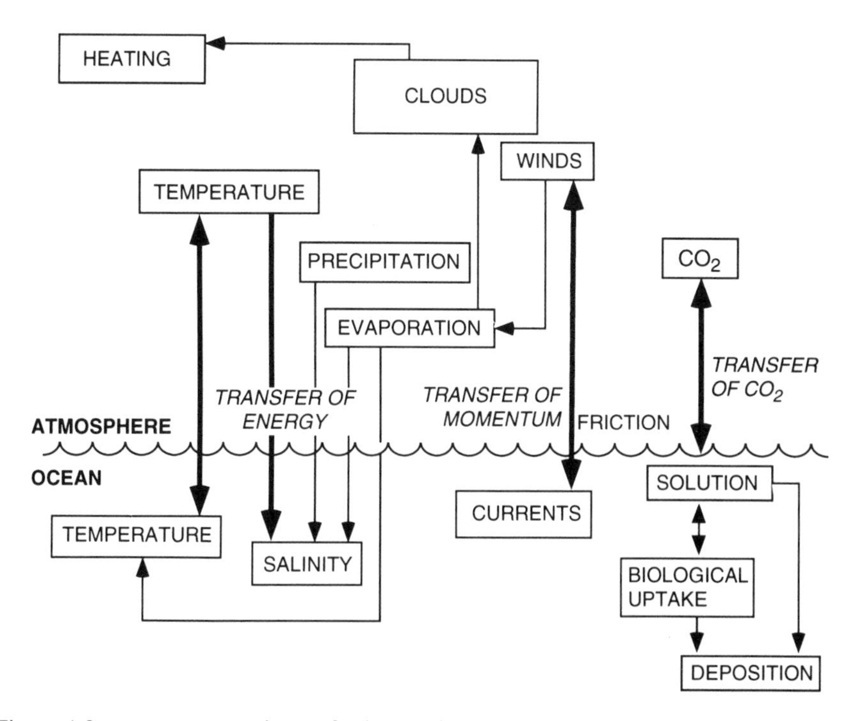

Figure 1.3 A representation of the major coupling mechanisms between the atmosphere and ocean subsystems. The relative importance of these coupling mechanisms varies with latitude. The feedback between atmospheric temperature and oceanic salinity is interesting because it is strong only in the sense of the atmosphere forcing the ocean

Inclusion of these subsystems is still in its early stages, but it is already clear that 'fully coupled' climate models will have to incorporate atmospheric and marine chemistry and the transient changes in the world's biota. The human component of the climate system, manifested particularly in trace gas and aerosol emissions and land use change, is perhaps its most difficult and challenging aspect. Human activities have only very recently begun to be parameterized in climate and 'integrated assessment' models.

In this rather clumsy fashion, and from mixed parentage, the discipline of climate modelling has evolved. Climate modellers have discovered that the system that they had summarized so neatly in 1975 is exceedingly complex, containing links and feedbacks which are highly non-linear and hence difficult to identify and reproduce.

1.2 CLIMATE CHANGE ASSESSMENT

Over the last 10 years, with increased awareness of the potential impacts of changes in atmospheric concentrations of trace gases and aerosols, there has been an evolving demand by policymakers for the results from climate models. In 1988, the Intergovernmental Panel on Climate Change (IPCC) was formed by the United Nations Environment Programme and the World Meteorological Organization. The IPCC was directed to produce assessments of available scientific information on climate change, written in such a way as to address the needs of policymakers and non-specialists. The First Scientific Assessment was published in 1990 in three volumes, encompassing science, impacts and response. There was a scientific update in 1992 and two further volumes were produced as input to the First Conference of the Parties to the FCCC in March 1995. The Second Scientific Assessment was published in 1996.

The IPCC process aims to determine the current level of confidence in our understanding of the forcings and mechanisms of climate change, to find out how trustworthy the assessments are, and to ask whether we have yet seen any human-induced climate change. Through an exhaustive review process, the IPCC aims to provide assessments which discuss climate change on a global scale and represent the international consensus of current understanding. Throughout the process, the aim is to include only information which has been subjected to rigorous review, although this is balanced by a desire to include the latest information in order that the best possible assessment can be made. These two competing goals mean that the development of the IPCC documents is an extremely time-consuming process, but ensure that the final result is a useful statement of the state of current knowledge of the climate system. The IPCC assessment covers three areas, which are handled by three 'working groups'. For the Second Assessment, Working Group I deals with the assessment of the science associated with climate modelling, climate observations and climate predictions. Working Group II deals with the impacts of, adaptations to, and mitigation of, climate change, while Working Group III deals with cross-cutting economic and technology issues.

1.2.1 The scientific perspective

It is generally accepted that physically based computer-modelling offers the most effective means of answering questions requiring predictions of the future climate and of potential impacts of climatic changes. Although there have been great advances in such modelling over the past 30–40 years, even the most sophisticated models are still far removed in complexity from the full climate system. Further advances are possible, but they need to be associated with increased understanding of the nature of interactions within the real climate system and translated to those within models. Perturbations caused by everything from industrial aerosols to volcanoes, and from solar luminosity to climatically induced variation in surface character, must be considered. Modelling in such a widely ranging subject is a formidable task and it requires co-operation between many disciplines if reliable conclusions are to be drawn.

Available computing power has increased greatly over the past 30–40 years (Figure 1.4). This rapid increase has meant that climate models have expanded both in terms of complexity and in total simulation time. Multi-decadal, seasonal simulations, with full diurnal and seasonal cycles, are now expected in climate experiments, and transient changes in, for example, the atmospheric CO_2 amount are becoming commonplace. As our knowledge increases, more aspects of the climate system are being, and will be, incorporated into climate models, and the resolution and length of integrations will further increase. Figure 1.5 shows how the demand for computer power varies with resolution of three-dimensional models and with the number and type of components included. The general trend has been that, as computer power increases, so do the complexity, resolution and length of simulations with climate models. Atmospheric modellers have tended to favour increasing the number of components in the models, while the ocean modellers, with fewer interactions (e.g. no clouds), have driven the resolution of their models higher, although both groups have done both.

Figure 1.6 shows the performance of a range of atmosphere-only global climate models between 1974 and 1984. Over this period there was considerable improvement in model simulation of observed characteristics of the

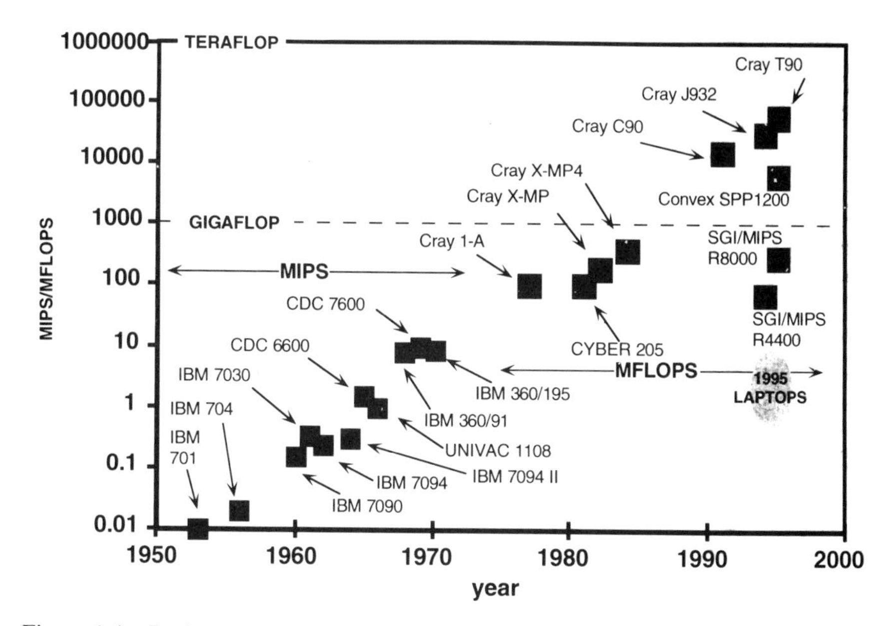

Figure 1.4 Peak performance of some of the most powerful computers between 1953 and 1995. The power is given in millions of instructions per second (MIPS) up to 1975 and, more recently, in millions of floating point operations per second (MFLOPS). Note that the vertical scale is logarithmic. The growth has been exponential, and shows no signs of levelling off (adapted from Simmons and Bengtsson, 1988)

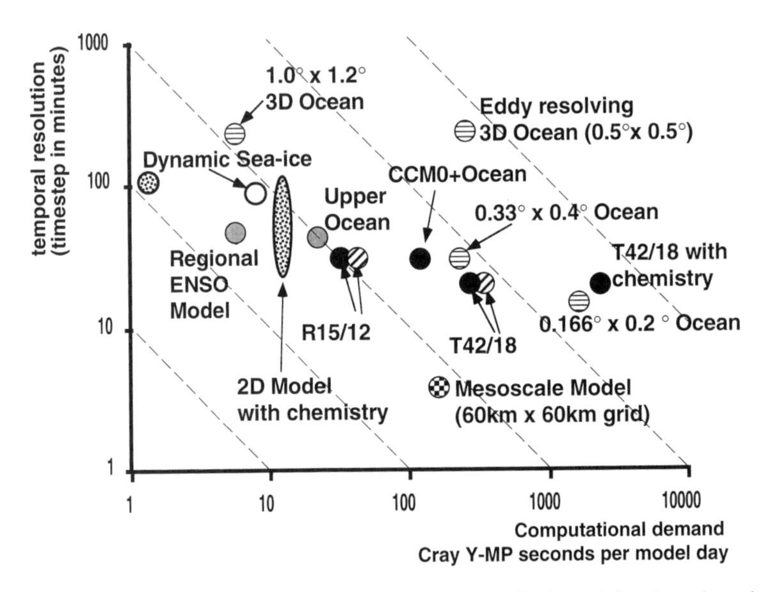

Figure 1.5 Temporal resolution (timestep) of some typical models plotted against the amount of dedicated Cray YMP processor time required to run the model (seconds required to compute one model day). The complexity of the models is indicated. The pairs for R15 and T42 are for CCM1 and CCM2 with and without the BATS land-surface scheme; the slightly greater computational time being with BATS. CCM0+Ocean is an R15 AGCM coupled to a $1° \times 1°$ global ocean. A range of timesteps is used by the 2D chemistry model to deal with different reaction rates for different chemical species. A range of timesteps is used by the 2D chemistry model to deal with different reaction rates for different chemical species. It is worth noting that the number of vertical layers employed by a model and the nature of the algorithms employed also influence the computational demands of these models (Terms and abbreviations are explained in Chapter 5)

climate system. In Figure 1.7, the 1995 performance of the available coupled ocean–atmosphere models shows that this improvement has continued. Certainly, faster computers help to increase the possible size and complexity of models, but simple models may be sufficient to answer particular, well-specified problems.

Whether its predictions are correct is the ultimate test of any model. Weather-forecast models can be tested over a period of a few hours to a few days, but models of climate are required to predict decades or even centuries in advance, or periods of the Earth's history for which validation data are scant. More importantly, climate model 'predictions' offer only a general case of the response since the model climate loses its association with the initial conditions within a few weeks. Hence, testing of single simulations is virtually impossible, although ensembles of results should characterize the climate. Despite the limitations placed by chaos theory on our ability to predict the *exact* state of the atmosphere beyond about 10–15 days into the future, there is good reason

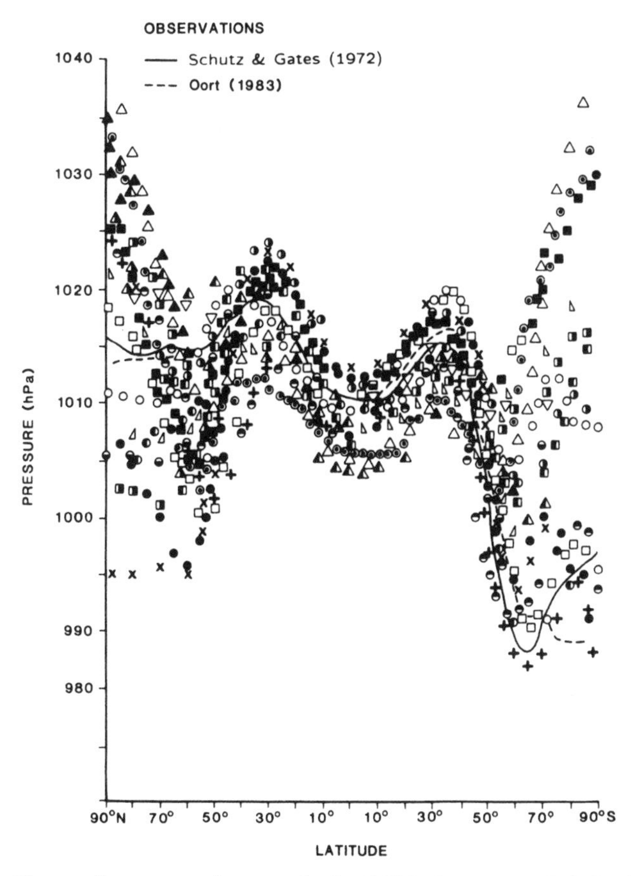

Figure 1.6 The performance of atmospheric GCMs in terms of their simulation of (a) sea-level pressure and (b) precipitation (both December–January–February). The range between different models is great, but reduces over the years. It is also important to recognize that these and other observational data are incomplete and, probably, somewhat inaccurate. Note also that the 'improving' trend can be dramatically, if temporarily, reversed in specific model simulations when only part of a process has been captured (reproduced by permission from Gates, 1985)

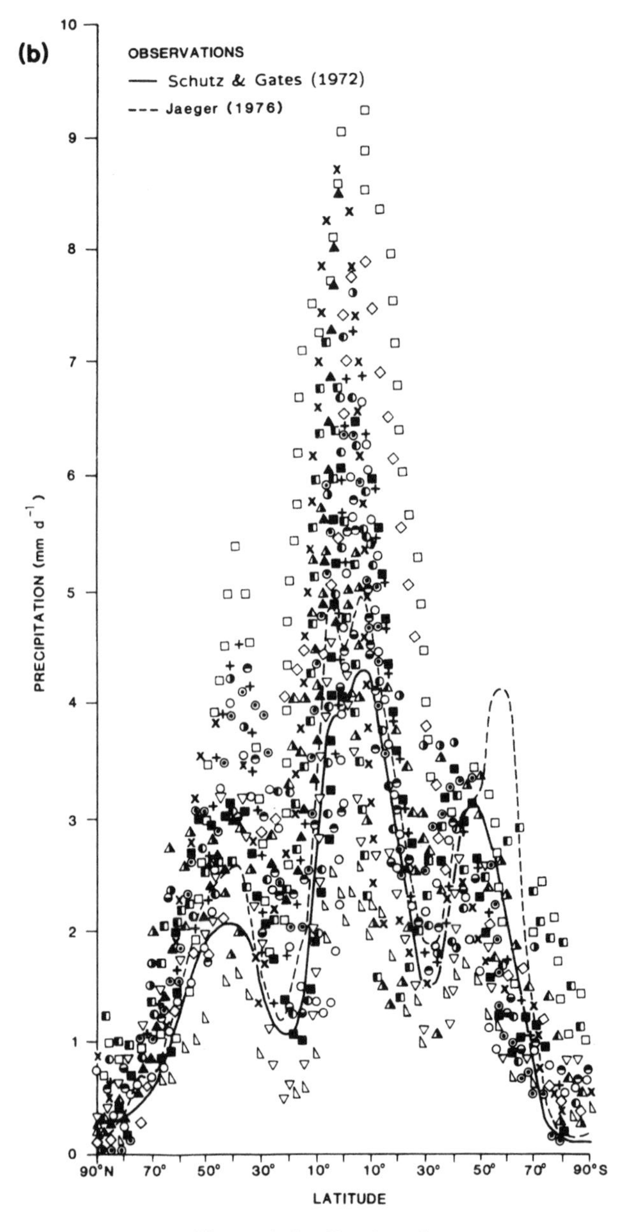

Figure 1.6 *Continued*

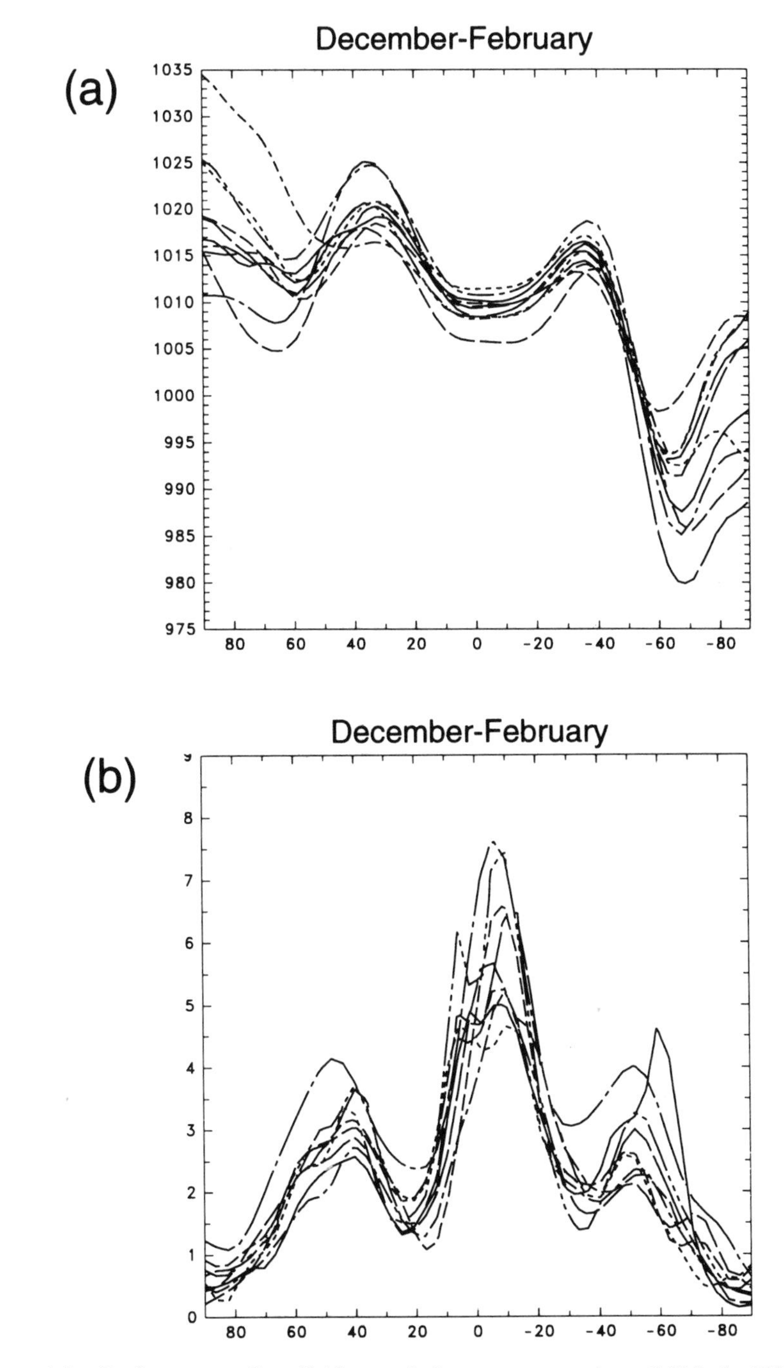

Figure 1.7 Performance of available coupled ocean–atmosphere GCMs in 1995 can be contrasted with the 1984 compilation in Figure 1.6. (a) Mean sea level pressure (hPa) during December–January–February. (b) Total precipitation in mm d^{-1} during December–January–February (reproduced by permission from Houghton *et al.*, 1996)

to believe that our ability to predict the nature of the ensemble state (the climate) is not significantly affected. There are also other model types. For example, even simple models can predict the effect on mean temperatures of volcanic eruptions such as that of Mount Pinatubo quite successfully on seasonal or longer time-scales.

The climate models which will be discussed in this book cover a wide range of space- and time-scales. These different types of climate models attract interest from many different disciplines. Long-period modelling may attract glaciologists, geologists or geophysicists. Atmospheric chemists, dealing with complex reactions that typically have very short time-scales, are successfully incorporating these processes into three-dimensional climate models. Implications of solar-system-scale phenomena attract planetary physicists and astronomers, while social and economic scientists are interested in the human component of the climate system. In this book we will attempt to show how these contributions fit together and jointly enhance the science of climate modelling.

1.2.2 The human perspective

Any changes in climate, whatever the cause, may have impacts on human activities. Crop-yield models have been used to quantify how food production depends on the weather. It might therefore be postulated, for example, that a change in climate could lead to consistently low or high yields in a particular area, which in turn would have led to a human response in terms of a change in agricultural practice. Such simple postulates can be misleading, since they conceal several problems which are inherent in relating climate change to human impact. These concern the nature of climatic changes themselves, the strength of the relationship between climate changes and human response, and the availability of (past) climatic and sociological data for evaluation.

It is possible to think of climatic changes as being represented by changes in the long-term mean values of a particular climatic variable. Superimposed on this changing mean value will be decadal fluctuations and year-to-year variations. Such short-period variations may, of course, be influenced by the change in the mean. On the human time-scale, changes in the mean value are likely to be so slow as to be almost imperceptible. For example, the changes over the last few decades can only be detected by careful analysis of instrument records. Much more noticeable will be variability, expressed, for example, as a 'run of bad winters'. Any human response will depend on such a perception, whether conscious or subconscious. A 'large' climate change may not lead to any response whereas a much smaller change in a particular feature, expressed as a perceived change in variability, may have a profound impact on human activity. Detection, for example, of climate change in response to increasing atmospheric trace 'greenhouse' gases is very difficult in the early stages if only one response is monitored. As an alternative, 'fingerprint' methods have been proposed which monitor a set of small changes in a number of variables, and require prespecified thresholds in all of them to be passed before a signal can be established.

Any attempt to establish the impact of past climatic changes must use historical information. Pre-instrumental historical records are qualitative and selective, and emphasize information about unusual conditions which were perceived as having an impact. Consequently they can tell us less about normal conditions than about abnormal ones. A great deal therefore tends to be inferred about the climate and its variability before any suggestions regarding its impacts can be made. Even if a change occurs which potentially has a significant impact on human activity, a societal response will not necessarily follow. Any response to a climate change is governed by a host of non-climatic factors which need to be considered. Consequently a direct link between climate change and human activity is often difficult to establish. This problem of 'attribution' following detection is also common to the issue of greenhouse warming.

It has also been suggested that as cold, damp winter conditions prevailed in some northern mid-latitudes during the Little Ice Age (*c*. 1450–1850), grain storage became impossible and famine susceptibility increased. However, population pressure and plague could have been equally important in creating the problems of this period, or could have exacerbated climatic stress. Indeed, trying to estimate widespread effects of the decrease of about 1 K in mid-latitude temperatures is not especially valuable as additional factors, such as the incidence of late spring frosts or destructive winds near harvest time, of which we have little or no information, may have been more significant.

Such an effect is typically seen on a local scale. One well documented example occurred in the Lammermuir Hills in Scotland. Careful study of the records of farming and settlement in this area gives credence to the suggestion that in marginal regions human response to climatic deterioration is identifiable. The combined isopleths of 1050 degree days plus 60 mm potential water surplus (at the end of the summer), which represents the approximate 'cultivation limit', expanded and moved downslope during the cooling period and were restored towards the end of the Little Ice Age. In the centre of an agricultural region, farming practices will be well adjusted to that particular climate and year-to-year variations will pose little threat. As the margins are approached, however, variability will become more significant. Usually overall production will be low, so that little surplus can be stored against the poor years that climate variability will inevitably bring. If a climate change occurs which alters the frequency of the poor years, some human response is very likely to follow.

The definition of 'marginality', of course, depends upon the climatic regime and agricultural practices considered. Three areas using different climatic indices for marginality are shown in Figure 1.8. Northern Europe is divided into agriculturally marginal and sub-marginal areas. The limits are given by a

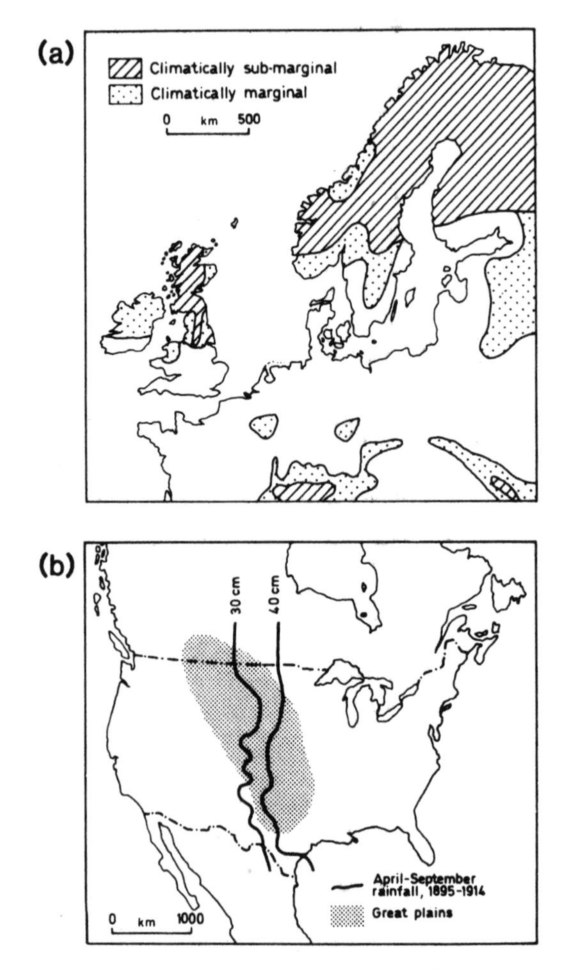

Figure 1.8 Climatologically marginal land in (a) northern Europe, (b) the Great Plains of the USA, and (c) eastern Australia. In (c) the shifts in climatic belts between 1881–1910 and 1911–1940 are seen (reproduced by permission from various sources including Gentilli, 1971, Elsevier Science Publications)

combined index which is a function of the number of months with a mean monthly temperature above 10°C and the degree of the precipitation deficit over evapotranspiration, if any, in the summer months. The region of marginal cultivation identified for the United States is based upon total rainfall in the period April–September rather than upon the combination of temperature and rainfall used for northern Europe. For Australia, three zones of marginality, for different climate regimes and agricultural enterprises, are shown in Figure 1.8(c). Here the limits are based on temperature and precipitation values and their ranges. The changes of these limits with time indicate the eastward encroachment of aridity and the establishment of new marginal areas.

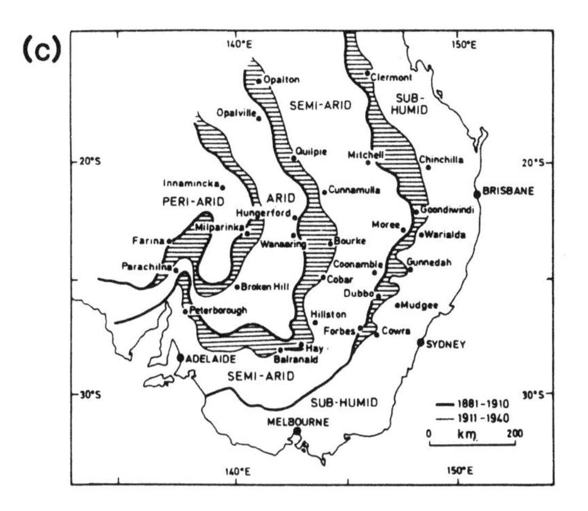

Figure 1.8 *Continued*

Future human responses to climatic change are likely to involve similarly complex webs of decisions. We will examine modelling the climatic part of these problems. Consideration of the nature of the interaction of elements of the climate system is an important, but by no means the only, prerequisite to consideration of potential human response. The remainder of this chapter is devoted to a description of the characteristics of the climate system which are of interest to modellers. The last chapter in the book returns to the issue of trying to simulate socio-economic interactions with climate.

1.3 CLIMATE FORCINGS

The climate system is a dynamic system in transient balance. This concept, which is vitally important in climate modelling, is easy to visualize in terms of vehicle movement. The heart of New York City, Manhattan Island, experiences a very large vehicular influx each morning and an equally large outflux in the evening. Over time periods greater than a few days, Manhattan has an (approximate) vehicular balance, while over time periods of a few hours there are large negative and positive fluxes of vehicles. If the authorities were either (i) to close all bridges and tunnels on only the east side of the island, or (ii) to close all the car parks and refuse to allow street parking, the fluxes of vehicles would alter considerably and the net flux budget would change in this part of the New York subsystem.

Fluxes are thus seen to be vectors (they are the movement of some quantity from one place to another, and the direction of flow is important), and net fluxes differ considerably as a function of the time period considered. Also, different budgets, the result of the net fluxes, are established when the imposed disturbance changes. The most important fluxes in the climate system are fluxes of radiant (solar and heat) energy, but the fluxes of water and, to a lesser extent, mass (matter) will also be found to affect climate dynamics.

A climate forcing is a change imposed on the planetary energy balance which, typically, causes a change in global temperature. Forcings imposed on the climate system may be considered as falling into two separate categories. External forcings are caused by variations in agents outside the climate system, such as solar radiation fluctuations. On the other hand internal forcing, such as that due to volcanic eruptions, ice-sheet changes, CO_2 increases and deforestation, are variations in components of the climate system. Longer-term internal forcings which occur as a result of continental drift and mountain building have an effect, and changes in the polarity of the Earth's magnetic field may influence the upper atmosphere and thus, perhaps, the whole climate.

1.3.1 External causes of climatic change

Milankovitch variations

The astronomical theory of climate variations, also called the *Milankovitch Theory*, is an attempt to relate climatic variations to the changing parameters of the Earth's orbit around the Sun. There are several different ways in which the orbital configuration can affect the received radiation, and thus possibly the climate. They are (Figure 1.9): (i) changes in eccentricity; (ii) changes in obliquity; (iii) changes in orbital precession. The Earth's orbit becomes more eccentric (elliptical) and then more circular in a pseudo-cyclic way, completing the cycle in about 110 000 years. The mean annual incident flux varies as a function of the eccentricity of the orbit, E. For a larger value of E there is smaller incident annual flux. The current value of E is 0.017. In the last 5 million years it has varied from 0.000 483 to 0.060 791. These variations would result in changes in the incident flux of +0.014 to −0.17% from the current value.

The obliquity, the tilt of the Earth's axis, is the angle between the Earth's axis and the plane of the ecliptic (the plane in which the bodies of the solar system lie). This tilt varies from about 22° to 24.5°, with a period of about 40 000 years. The current value is 23.5°. Seasonal variations depend upon the obliquity: if the obliquity is large, so is the range of seasonality. Although the total received radiation is not altered, a greater seasonal variation in received flux is accompanied by a smaller meridional gradient in the annual radiation.

The orbit of the Earth is an ellipse around the Sun, which lies at one of the foci. Owing to gravitational interaction with the other planets, primarily Jupiter, the perihelion (the point of the Earth's orbit closest to the Sun) moves in space so that the ellipse is moved around in space. This orbital precession will cause a progressive change in the time of the equinoxes. These changes occur in such a way that two periodicities are apparent: 23 000 years and 18 800 years. This change, like that of obliquity, does not alter the total

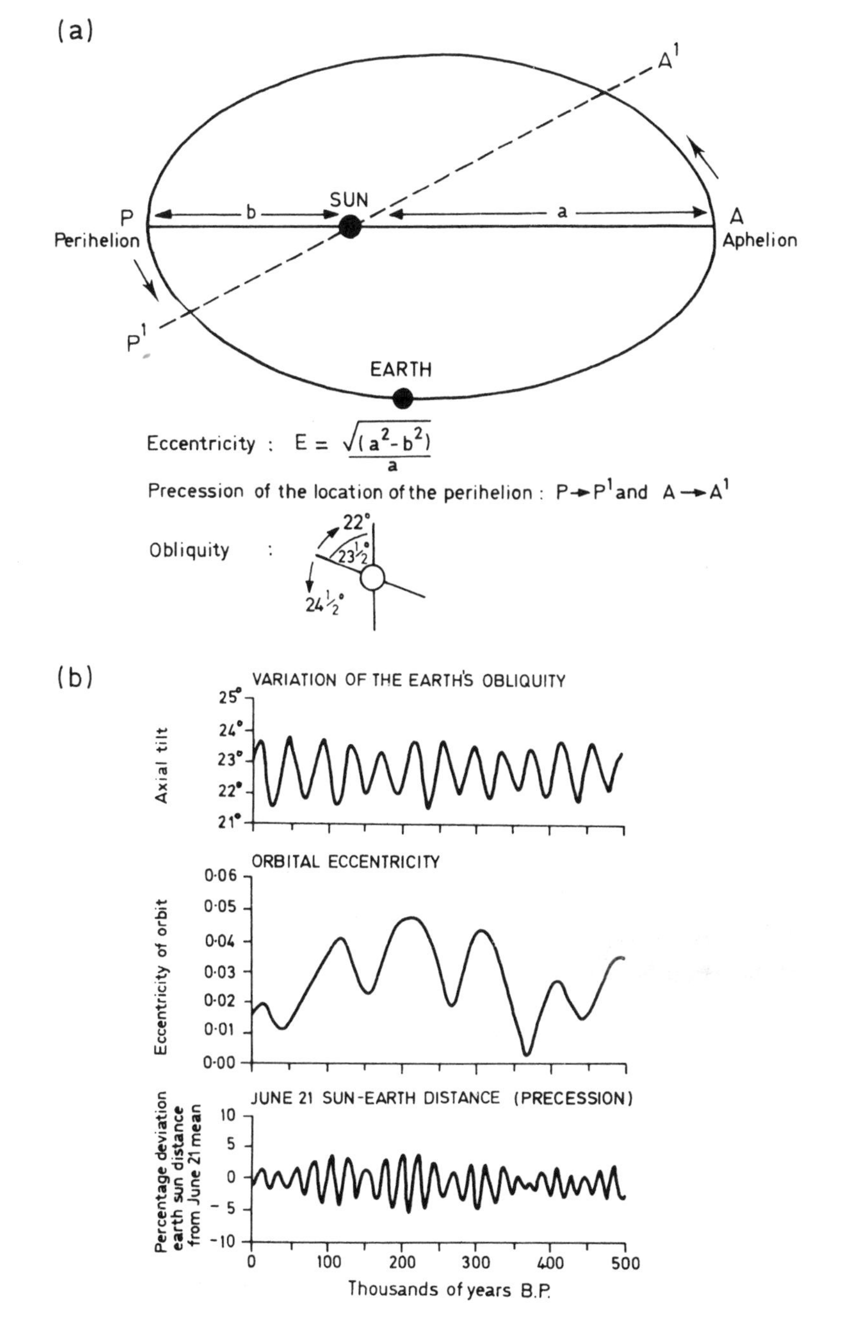

Figure 1.9 (a) Schematic diagram showing the variations in the three orbital components: obliquity (axial tilt), orbital eccentricity and precession of perihelion. (b) Variations in these three components, from 500 000 years ago to the present, as a function of time (reproduced by permission from Broecker and Van Donk (1970) *Rev. Geophys.*, **8**, 169–196. Copyright by the American Geophysical Union)

radiation received but does affect its temporal and spatial distribution. For example, perihelion is currently on 5 January, in the middle of the northern hemisphere winter, but 11 000–15 000 years from now it will occur in July. At the present-day value of eccentricity there is a range of ~6% in the solar constant between perihelion and aphelion (i.e. 1411–1329 $\mathrm{W\,m^{-2}}$).

Spectral analysis of long-term temperature data has shown the existence of cycles with periods of ~20 000, ~40 000 and ~100 000 years (Figure 1.10). These correspond closely with the Milankovitch cycles. The strongest signal in the observational data, however, is the 100 000-year cycle. This cycle corresponds to that of eccentricity variations in the Earth's orbit, but eccentricity variations produce the smallest insolation changes. Hence the exact effect of the Milankovitch cycles is far from clear. For example, modelling results have suggested that the present configuration of the land masses in the northern hemisphere may favour rapid development of ice caps when conditions favour cool northern hemisphere summers. While the Milankovitch forcing offers an interesting 'explanation' for long-term, cyclic climatic changes, the energy distributions within spectral analyses of climate and of orbital variations are interestingly different, and only recently have models begun to produce observed temperature changes from observed forcing. Almost certainly, these

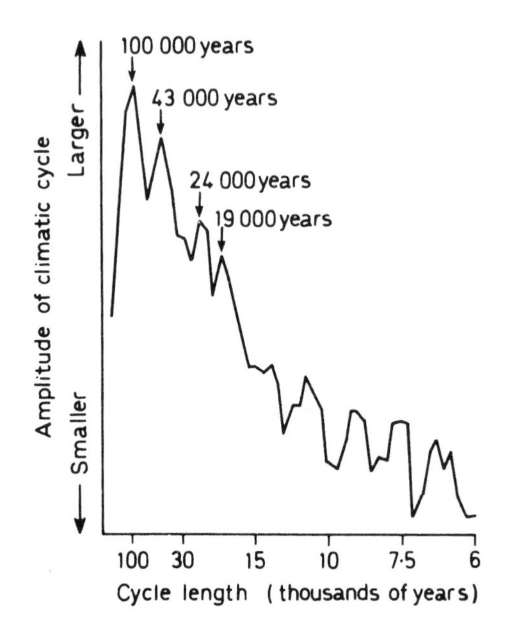

Figure 1.10 Spectrum of climatic variations over the past 500 000 years. The graph shows the importance of the climatic cycles of 100 000 years (eccentricity): 43 000 years (obliquity) and 24 000 and 19 000 (precession of the location of the perihelion). The curve is constructed from an isotopic record of two Indian Ocean cores (reproduced from Imbrie and Imbrie (1979) by permission of Macmillan, London and Basingstoke)

external changes trigger large feedback effects in the climate system which are yet to be fully understood.

Solar activity

Variations in the climate during historical times have been linked with the sunspot cycle, which is a second possible cause of solar-produced climatic change. This cycle occurs with a 22 year periodicity: the 'Hale' double sunspot cycle. The overall amplitude of the cycles seems to increase slowly and then fall rapidly, with a period of 80–100 years. There also appears to be a quasi-cyclic fluctuation of the order of 180 years. No mechanistic link between sunspot activity and surface conditions on the Earth has been demonstrated. Correlations between climate and sunspots usually fail when global conditions are considered. In particular, the Little Ice Age has been linked to the 'Maunder minimum' in sunspots, although it should be noted that the actual period of the Little Ice Age itself seems to vary according to the geographical area from which data are taken.

Recent studies have suggested that changes in the energy output from the Sun between the Maunder minimum (*c.* 1645–1715) and the 1980s was likely to be $0.4 \pm 0.2\%$. The magnitude of this forcing is very much less than the forcing due to enhanced CO_2 over that time, but short-term variability associated with the solar cycle can be comparable to short-term greenhouse forcing this century. For example, it has been proposed that the decrease in solar radiative output from 1980 to 1986 was approximately equal and opposite to the forcing by greenhouse gases over this same period.

Other external factors

Collisions of comets with the Earth and very large meteoritic impacts have also been proposed as causes of climatic fluctuations. Many of the disturbances that such impacts would cause, such as an increase in stratospheric and tropospheric aerosols, are similar to disturbances internal to the system, as described in the next subsection.

1.3.2 Internal factors: Human-induced changes

Most current concern is about the possible impacts of human activities, which could operate on the relatively short time-scales necessary to create noticeable changes within the next century. These include the emissions of greenhouse gases and aerosols, changes in land use and the depletion of stratospheric ozone. The only natural effects which are thought likely to be important on similar time-scales are volcanic activity and, possibly, oscillations in the deep ocean circulation.

Greenhouse gases

The increased concentrations of a number of greenhouse gases in the atmosphere is well documented, and simulating its potential effect is widely reported in the climate-modelling literature. Apart from water vapour, over which we have no control, CO_2 is the major component of both the natural greenhouse effect and of the greenhouse warming which is projected to occur as a result of continued burning of fossil fuels. The magnitude of the warming and the relative impacts on different regions of the world will depend on the nature of the feedbacks within the climate system. The 'greenhouse warming' literature is so widespread that we opt not to review it here.

The magnitudes of the forcings which act to perturb the climate system as determined by the IPCC Second Scientific Assessment are shown in Figure 1.11(a). There is significant uncertainty in the magnitude of many of these forcings, but it is worth noting that the combined effect of cloud and aerosol forcings is potentially comparable to the forcing due to carbon dioxide but in the opposite direction. Figure 1.11(b) shows a 1981 assessment which compares the effect of many other internal and external forcing agents.

Tropospheric aerosols and clouds

The influence of volcanic aerosols has long been recognized, but the influence of tropospheric aerosols associated with industrial pollution and fossil fuel and biomass burning has only recently been identified and, to some extent, quantified. Solid sulphate particles result from the oxidation of SO_2 emitted when fossil fuels are burned. Other industrial processes and natural and human initiated biomass burning also contribute particulates, often termed aerosols, to the troposphere. These aerosols are localized and have two effects on the climate system. The 'direct' effect of most aerosols is to reflect some solar radiation back into space and so act to cool the affected area. Some particulates, such as soot, are dark in colour and have the opposite effect, causing local warming. The magnitude of the cooling or warming depends on the nature of the aerosols and their distribution in the atmosphere.

There is an important 'indirect' effect of tropospheric aerosols. They act as additional cloud condensation nuclei and cause more, smaller, drops to form in clouds, increasing the reflectivity of the clouds, further cooling the planet (Figure 1.11(a)). The effect of changes in cloud character can have complex repercussions, since the clouds also affect the amount of radiation which escapes from the Earth system.

Stratospheric ozone

The discovery of the 'ozone hole' in 1986 and, more recently, a similar, but less intense, ozone depletion over the Arctic has focused attention on the need for interactive chemical sub-models in global climate models. The ozone 'destruction', which is observed, now appears to be due to the disturbance of the natural balance of destruction and production which previously existed in the stratosphere. Paul Crutzen, Mario Molina and Sherwood Rowland were awarded the 1995 Nobel prize for Chemistry for their role in identifying the threat to stratospheric ozone from anthropogenic compounds.

The presence of free chlorine atoms in the stratosphere can now be traced to the photochemical disruption of chlorofluorocarbons (CFCs) and hydrochlorofluorocarbons (HCFCs) when these 'inert' gases migrate from the troposphere.

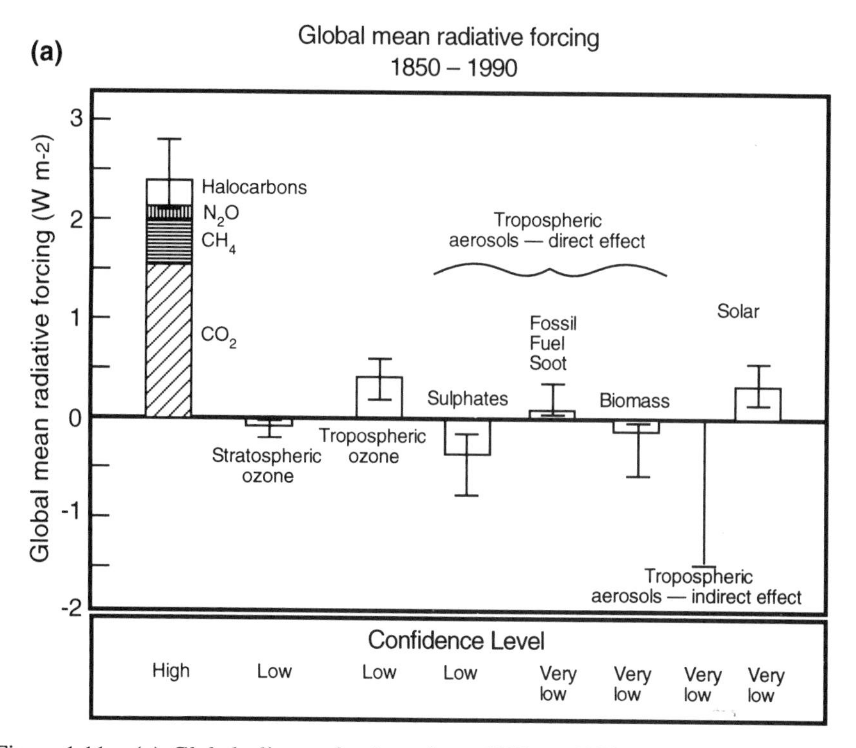

Figure 1.11 (a) Global climate forcings from 1850 to 1990 according to the IPCC 1995 scientific assessment (reproduced by permission from Houghton *et al.*, 1996). (b) The effect of various climatological perturbations upon surface temperatures as predicted by a one-dimensional RC climate model. One of the major perturbations causing cooling is seen to be increased loading of stratospheric aerosols. Column I shows that increasing cirrus cloud operates to warm, rather than to cool as do increases in middle- and low-level clouds (Columns G and H). Warming by a 2% increase in high clouds is greater than the cooling caused by a 2% increase in middle-level clouds (reproduced by permission from Hansen *et al.*, *Science*, **213**, 957–966. Copyright 1981 by the AAAS)

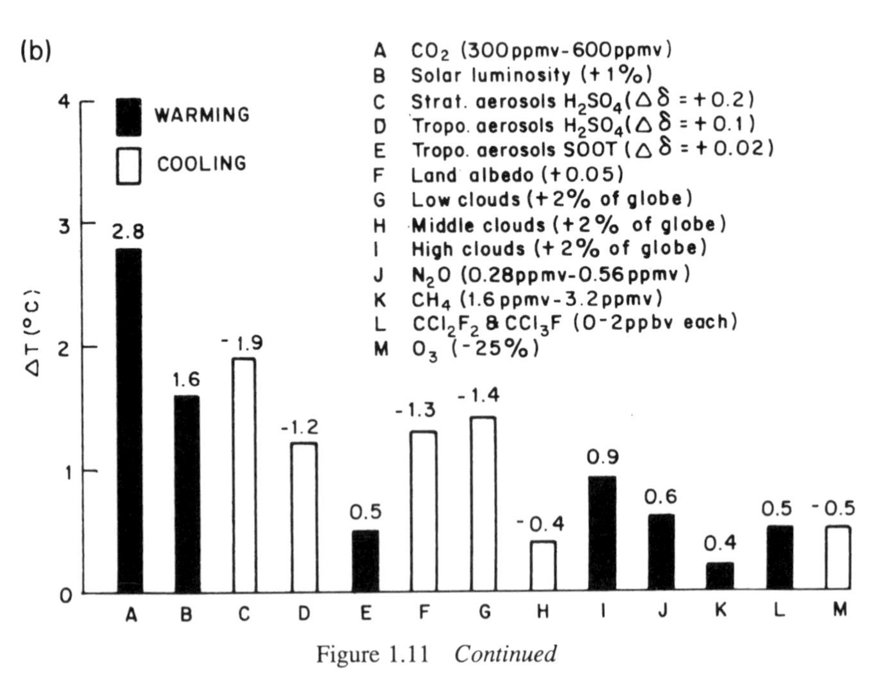

Figure 1.11 *Continued*

Chlorine is the principal cause of the disturbance in ozone chemistry which produces the 'ozone hole'. Although the build-up of CFCs, at least, in the atmosphere is levelling off as a result of the Vienna Convention and the Montréal Protocol, the very long lifetimes of these gases means that they will persist in the atmosphere for hundreds of years.

The particular reactions which act to accelerate the ozone destruction rely on the presence of free chlorine atoms and a solid surface, provided by stratospheric ice clouds. Suitable conditions exist over the Antarctic continent during the winter, and to a lesser extent over the Arctic Ocean in winter. It is possible that, in addition to the role played by ice crystals in the chemistry of the ozone breakdown, volcanic aerosols may also provide a suitable surface upon which the chemistry can take place.

Since CFCs, HCFCs and the hydrofluorocarbons (HFCs), which are replacing them, are radiatively active (they are much more effective greenhouse gases than CO_2), they also act to change the atmospheric temperature, and this alters the rate of the chemical reactions. CFCs that remain in the troposphere are effective absorbers of infrared radiation which would otherwise escape to space. These gases therefore act to enhance the atmospheric greenhouse and to provide a warming influence for the planet. The radiative effect of the reduced stratospheric ozone is to cool the planet. The enhanced levels of tropospheric ozone which have been observed result in a warming (Figure 1.11(a)).

Land-surface changes

Humans are beginning to make what can be regarded as regional-scale changes to the character of the Earth's surface. These include desertification, re- and deforestation and urbanization. Climate modellers have investigated the climatic effect of such changes in the nature of the Earth's continental surface.

Desertification is a problem which affects millions of people. The sparse vegetation natural to arid and semi-arid areas can be easily removed as a result of relatively minor changes in the climate, or by the direct influence of human activity such as overgrazing or poor agricultural practices. Removal of vegetation and exposure of bare soil decrease soil water storage because of increased runoff and increased albedo. Less moisture available at the surface means decreased latent heat flux, leading to an increase in surface temperature. On the other hand, the increased albedo produces a net radiative loss. In climate model calculations the latter effect appears to dominate, and the radiation deficit causes large-scale subsidence. In this descending air, cloud and precipitation formation would be impossible and aridity would increase. The result of a relevant model simulation is shown in Figure 1.12(a). This global simulation involves a surface albedo change for a group of semi-arid areas. It can be seen that an increase in surface albedo does seem to decrease rainfall. Use of a global model emphasizes that all parts of the climate system are interlinked. Although this particular model includes many simplifications, the results are illustrative of the types of surface-induced climatic effects which are currently captured by models.

At present around 30% of the land surface of the Earth is forested and about a third as much is cultivated. However, the amount of forest land, particularly in the tropics, is rapidly being reduced, while reforesting is prevalent in mid-latitudes. As a consequence the surface characteristics of large areas are being greatly modified. Modellers have attempted to examine the climatic effects of forest planting and clearance. The change in surface character can be especially noticeable when forests are replaced by cropland. One area which is undergoing deforestation is the Amazon Basin in South America. The important change after deforestation is in the surface hydrological characteristics, since the evapotranspiration from a forested area can be many times greater than that from adjacent open ground. Most climate model simulations of Amazonian deforestation show a reduction in moisture recycling (because of the lack of the moist forest canopy), which reduces precipitation markedly (Figure 1.12(b)). However, the available global model experiments do not agree on whether an increase in surface temperature occurs. The largest impacts are the local and regional effects on the climate, which could exacerbate the effects of soil impoverishment and reduced biodiversity accompanying the deforestation. It has proved possible to detect impacts resulting from tropical deforestation propagating to the global scale by increasing the length of the integrations to improve the statistics.

DESERTIFICATION ALBEDO DISTRIBUTION AND SAHEL RAINFALL

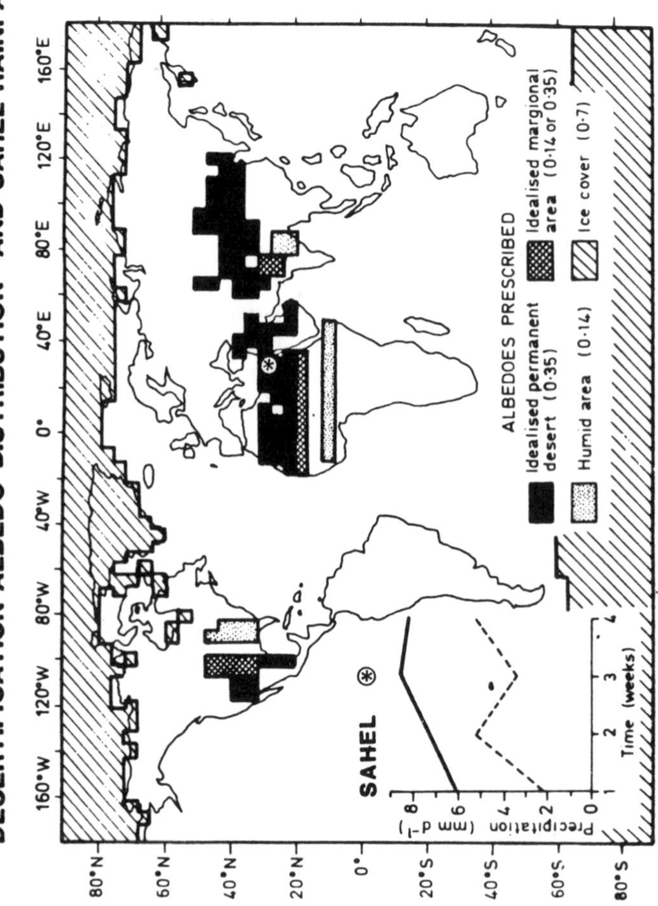
160°W
120°W
80°W
40°W
0°
40°E
80°E
120°E
160°E
80°N
60°N
40°N
20°N
0°
20°S
40°S
60°S
80°S
SAHEL
Precipitation (mm d⁻¹)
8
6
4
2
0
1
2
3
4
Time (weeks)
ALBEDOES PRESCRIBED
Idealised permanent desert (0·35)
Idealised margional area (0·14 or 0·35)
Humid area (0·14)
Ice cover (0·7)

(a)

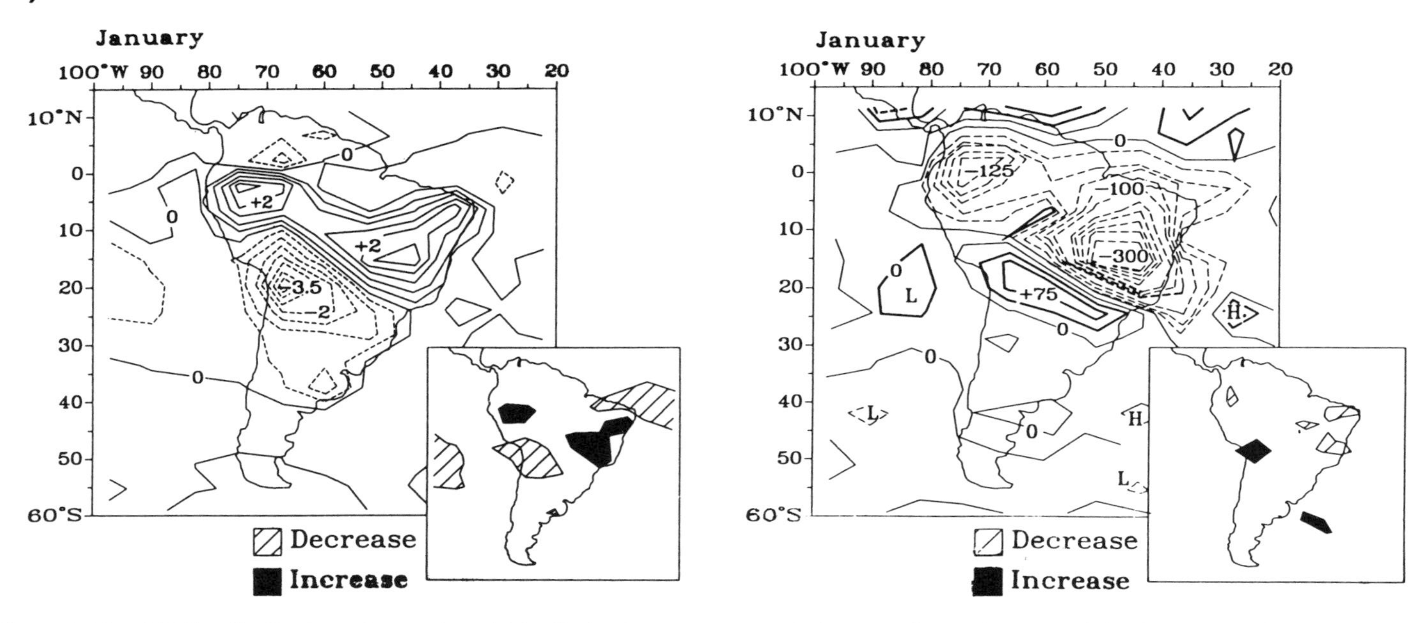

Figure 1.12 (a) The distribution of areas for which albedo changes were made in a set of experiments, originally conducted by Charney (1975), designed to examine desertification. The inset graph shows the rainfall resulting from increasing the surface albedo from 0.14 to 0.35 in the Sahel region when free evaporation was permitted (redrawn by permission from Henderson-Sellers and Wilson (1983) *Rev. Geophys. Space Phys.*, **21**, 1743–1778. Copyright by the American Geophysical Union). (b) Simulated temperature (°C) and total precipitation (mm) changes (left and right, respectively) following replacement of the Amazon tropical moist forest by a scrub grassland in a GCM. These are 5-year means from the end of a 6-year deforestation experiment. The insets show changes which are significant at the 95% level using Student's *t*-test

1.3.3 Internal factors: Natural changes

Volcanic eruptions

Volcanoes influence climate by projecting large quantities of particulates and gases into the atmosphere. Volcanic eruptions can thereby produce measurable temperature anomalies of at least a few tenths of a degree. The effect of the injected aerosol upon the radiation balance, and whether heating or cooling ensues, will depend largely on the height of injection into the atmosphere.

Most eruptions inject particulates into the troposphere at heights between 5 and 8 km. These are rapidly removed by either gravitational fall-out or rain-out, and the resultant climatic effect is minimal. More violent eruptions hurl debris into the upper troposphere or even into the lower stratosphere (15–25 km) (e.g. Mount Agung in 1963, El Chichón in 1982 and Mount Pinatubo in 1991). Eruptions such as these are much less frequent but are likely to have more extensive climatic effects. The particulates have a long residence time in the stratosphere: of the order of a year for aerosols of radii 2–5 μm but as long as 12 years for smaller aerosols of radii 0.5–1.0 μm. Mount Pinatubo injected around 20 million tonnes of SO_2 to heights of 25 km. As it was dispersed by the stratospheric winds, the SO_2 was photochemically transformed into sulphate aerosols. These non-absorbing aerosols increase the albedo of the atmosphere and reduce the amount of solar radiation that reaches the surface. If the aerosol absorbs in the visible part of the spectrum, energy is transferred directly to the atmosphere. If the aerosol absorbs and emits in the infrared, the greenhouse effect is increased.

Immediately following an eruption the stratosphere is dominated by dust particles which scatter radiation of wavelengths up to 10 μm roughly 10 times as efficiently as normal stratospheric particles. The 'clear sky' optical thickness can rise to 0.1 (20 times the normal value) after large eruptions since the particles also absorb visible radiation. Sulphate production is increased a few months later, and a further increase in the visible scattering occurs along with a slight increase in the infrared absorption. These changes will affect the atmospheric heating rates. The enhanced absorption of visible radiation is typically not sufficient to compensate for enhanced cooling by emission of infrared radiation. The aerosols generated by the eruption of Mount Pinatubo have been estimated to have resulted in a forcing on the climate system of around $-0.4\ \mathrm{W\,m^{-2}}$, with a resultant temporary global cooling of about 0.5°C. Since the eruption of Mount Pinatubo occurred at a time when the global observing network was extraordinarily well equipped to gather appropriate data, it prompted enormous advances in our knowledge of the effects of volcanic eruptions on atmospheric processes.

Model simulations suggest that radiative effects will, overall, produce a global cooling when large-scale volcanic eruptions occur. Figure 1.13(a)

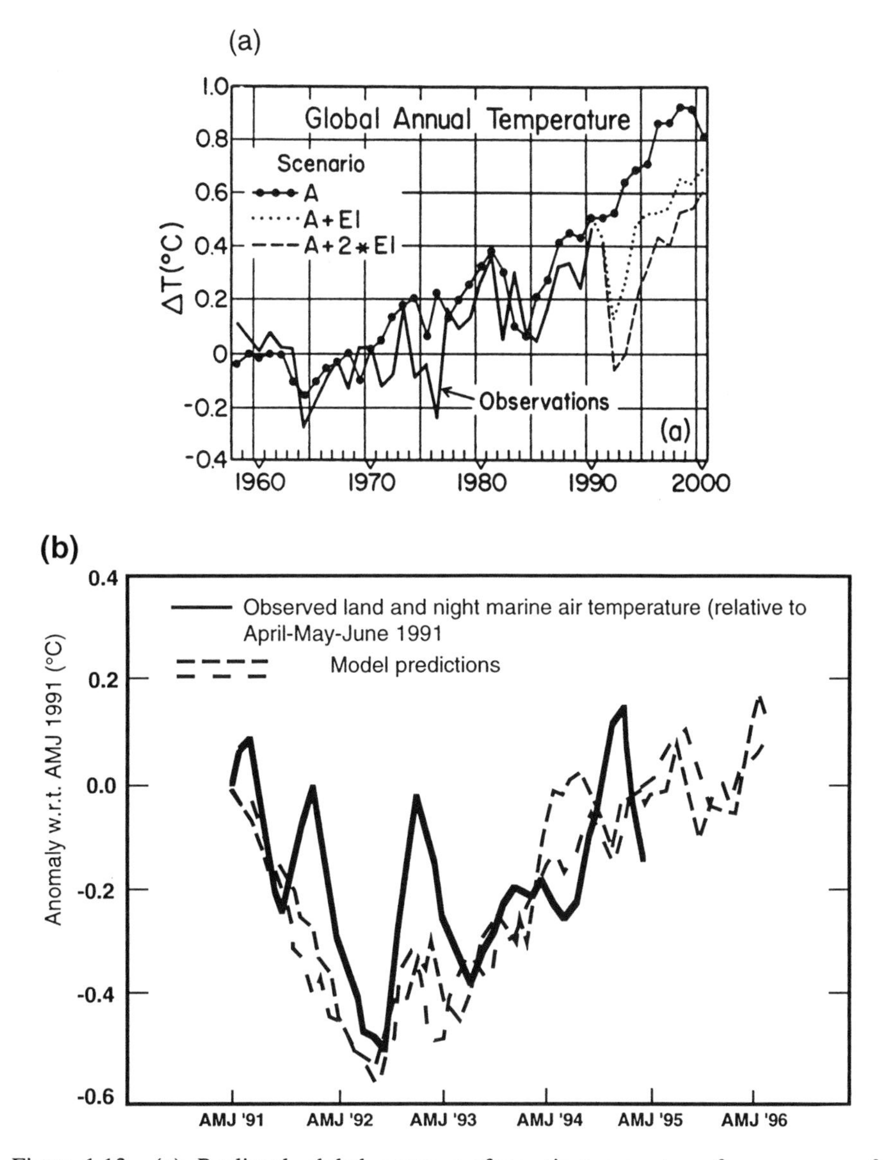

Figure 1.13 (a) Predicted global mean surface air temperature for a range of scenarios following the 1991 eruption of Mount Pinatubo. The upper frame shows a situation where there is an exponentially increasing concentration of greenhouse gases. Scenario A has no volcanic aerosol after 1985, A + El supposes that the aerosol load from Mount Pinatubo is comparable to that from El Chichón. A + 2∗El supposes an aerosol load twice that from El Chichón, which corresponded to early estimates. The lower panel shows the same aerosol scenarios imposed on a linear growth in greenhouse gas concentrations (after Hansen *et al.*, 1992). (b) Predicted and observed changes in land and ocean surface air temperature following the eruption of Mount Pinatubo (reproduced by permission from Houghton *et al.*, 1996)

shows results of a set of simulations in which an attempt was made to predict temperature anomalies resulting from the eruption of Mount Pinatubo prior to observations being available. The results of the 'prediction', and the results of two other simulations, agree well with the observed temperature anomalies in Figure 1.13(b). Complications arise because volcanic particles can serve as cloud condensation nuclei and consequently feedbacks which involve cloudiness changes may be invoked. Eruptions like Mount Pinatubo and their effects on the atmosphere are very short-lived compared with the time needed to influence the heat storage of the oceans. Hence temperature anomalies do not persist, nor are they likely to initiate significant long-term climatic changes.

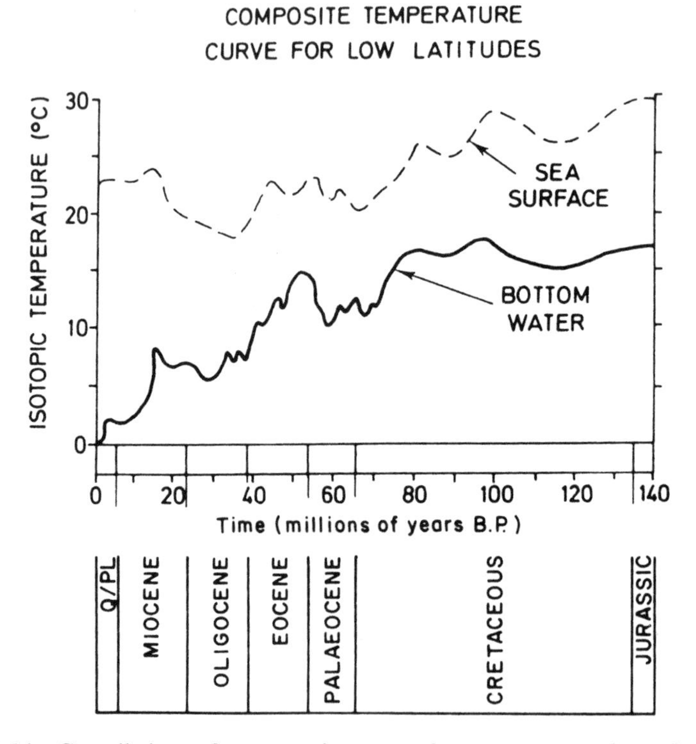

Figure 1.14 Compilation of oxygen isotope palaeotemperature data obtained by analysis of benthic and planktonic *foraminifera* from Deep Sea Drilling Project cores. (Q/PL denotes the Pliocene epoch and the Quaternary period, the latter beginning about 2 million years before present (BP).) The upper curve is drawn through tropical sea-surface temperatures and the lower curve through bottom water temperatures (reproduced by permission from Douglas and Woodruff, 1981)

Ocean circulation changes

The natural variability of the ocean circulation is an important factor for climate. The ocean circulation varies on glacial time-scales, when the circulation is known to change markedly, and on interannual time-scales where the El Niño Southern Oscillation (ENSO) phenomenon is important. Modellers have recently had some success in developing predictive models of ENSO events in the Equatorial Pacific on seasonal time-scales using spatially restricted ocean models.

Another challenge which faces ocean scientists is trying to explain the sudden changes which may occur in the circulation of the North Atlantic. The relative warmth of Europe (palms in Western Scotland) in our present era is attributable to the formation of North Atlantic Deep Water (NADW), which maintains the flow of warm surface water from the south. However, geological evidence from mid-Atlantic ocean drilling shows that NADW production has varied greatly over the last 25 000 years, seeming to be tied closely to stages of the last glaciation. The mechanisms which trigger changes in NADW production are, however, not well understood.

Figure 1.14 shows a compilation of tropical sea-surface and bottom-water temperatures. The tropical sea-surface temperatures vary very little over the 140 million years shown, while the bottom-water temperatures have decreased by more than 10 K since the Cretaceous period. How these high, deep ocean temperatures could have occurred and what effects the changes which caused them had on other components of the climate are questions which models can, perhaps, be used to answer.

1.4 CLIMATE FEEDBACKS AND SENSITIVITY

In the broadest sense, a feedback occurs when a portion of the output from the action of a system is added to the input and subsequently alters the output. The result of such a loop system can either be an amplification of the process or a dampening. These feedbacks are labelled positive and negative, respectively. Positive feedbacks enhance a perturbation, whereas negative feedbacks oppose the perturbation (Figure 1.15(a)).

The importance of the direction of a feedback can be simply illustrated by considering the impact of self-image on diet. Someone slightly overweight who eats for consolation can become depressed by their increased food intake and so eat more and rapidly become enmeshed in a detrimental positive-feedback effect. On the other hand, perception of a different kind can be used to illustrate negative feedback. As a city grows there is a tendency for immigration, but the additional influx of industry, cars and people is often detrimental to the environment so that it may be balanced, or even exceeded, by an outflux of wealthier inhabitants, with a potentially negative impact on the economy. In this section some of the feedback mechanisms inherent in the climate system are described.

1.4.1 The ice-albedo feedback mechanism

If some external or internal perturbation acts to decrease the global surface temperature, then the formation of additional areas of snow and ice is likely. These cryospheric elements are bright and white, reflecting almost all the solar

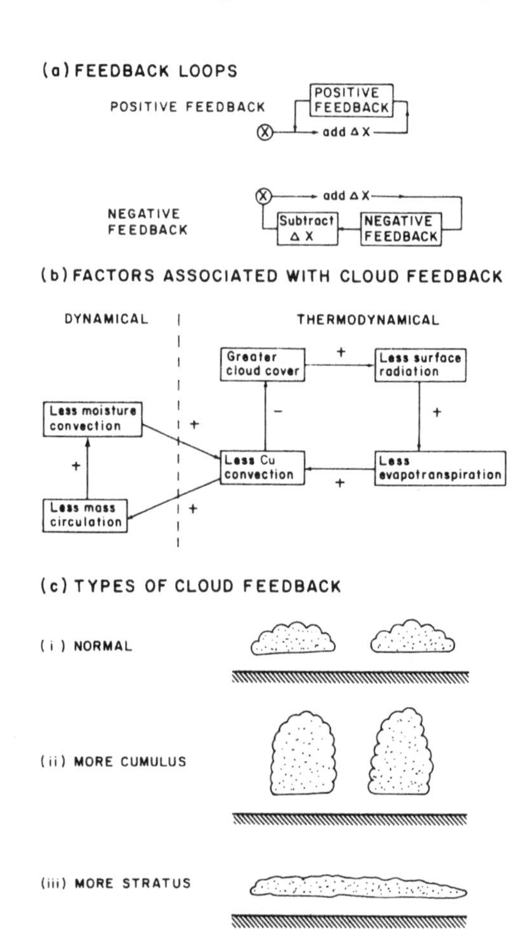

Figure 1.15 Flow diagram illustrating positive and negative feedbacks; (b) specific examples of dynamical and thermodynamical feedbacks and their directions in the case of a change in the amount of cumulus convection; (c) the exact nature of an increase in cloud amount is unclear. The cloud could either be more extensive vertically or more extensive horizontally (reproduced by permission of Longman from Henderson-Sellers and Robinson, *Contemporary Climatology*, 1986)

radiation incident upon them. Their albedo (ratio of reflected to incident radiation) is therefore high. The surface albedo, and probably the planetary albedo (the reflectivity of the whole atmosphere plus surface system as seen from 'outside' the planet), increases. Thus a greater amount of solar radiation is reflected away from the planet and temperatures decrease further. A further increase in snow and ice results from this decreased temperature and the process continues. This positive feedback mechanism is known as the ice-albedo feedback mechanism. Of course, this ice-albedo feedback mechanism is also positive if the initial perturbation causes an increase in global surface temperatures. With higher temperatures, the areas of snow and ice are likely to be reduced, thus reducing the albedo and leading to further enhancement of temperatures. The existence of clouds over regions of snow and ice can greatly modulate the shortwave feedbacks associated with the cryosphere. The presence of a snow or ice surface also affects the temperature structure of the atmosphere, introducing feedbacks associated with longwave radiation.

1.4.2 The water-vapour 'greenhouse'

Another positive feedback mechanism occurs with the increase of atmospheric water vapour resulting from increased evaporation as temperatures are raised. Many atmospheric gases contribute to the greenhouse warming of the surface as a result of their absorption of infrared radiation emitted from the surface. The dominant greenhouse gas in the Earth's atmosphere is water vapour, although carbon dioxide and other trace gases, such as chlorofluorocarbons, are becoming increasingly important. The additional greenhouse effect of the extra water vapour enhances the temperature increase. Similarly, if temperatures fall, there will be less water vapour in the atmosphere and the greenhouse effect is reduced.

1.4.3 Cloud feedbacks

To establish even the direction of the feedback associated with clouds is difficult, since they are both highly reflective (thus contributing to the albedo) and composed of water and water vapour (thus contributing to the greenhouse effect, because of their control of the longwave radiation). It has been suggested that for low- and middle-level clouds the albedo effect will dominate over the greenhouse effect, so that increased cloudiness will result in an overall cooling. On the other hand, cirrus clouds, which are fairly transparent at visible wavelengths, have a smaller impact upon the albedo so that their overall effect is to warm the system by enhancing the greenhouse effect.

Cloud feedback, however, is not this straightforward. There are dynamical and thermodynamical factors to be considered (Figure 1.15(b)), so that it is uncertain whether an increased temperature will lead to increased or decreased cloud cover (as opposed to cloud amounts). Although it is generally agreed that increased temperatures will cause higher rates of evaporation and hence make more water vapour available for cloud formation, the form which these additional clouds will take is much less certain. For the same 'volume' of new cloud, an increased dominance of cumuliform clouds probably reduces the percentage of the surface covered by clouds. Stratiform clouds, on the other hand, would increase the area covered (Figure 1.15(c)). Thus using the simplest reasoning it might be claimed that an increase in cumuliform clouds implies positive feedback, whereas an increase in stratiform clouds implies a negative feedback.

Another unknown factor about how clouds change in response to a climate perturbation is their height of formation. The situation is still further complicated by the lack of understanding of how the radiative properties of clouds may change. The sizes of the droplets in a cloud have an important influence on how the clouds interact with radiation, and the amount of water in the clouds also changes the way the clouds interact with the radiation. Clouds with larger drops have a lower albedo than clouds composed of smaller drops but with the same amount of liquid water (usually described in terms of the 'liquid water path'). Successful modelling of cloud liquid water must account for the competing effects of changing drop size and liquid water path which will ultimately affect the nature of the interaction with the solar and terrestrial radiation streams.

1.4.4 Combining feedback effects

Since more than one feedback effect is likely to operate within the climate system in response to any given perturbation, it is important to understand the way in which these feedbacks are combined. For example, consider a system in which a change of surface temperature of magnitude ΔT is introduced. Given no internal feedbacks, this temperature increment will represent the change in the surface temperature. If feedbacks occur then there will be an additional surface temperature change and the new value of the surface temperature change will be

$$\Delta T_{\text{final}} = \Delta T + \Delta T_{\text{feedbacks}} \tag{1.1}$$

where $\Delta T_{\text{feedbacks}}$ can be either positive or negative. The value of ΔT_{final} (i.e. whether it is large or small) is usually related to the perturbation which caused it by a measure of the sensitivity of the climate system to disturbance. There are a number of such sensitivity parameters in the climate modelling literature. An early measure of a model's sensitivity was termed the β parameter, where β is equal to the ratio of the calculated surface temperature change to an incremental change in the prescribed incident solar radiation. More recently, equation (1.1) has been rewritten in terms of a feedback factor, f, so that

$$\Delta T_{\text{final}} = f\,\Delta T \tag{1.2}$$

This feedback factor has, in turn, been related to the amplification or gain, g, of the system which is defined, using the analogy of gain in an electronic system, by

$$f = 1/(1-g) \tag{1.3}$$

The f factor is neither additive nor multiplicative, and is thus not an especially useful parameter. Gain factors, g, are additive but depend, as does the β parameter, on knowing the present climate system albedo or outgoing fluxes. A much more convenient climate sensitivity parameter is given in terms of a perturbation in the global surface temperature, ΔT, which occurs in response to an externally prescribed change in the net radiative flux crossing the tropopause, ΔQ,

$$C[\delta(\Delta T)/\delta t] + \lambda\Delta T = \Delta Q \tag{1.4}$$

Here $\lambda\Delta T$ is the net radiation change at the tropopause resulting from the internal dynamics of the climate system, t is time and C represents the system heat capacity. Although equation (1.4) represents an extreme simplification of the system, it is useful in interpreting and summarizing the sensitivity of more complex climate models. A convenient reference value for λ is the value λ_B which λ would have if the Earth were a simple black body with its present-day albedo,

$$\lambda_B = 4\sigma T_e^3 = 3.75\ \text{W m}^{-2}\ \text{K}^{-1} \tag{1.5}$$

where σ is the Stefan–Boltzmann constant and T_e is the Earth's effective temperature, both of which are defined in the glossary and explained in Chapter 3. The overall climate system sensitivity parameter λ_{TOTAL} is composed of the summation of λ_B and all contributing feedback factors λ_i such that, for example,

$$\lambda_{\text{TOTAL}} = \lambda_B + \lambda_{\text{water vapour}} + \lambda_{\text{ice-albedo}} \tag{1.6}$$

Thus for a given system heat capacity, a positive value of feedback factors (λ_i) implies stability or negative feedback, and a negative value implies positive feedback and possibly growing instability. It is worth noting that, as discussed in relation to the ice-albedo feedback, the feedback factors are not necessarily independent. To establish the resulting temperature change, the inverse of the value of λ_{TOTAL} is multiplied by the value of ΔQ for the perturbation considered. The relationship (derived directly from equation (1.4) for the case of zero temperature change) that $\Delta T = \Delta Q/\lambda$ has given rise to another definition of a feedback factor as $1/\lambda$ or λ'. It is this sensitivity parameter which has been used as a measure of the sensitivity of climate models in the IPCC assessment and in some recent Global Climate Model (GCM) intercomparisons (see Chapter 6).

For doubling atmospheric CO_2 it has been shown that $\Delta Q \approx 4.2$ W m^{-2}. If we take $\lambda_B = 3.75$ W m^{-2} K^{-1}, $\lambda_{\text{water vapour}} = -1.7$ W m^{-2} K^{-1} and $\lambda_{\text{ice-albedo}} = -0.6$ W m^{-2} K^{-1}, then we have $\lambda_{TOTAL} = 1.45$ W m^{-2} K^{-1}, so that the globally averaged temperature rise due to doubling atmospheric CO_2 is found to be about 2.9 K, whereas if we had neglected the ice-albedo feedback the temperature increase would have been only about 2.0 K. Various estimates have been made of the feedback effects likely to be caused by changes in cloud amount and cloud type. These estimates range from λ_{cloud} is zero (i.e. the effects cancel) to results from GCMs which suggest that λ_{cloud} could be as large as -0.8 W m^{-2} K^{-1}. The addition of this feedback effect to those considered above would raise the surface temperature increase due to doubling CO_2 to about 6.5 K. This example demonstrates clearly how powerful a combination of positive feedback effects can be for the predicted surface temperature change.

Great care must be taken in interpretation of quoted values for feedback factors since several different definitions can be used. Three of those used are shown in Figure 1.16 plotted as functions of the feedback factor, λ_{TOTAL}. Note the areas of the graph (and the values of the feedback factors) which represent positive and negative feedback. As the term λ is often used synonymously with λ_{TOTAL}, it is important to establish which is meant by careful contextual reading.

The climatic system is clearly extremely heterogeneous. There are many subsystems which interact with one another producing feedback effects. Climate dynamics is not unique in being controlled by changing feedback effects. If we reconsider the feedbacks affecting the population of a city, mentioned earlier, it is easy to imagine a range of other feedbacks operating in the same, and opposite, direction as the negative feedback on population induced by the perception of declining environmental character. As the city grows there are greater profits to be made in centrally located businesses (a positive feedback on immigration) while land prices, rents etc. increase (a negative feedback), but street crime probably increases and long journey-to-work times are detrimental to family life. All these and many other feedback effects operate in a dynamically changing 'control' of the city size. Climatic feedbacks can be thought of as analogous to these geographical and economic controls.

Often the importance of feedback effects depends upon the time-scale of behaviour of the subsystems they affect. This concept of time-scale of response is crucially important to all aspects of climate modelling. This time-scale is variously referred to as the equilibration time, the response time, the relaxation time or the adjustment time. It is a measure of the time the subsystem takes to re-equilibrate following a small perturbation to it. A short equilibration time-scale indicates that the subsystem responds very quickly to perturbations, and can therefore be viewed as being quasi-instantaneously

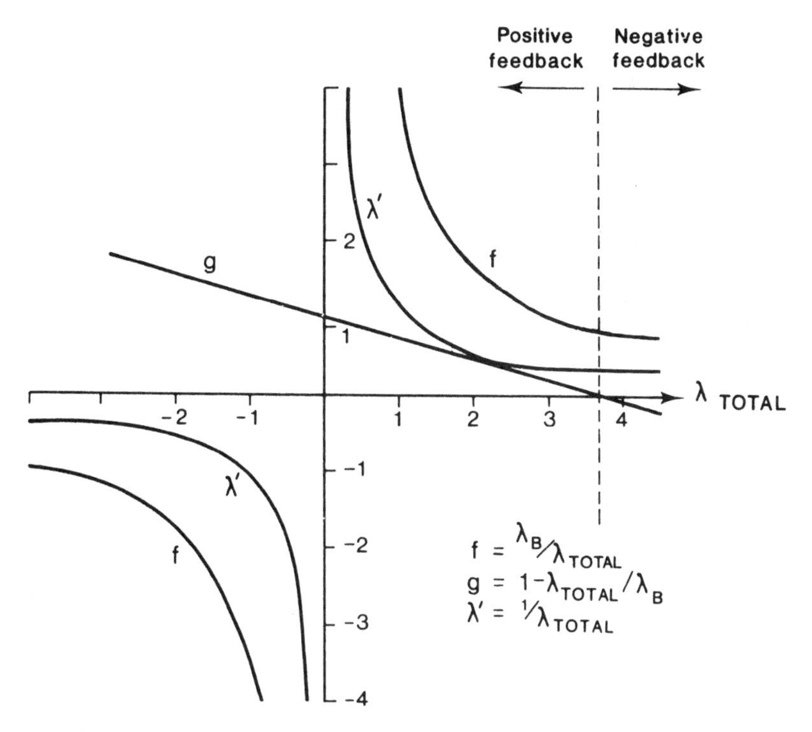

Figure 1.16 The feedback factor, f, the gain of the same system, g, and a third climate system feedback parameter λ' used by the IPCC, all plotted as a function of λ_{TOTAL} (probably the most useful measure of climate system sensitivity), the sum of all the contributing λ_is; λ_B is the value of λ_{TOTAL} for zero feedback. The areas of the diagram signifying overall positive and negative climatic feedback are shown. (The older sensitivity parameter to be found in the literature, the β parameter, not shown here, can be written as $\beta = S_0/\lambda_{\text{TOTAL}}$, where S_0 is the global average incoming solar radiation ($\simeq$340 W m^{-2}: one-fourth of the solar constant). Consequently for $\lambda_{\text{TOTAL}} = 3.75$ W m^{-2} K^{-1}, $\beta = 91.33$ K)

equilibrated with an adjacent subsystem which possesses a much longer equilibration time. It is common to express equilibration times in terms of the time (termed the *e*-folding time) it would take a system or subsystem to reduce an imposed displacement to 1/*e*th of the displaced value. For example, a pot of water removed from a stove will re-equilibrate with the room environment with an *e*-folding time depending upon the difference in temperature of the pot contents and the room, and the size and shape of the pot. A smaller temperature difference, a smaller pot or a larger surface-to-volume ratio of the container will result in relatively shorter *e*-folding times. Large *e*-folding times are possessed by subsystems which respond only very slowly. The response time is generally assessed in terms of the thermal response time. Table 1.1 lists equilibration times for a range of subsystems of the climate system. The

Table 1.1 Equilibrium times for several subsystems of the climate system

Climatic domain	Seconds	Equivalent
Atmosphere		
Free	10^6	11 days
Boundary layer	10^5	24 hours
Ocean		
Mixed layer	10^6–10^7	Months to years
Deep	10^{10}	300 years
Sea-ice	10^6–10^{10}	Days to 100s of years
Continents		
Snow and surface ice layer	10^5	24 hours
Lakes and rivers	10^6	11 days
Soil/vegetation	10^6–10^{10}	11 days to 100s of years
Mountain glaciers	10^{10}	300 years
Ice sheets	10^{12}	3000 years
Earth's mantle	10^{15}	30 million years

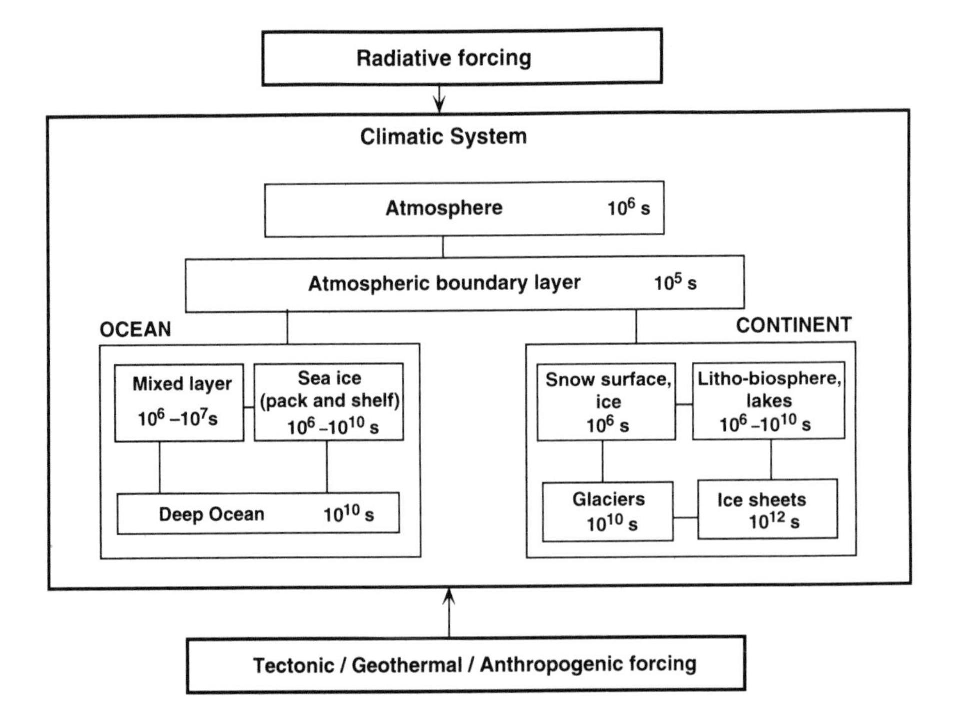

Figure 1.17 Schematic representation of the domains of the climatic system showing estimated equilibration times (reproduced by permission of Academic Press from Saltzman, 1983)

longest equilibration times are those for the deep ocean, the glaciers and ice sheets ($10^{10}-10^{12}$ s), while the remaining elements of the climate system have equilibration times nearer $10^{5}-10^{7}$ s.

Thus the climatic system can be pictured as in Figure 1.17 in terms not only of subsystems and their directions and types of interactions, but also in terms of their approximate equilibration times. The very long equilibration time of the deep ocean poses a particularly difficult problem for climate modellers. The methods by which the short response time of the atmospheric features can be linked to the much slower response time of the ocean system are discussed in Chapter 5. It must be noted, however, that some elements of the cryosphere have equally long response times and have, so far, not been included in the parameterizations of GCMs.

Clearly, modelling of climatic feedbacks (i.e. the processes and interactions) will be crucial to the results of the modelling experiment. The inclusion or exclusion of a feedback mechanism could radically alter results. Understanding feedbacks can only come through careful examination of the action of likely perturbations and the relative equilibration times of various parts of the climate system. Such a wide range of time-scales is reflected in the wide range of model types currently in use.

1.5 RANGE OF QUESTIONS FOR CLIMATE MODELLING

The type of question asked of climate modellers has changed over time. In the early years of climate modelling the question was 'Can the model capture the fundamental characteristics of the atmosphere?', so that particular attention was paid to reproducing the atmospheric time mean state adequately. Over the last twenty or so years climate models of all types have been applied to questions which are strictly demands for predictions: for example 'What is the impact on the climate of doubling or tripling atmospheric CO_2?', 'Will removing the tropical forests of the world affect the climate at locations distant from the deforested area?' and 'Is the North Atlantic ocean circulation likely to change rapidly?'. As will become clear in the rest of this book, many 'predictions' have been made with models that have not yet been fully tested. Sensitivity testing, intercomparison and careful evaluation of climate models has only become widespread since the models have been shown to be doing a good basic job.

The next stage in the evolution of climate modelling seems likely to be an attempt to answer still more difficult questions about the climate system. One such question, which could not even be considered by modellers until they felt confident of their predictions of greenhouse warming, is the likely social and economic implications of this future climate change.

The rest of this book is intended to give the reader a basic understanding of the types and complexity of climate models. We have not tried to answer specific questions such as those outlined above or describe, in detail, particular models or experiments. In writing this primer our aim has been to help those new to climate modelling to a quicker and fuller understanding of the available literature.

RECOMMENDED READING

Charlson, R.J. and Heintzenberg, J. (1995) *Aerosol Forcing of Climate.* John Wiley & Sons, Chichester, 399 pp.

Dickinson, R.E. (1985) Climate sensitivity. *In* S. Manabe (ed.), *Issues in Atmospheric and Oceanic Modelling. Part A. Climate Dynamics.* Advances in Geophysics, Vol. 2. Academic Press, New York, pp. 99–129.

GARP (1975) *The Physical Basis of Climate and Climate Modelling.* GARP Publication Series No. 16, WMO/ICSU, Geneva.

Hansen, J.E. and Takahashi, T. (eds) (1984) *Climate Processes and Climate Sensitivity.* Geophysical Monograph No. 29, Maurice Ewing Vol. 5. American Geophysical Union, Washington, DC, 368 pp.

Hansen, J.E., Lacis, A., Ruedy, R., Sato, M. and Wilson, H. (1993) How sensitive is the world's climate. *National Geographic Research and Exploration* **9**, 142–158.

Henderson-Sellers, A. (ed.) (1995) *Future Climates of the World: A Modelling Perspective.* World Survey of Climatology Series, Vol 16. Elsevier, Amsterdam, 568 pp.

Houghton, J.T. (ed.) (1984) *The Global Climate.* Cambridge University Press, Cambridge, 233 pp.

Houghton, J.T., Jenkins, G.J. and Ephraums, J.J. (1990) *Climate Change: The IPCC Scientific Assessment.* Cambridge University Press, Cambridge, 403 pp.

Houghton, J.T., Callander, B.A. and Varney, S.K. (1992) *Climate Change 1992: The Supplementary Report to the IPCC Scientific Assessment.* Cambridge University Press, Cambridge, 150 pp.

Houghton, J.T., Mera Filho, L.G., Callander, B.A., Harris, N., Kattenberg, A. and Maskell, K. (eds) (1996) *Climate Change 1995. Contribution of Working Group I to the Second Assessment Report of the Intergovernmental Panel on Climate Change.* Cambridge University Press, Cambridge, 572 pp.

Manabe, S. (ed.) (1985) *Issues in Atmospheric and Oceanic Modelling. Part A. Climate Dynamics.* Advances in Geophysics, Vol. 28, Academic Press. New York, 591 pp.

Oort, A.H. and Peixoto, J.P. (1983) Global angular momentum and energy balance requirements from observations. *Advances in Geophysics* **26**, 355–490.

Peixoto, J.P. and Oort, A.H. (1991) *Physics of Climate.* American Institute of Physics, Woodbury, New York, 520 pp.

Schlesinger, M.E. (ed.) (1988) *Physically Based Modelling of Climate and Climatic Change. Parts 1 and 2.* NATO ASI Series C: No. 243. Kluwer Academic Publishers, Dordrecht, 990 pp.

Trenberth, K.E. (1992) *Coupled Climate System Modelling.* Cambridge University Press, Cambridge, 600 pp.

Wang, W.-C. and Isaksen, I.S. (1995) *Atmospheric Ozone as a Climate Gas: General Circulation Model Simulations*, NATO ASI Series I: Volume 32, Springer, New York, 459 pp.

CHAPTER 2

A History of and Introduction to Climate Models

There has been a renaissance in climate studies over the last decade. Scientists in different disciplines concerned with the climate system have grown increasingly appreciative of the connections between the various components of the climate system and the hazards of overly narrow viewpoints.

R. E. Dickinson (1983)

2.1 INTRODUCING CLIMATE MODELLING

Any climate model is an attempt to simulate the many processes that produce climate. The objective is to understand these processes and to predict the effects of changes and interactions. The simulation is accomplished by describing the climate system in terms of basic physical, chemical and biological principles. Hence a model can be considered as being comprised of a series of equations expressing these laws. Climate models can be slow and costly to use, even on the fastest computer, and the results can only be approximations.

The need for simplification

For several reasons, a model must be a simplification of the real world. The processes of the climate system are not fully understood, but they are known to be complex. Furthermore, they interact with each other, producing feedbacks (Section 1.4), so that any solution of the governing equations must involve a great deal of computation. The solutions that are produced start from some initialized state and investigate the effects of changes in a particular component of the climate system. The boundary conditions, for example the solar radiation, sea-surface temperatures or vegetation distribution, are set from observational data or other simulations. These data are rarely complete or of adequate accuracy to specify completely the environmental conditions, so that there is inherent uncertainty in the results.

Large-scale global climate models, designed to simulate the climate of the planet, must take into account the whole climate system (see Figure 1.2). All of the interactions between components must be integrated in order to develop a climate model. This presents great problems, because the various interactions operate on different time-scales. For example, the effects of deep ocean overturning may be very important when considering climate averaged over decades to centuries, while local changes in wind direction may be unimportant on this time-scale. If, on the other hand, monthly time-scales are of concern, the relative importance would be reversed.

Early global models were used to generate average conditions for January and July. This was usually done by maintaining forcing appropriate to one particular month and running the model for hundreds of days. These simulations were termed, for example, 'perpetual January' or 'perpetual July' (depending on forcing). Climate models are now usually run for multi-year seasonal cycles, and these are used to produce ensemble averages. This is not to imply that a particular January in the period for which a climate model prediction is made would have these conditions, only that the conditions apply to an average January. More recent simulations average over many years or produce ensemble series by perturbing initial conditions for each of an ensemble set. Indeed, it is always implied that any 'new' climate predicted will have variation about the mean, just as with the present climate. The amount of this new variation is often of concern when the results of global-scale models are used to estimate the possible impact of climatic change in a local or regional area.

The simplifications that must be made to the laws governing climatic processes can be approached in several ways. Consequently, there are numerous different global-scale climate models available. In general, two sets of simplifications need to be made. The first involves the processes themselves. It is usually possible to treat in detail some of the processes, specifying their governing equations fully. However, other processes must be treated in an approximate way, either because of our lack of exact information or of understanding, or because there are inadequate computer resources to deal with them. For example, it might be decided to treat the radiation processes in great detail, but only approximate the horizontal energy flows associated with regional-scale winds. The approximation may be approached either by using available observational data, the empirical approach, or through specification of the physical laws involved, the theoretical (or conceptual) approach.

Resolution in time and space

The second set of simplifications involves the resolution of the model in both time and space. While it is generally assumed that finer spatial resolutions produce more reliable results, constraints of both data availability and computational time may dictate that a model may have to have, for example, latitudinally averaged values as the basic input. In addition, too fine a resolution may be inappropriate because processes acting on a smaller scale than the model is designed to resolve may be inadvertently incorporated. Similar considerations are involved in the choice of temporal resolution. Most computational procedures require a 'timestep' approach to calculations. The processes are allowed to act for a certain length of time and the new conditions are calculated. The process is then repeated using these new values. This continues until the conditions at the required time have been established. Timestepping is a natural consequence of there not being a steady-state solution to the model equations. Although accuracy potentially increases as the timestep decreases, there are constraints imposed by data, computational ability and the design of the model. The time and space resolutions of the model are also linked, as explained in Chapter 5.

Although models are designed to aid in predicting future climates, performance can only be tested against the past or present climate. Usually when a model is developed an initial objective is to test the sensitivity of the model and to ascertain how well its results compare with the present climate. Thereafter it may be used to simulate past climates, not only to see how well it performs, but also to gain insight into the causes of these climates. Although such past climates are by no means well known, this comparison provides a very useful step in establishing the validity of the modelling approach. After such tests, the model may be used to gain insight into possible future climates.

2.2 TYPES OF CLIMATE MODELS

The important components to be considered in constructing or understanding a model of the climate system are:

1. Radiation – the way in which the input and absorption of solar radiation and the emission of infrared radiation are handled;
2. Dynamics – the movement of energy around the globe by winds and ocean currents (specifically from low to high latitudes) and vertical movements (e.g. small-scale turbulence, convection and deep-water formation);
3. Surface processes – inclusion of the effects of sea- and land-ice, snow, vegetation and the resultant change in albedo, emissivity and surface–atmosphere energy and moisture interchanges;
4. Chemistry – the chemical composition of the atmosphere and the interactions with other components (e.g. carbon exchanges between ocean, land and atmosphere);
5. Resolution in both time and space – the timestep of the model and the horizontal and vertical scales resolved.

The relative importance of these processes and the theoretical (as opposed to empirical) basis for parameterizations employed in their incorporation can be discussed using the 'climate modelling pyramid' (Figure 2.1). The edges represent the basic elements of the models, and complexity is shown increasing

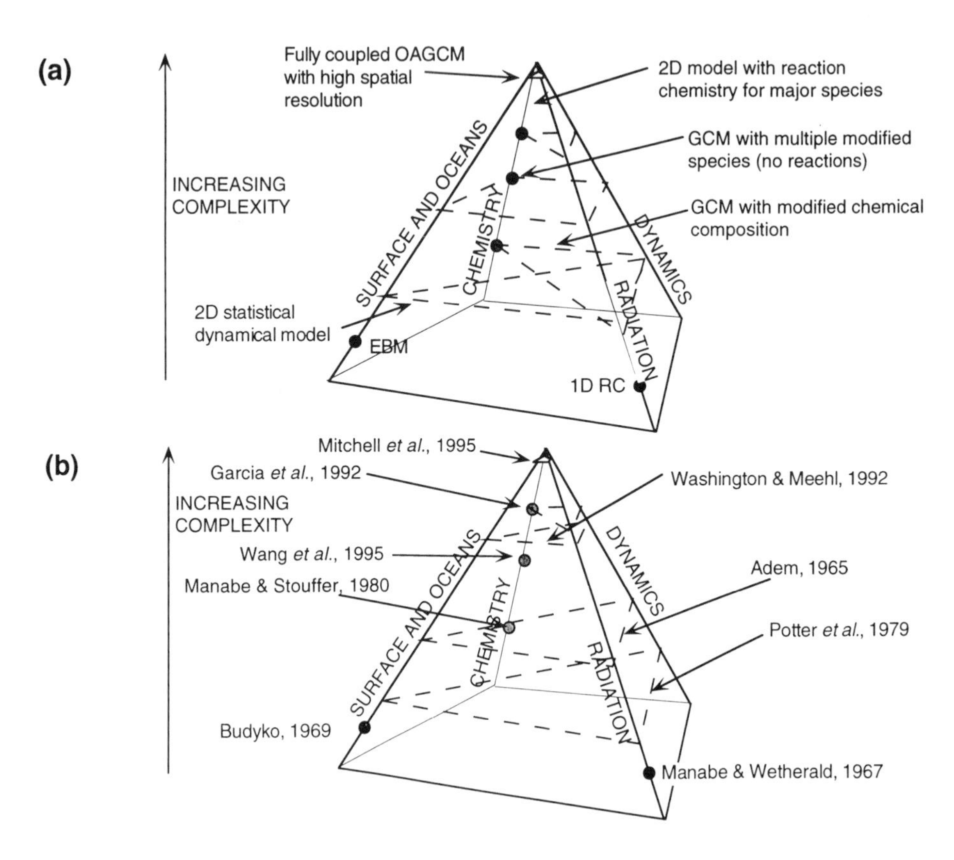

Figure 2.1 The climate modelling pyramid. The position of a model on the pyramid indicates the complexity with which the four primary processes (dynamics, radiation, surface and oceans, and chemistry) interact. Progression up the pyramid leads to greater interaction between each primary process. The vertical axis is not intended to be quantitative. (a) The positions of the various model types; (b) examples from the literature and their positions on the pyramid

upwards. Around the base of the pyramid are the simpler climate models which incorporate only one primary process. As indicated in Figure 2.1, there are four basic types of model.

1. Energy balance models (EBMs) are one-dimensional models predicting the variation of the surface (strictly the sea-level) temperature with latitude. Simplified relationships are used to calculate the terms contributing to the energy balance in each latitude zone.
2. One-dimensional radiative–convective (RC) models compute the vertical (usually globally averaged) temperature profile by explicit modelling of radiative processes and a 'convective adjustment' which re-establishes a predetermined lapse rate.

3. Two-dimensional statistical dynamical (SD) models deal explicitly with surface processes and dynamics in a zonally averaged framework and have a vertically resolved atmosphere. These models have been the starting point for the incorporation of reaction chemistry in global models.
4. Global circulation models (GCMs), where the three-dimensional nature of the atmosphere and/or ocean is incorporated. In these models an attempt is made to represent most climatic processes believed to be important.

The vertical axis in Figure 2.1 shows increasing complexity (i.e. more processes included and linked together) and also indicates increasing resolution: models appearing higher up the pyramid have higher spatial and temporal resolutions.

There is some ambiguity concerning the expansion of 'GCM'. Two possible terms are the more recent 'global climate model' and the older 'general circulation model'. The latter also refers to a weather forecast model, so that in climate studies 'GCM' is understood to mean 'general circulation climate model'. A further distinction is often drawn between oceanic general circulation models and atmospheric general circulation models by terming them OGCMs and AGCMs, respectively. As the pyramid is ascended, more processes are coupled to develop a coupled ocean–atmosphere global model (OAGCM or CGCM). It has been suggested that as processes which are currently fixed come to be incorporated into GCMs, the coupling will be more complete, say including changing biomes (an AOBGCM) or changes in atmospheric, ocean and even soil chemistry. In this book, the generic term GCM is used to mean a complex three-dimensional model of the atmosphere and/or ocean, and possibly other components, used for climate simulation. The meaning will be clear from the context.

2.2.1 Energy balance climate models

These models have been instrumental in increasing our understanding of the climate system and in the development of new parameterizations and methods of evaluating sensitivity for more complex and realistic models. This type of model can be readily programmed and implemented on most small computers, and the inherent simplicity of EBMs combined with the ease of interpreting results make them ideal instructional tools. They are widely used to investigate the sensitivity of the climate system to external changes and to interpret the results of more complex models. Energy balance models are discussed more fully in Chapter 3.

Energy balance models are generally 'one-dimensional', the dimension in which they vary being latitude. Vertical variations are ignored and the models are used with surface temperature as the predicted variable. Since the energy balance is allowed to vary from latitude to latitude, a horizontal energy transfer term must be introduced, so that the basic equation for the energy balance at each latitude, ϕ, is

$$C_m[\Delta T(\phi)/\Delta t] = R\downarrow(\phi) - R\uparrow(\phi) + \text{transport into zone } \phi \qquad (2.1)$$

where C_m is the heat capacity of the system and can be thought of as the system's 'thermal inertia', and $R\downarrow$ and $R\uparrow$ are the incoming and outgoing radiation fluxes, respectively.

The radiation fluxes at the Earth's surface must be parameterized with care, since conditions in the vertical are not considered in this type of model. To a large extent the effects of vertical temperature changes are treated implicitly. In a clear atmosphere, convective effects tend to ensure that the lapse rate remains fairly constant. However, cloud amount depends only weakly on surface temperature, so that cloud albedo is only partially incorporated in the model. In particular, clouds in regions of high temperatures, such as the intertropical convergence zone, are ignored in the parameterization of albedos in EBMs.

When using the model for annual average calculations, the surface albedo can be regarded as constant for a given latitude. This type of model, however, can also be used for seasonal calculations. In this case it is usual to allow the albedo to vary with temperature to simulate the effects of changes in sea-ice and snow extent.

Atmospheric dynamics are not modelled in an EBM; rather it is assumed that a 'diffusion' approximation is adequate for including heat transport. This approximation relates energy flow directly to the latitudinal temperature gradient. This flow is usually expressed as being proportional to the deviation of the zonal temperature, T, from the global mean, $\bar{T}$.

Early EBMs were originally found to be stable only for small perturbations away from present-day conditions. For instance, they predicted the existence of an ice-covered state for the Earth for only slight reductions in the present solar constant. This result prompted studies of the sensitivity of various climate model types to perturbations (see Section 2.4).

2.2.2 One-dimensional radiative–convective climate models

One-dimensional RC models represent an alternative approach to relatively simple modelling of the climate and they also occur at the bottom of the modelling pyramid (Figure 2.1). In this case the 'one dimension' in the name refers to altitude. One-dimensional RC models are designed with an emphasis on the global average surface temperature, although temperatures at various levels in the atmosphere can be obtained. The models can also be applied to zonally averaged conditions by including a description of the horizontal energy transport.

The main emphasis in these models is on the explicit calculation of the fluxes of solar and terrestrial radiation (the radiation streams). Given an initially isothermal atmosphere the heating rates for a number of layers in the atmosphere are calculated, although the cloud amount, optical properties and the albedo of the surface generally need to be specified. The temperature change in each layer, which results from an imbalance between the net radiation at the top and bottom of the layer, is calculated. At the end of each timestep a revised radiative temperature profile is produced. If the calculated lapse rate exceeds some predetermined 'critical' lapse rate, the atmosphere is presumed to be convectively unstable. An amount of vertical mixing, sufficient to re-establish the prescribed lapse rate, is carried out and the model proceeds to calculate the next radiative timestep. This procedure continues until convective readjustment is no longer required and the net fluxes for each layer approach zero. One-dimensional RC models operate under the constraints that at the top of the atmosphere there must be a balance of shortwave and longwave fluxes, and that surface energy gained by radiation equals that lost by convection. However, they vary in the way they incorporate the critical lapse rate. Some use the dry adiabatic lapse rate, some the saturated one, while many use a value of 6.5 $\mathrm{K\,km^{-1}}$, which is the value in an observed standard atmospheric profile. Similarly, different humidity and cloud formulations are possible.

Radiative–convective models (discussed more fully in Chapter 4) can be constructed either as equilibrium models or in a time-dependent form. FORTRAN code for the latter type is included on the CD – see Appendix B. The main use of radiative–convective models is to study the effects of changing atmospheric composition and to investigate the likely relative influences of different external and internal forcings. They are the basis for the 'column' models that have recently begun to be used to evaluate aspects of the parameterizations of the atmospheric (and surface) 'columns' in more complex GCMs. Column models are, in effect, single columns from a GCM, and include the sophisticated physics usually found in these models.

2.2.3 Two-dimensional climate models

Two-dimensional climate models can represent either the two horizontal dimensions or the vertical and one horizontal dimension. It is the latter which are more common, combining the latitudinal dimension of the energy balance models with the vertical one of the radiative–convective models. These models also include a more realistic parameterization of the latitudinal energy transports. In such models the general circulation is assumed to be composed mainly of a cellular flow between latitudes, which is characterized using a combination of empirical and theoretical formulations. A set of statistics summarizes the wind speeds and directions, while an eddy diffusion coefficient of the type used in EBMs governs energy transport. As a consequence of this approach, these models are called 'statistical dynamical' (SD) models.

Statistical dynamical models are about one-third of the way up the modelling pyramid (Figure 2.1), being more complicated than the vertically or latitudinally resolved one-dimensional models. Their use has provided insight into the operation of the present climate system, in particular showing that the relatively simple diffusion coefficient approach for poleward energy transports is appropriate provided that the coefficient, as well as the transport, is allowed to vary with the latitudinal temperature gradient. Similarly, advances in the understanding of baroclinic waves have been achieved from studies of the results of these SD models. Traditional two-dimensional models are insensitive to changes within a latitude band. Changes, such as new land–sea temperature contrasts and cloud regimes, which might result from alterations of land-surface albedo are impossible to investigate. Some compromise may be obtained by considering each zone as being divided into a land part and an ocean part. This type of 'two-channel' approach is discussed with reference to a more complex EBM in Section 4.8. Two-dimensional models have been employed to make simulations of the chemistry of the stratosphere and mesosphere. These models typically involve the modelling of tens to hundreds of chemical species and many hundreds of different reactions, and are much more demanding of computer time than 'atmosphere-only' 2D models.

As a result of the lack of zonal resolution, two-dimensional SD models were largely been superseded by GCMs for consideration of the effect of perturbations on the present climate. However, recent developments have pointed to the value of 'stripped-down' or 'computationally efficient' GCMs. In these models, some of the complexity and resolution of a full GCM is intentionally removed, producing a global model which is lower on the pyramid and has much in common with 2D SD models.

2.2.4 General circulation models

The aim of GCMs is the calculation of the full three-dimensional character of the climate comprising at least the global atmosphere and the oceans (Figure 2.2). The solution of a series of equations (Table 2.1) that describe the movement of energy, momentum and various tracers (water vapour in the atmosphere and salt in the oceans) and the conservation of mass is therefore required. Generally the equations are solved to give the mass movement (i.e. wind field or ocean currents) at the next timestep, but models must also include processes such as cloud and sea-ice formation, and heat, moisture and salt transport. The first step in obtaining a solution is to specify the atmospheric and oceanic conditions at a number of 'grid points', obtained by dividing the Earth's surface into a series of rectangles so that a regular grid results (Figure 2.3). Conditions are specified at each grid point for the surface and several layers in the atmosphere and ocean. The resulting set of coupled non-linear equations is then solved at each grid point using numerical techniques. Various techniques are available, but all use a timestep approach.

Although GCMs formulated in this way have the potential to approach the real oceanic and atmospheric situation closely, at present there are a number of practical and theoretical limitations. The prime practical consideration is one of the time needed for the calculations. For example, one particular low-resolution AGCM requires around 48 Mbytes of memory, whereas a more recent, higher-resolution version of the model requires over 160 Mbytes. Much of this stored information must be accessed and updated at each model timestep, and this places a strain on the resources of even the largest and

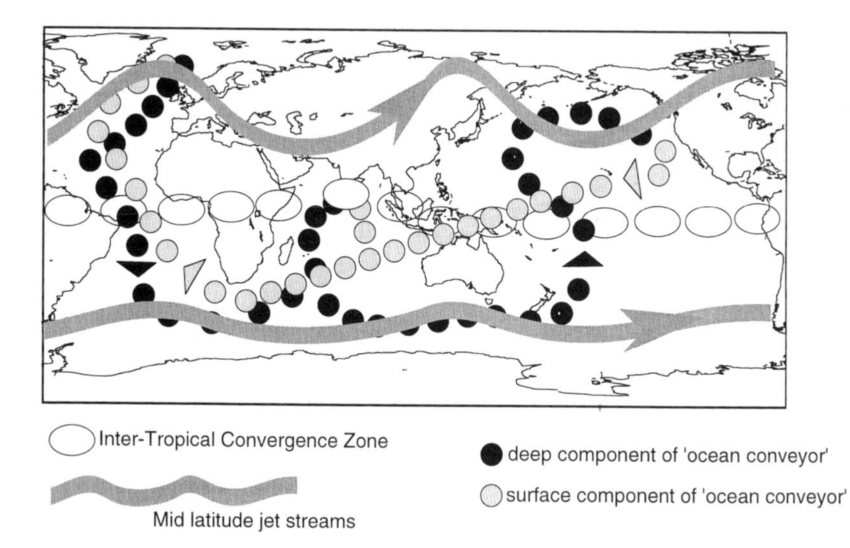

Figure 2.2 Illustration of the main features of the atmospheric and oceanic circulation. The thermohaline circulation of the ocean (circles), often referred to as the 'ocean conveyor', results in the movement of water through the major ocean basins of the world over periods of hundreds of years to thousands of years. The inter-tropical convergence zone (ITCZ), shown by ovals, and the upper tropospheric, mid-latitude waves (broad, solid lines) represent the dominant features of the atmospheric circulation which operate over periods of a few days

Table 2.1 Fundamental equations solved in GCMs

1. Conservation of energy (the first law of thermodynamics)
   i.e. Input energy = increase in internal energy plus work done
2. Conservation of momentum (Newton's second law of motion)
   i.e. Force = mass × acceleration
3. Conservation of mass (the continuity equation)
   i.e. The sum of the gradients of the product of density and flow-speed in the three orthogonal directions is zero. This must be applied to air and moisture for the atmosphere and to water and salt for the oceans, but can also be applied to other oceanic 'tracers' and to cloud liquid water.
4. Ideal gas law (an approximation to the equation of state – atmosphere only)
   i.e. Pressure × volume = gas constant × absolute temperature

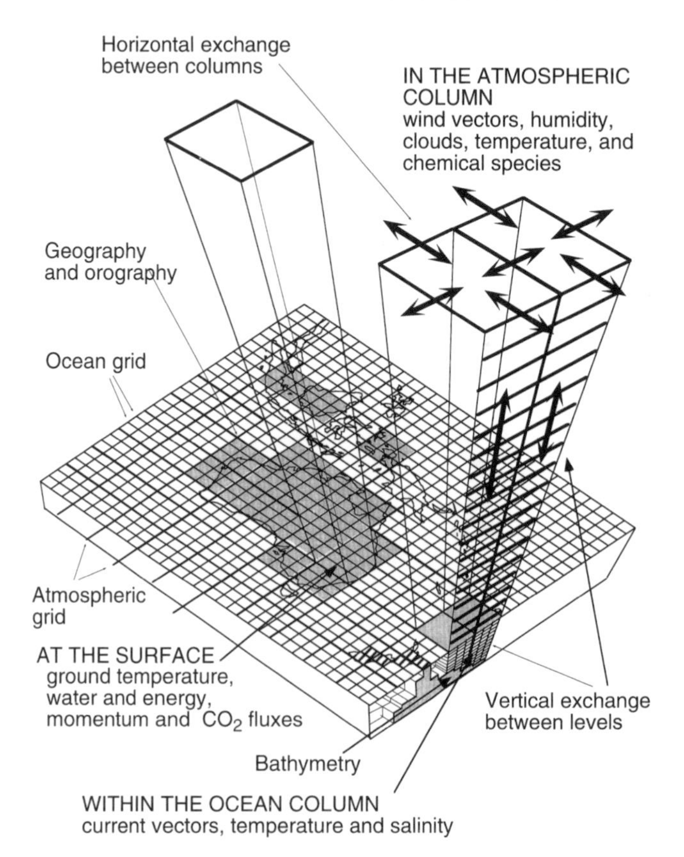

Figure 2.3 Illustration of the basic characteristics of a general circulation model, showing the manner in which the atmosphere and ocean are split into columns. Both atmosphere and ocean are modelled as a set of interacting columns distributed across the Earth's surface. The resolutions of the atmosphere and ocean models are often different

fastest computers (cf. Figure 1.5). Since the accuracy of the model partly depends on the spatial resolution of the grid points and the length of the timestep, a compromise must be made between the resolution desired, the length of integration and the computational facilities available. At present, atmospheric grid points are typically spaced between 2° and 5° of latitude and longitude apart, and timesteps of approximately 20–30 min are used. Vertical resolution is obtained by dividing the atmosphere into between 6 and 50 levels, with about 20 levels being typical.

The ocean is a three-dimensional fluid which must be modelled using the same principles as for the atmosphere. As well as acting as a thermal 'fly-wheel' for the climate system, the ocean also plays a central role in the carbon cycle, absorbing approximately half of the carbon which is released into the atmosphere every year. The dynamics of the ocean are governed by the amount of radiation which is available at the surface and by the wind stresses imposed by the atmosphere. The flow of ocean currents is also constrained by the positions and shapes of the continents (Figure 2.2). Ocean GCMs calculate the temporal evolution of oceanic variables (velocity, temperature and salinity) on a three-dimensional grid of points spanning the global ocean domain. Many climate model simulations incorporate very simple models of the ocean, which do not explicitly include ocean dynamics.

Modelling the full three-dimensional nature of the ocean is made difficult by the fact that the scale of motions which exist in the oceans is much smaller than those in the atmosphere (ocean eddies are around 10–50 km compared with around 1000 km for atmospheric eddies), and that the ocean also takes very much longer to respond to external changes (cf. Table 1.1). The deep water circulation of the ocean (Figure 2.2) can take hundreds or even thousands of years to complete. Ocean models which include these dynamic processes are now being coupled with atmospheric GCMs to provide our most detailed models of the climate system. The formation of oceanic deep water is closely coupled to the formation and growth of sea-ice, so that ocean dynamics demands effective inclusion of sea-ice dynamics and thermodynamics.

Originally, computational constraints dictated that global circulation models could only run for very short periods. For the atmosphere this meant only simulating a particular month or season, rather than a full seasonal cycle, although now all models include a seasonal cycle and most include a diurnal cycle. For the oceans, restrictions of computer power meant that the models were used before they had fully equilibrated. This could result in the ‘drift’ of the ocean climate away from present-day conditions, which was often corrected by applying adjusting fluxes at the ocean surface to compensate for systematic errors which persist at equilibrium. This is a particular problem for coupled OAGCMs. The importance of reducing (or preferably removing) such arbitrary adjustments and of including realistic time-dependent phenomena is now well established, and modellers have striven to include increasing numbers of these phenomena as well as using the increased computer power to provide higher resolution (cf. Figure 2.1).

It is important to identify the very different aims of those developing and using GCMs as compared with the designers of numerical weather forecast models. The latter are prediction tools, while GCMs can represent only probable conditions. For this reason many GCM integrations must be performed and their results averaged to generate an ‘ensemble’ before a climate ‘prediction’ can be made.

Computational constraints lead to problems of a more theoretical nature. With a coarse grid spacing, small-scale atmospheric motions (termed sub-gridscale), such as thundercloud formation, cannot be modelled, however important they may be for real atmospheric dynamics. Fine grid models can be used for weather prediction because the integration time is short. In contrast,

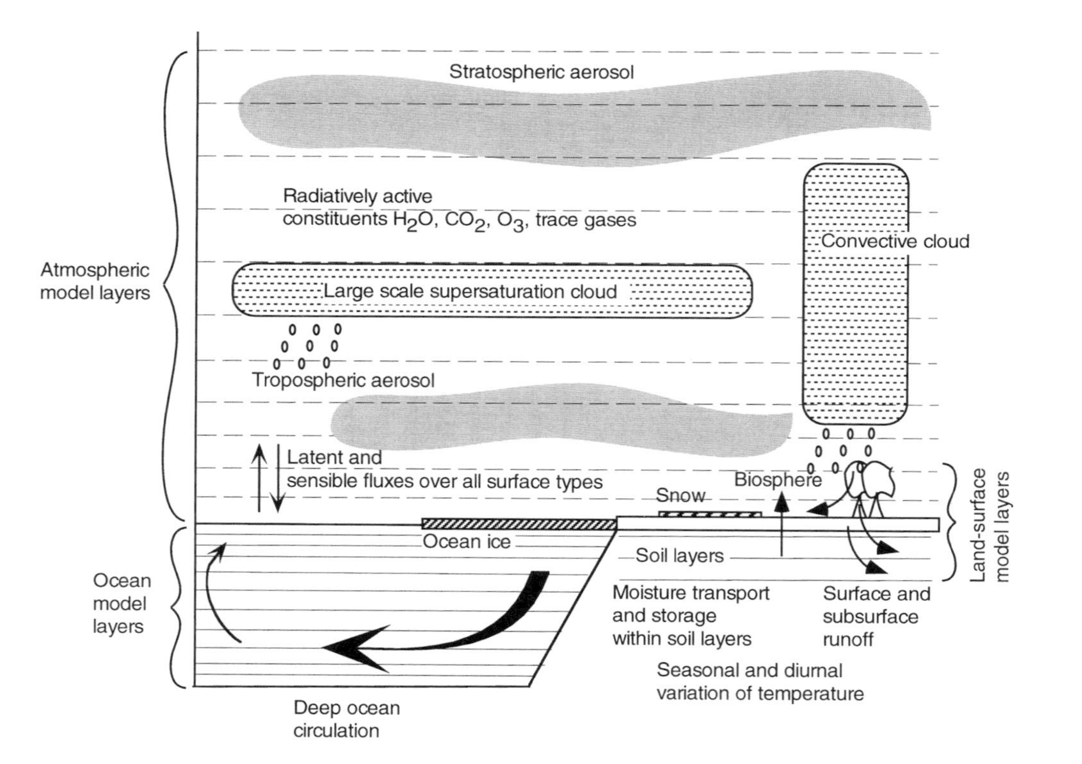

Figure 2.4 Schematic illustration of the processes in a global circulation climate model. In most models two types of cloud are treated. In this example, soil moisture is modelled in a number of layers, and tropospheric and stratospheric aerosols are included (adapted from Hansen *et al.*, 1983)

climate models must rely on some form of parameterization of sub-gridscale processes (e.g. see Section 5.2.3).

Some of the processes usually incorporated into global circulation climate models are shown in Figure 2.4. Within the atmosphere the modellers adopt an approach similar to that used for the RC models in calculating heating rates (although often computationally simpler), but also often include cloud formation processes as part of the convection and consider in detail the effects of horizontal transport. The interaction between the surface and the near-surface layer of the atmosphere, however, must be parameterized. Detailed consideration of the transfer processes at the surface of the Earth are computationally too demanding for explicit inclusion. Commonly, the surface fluxes of momentum, sensible heat and moisture are taken to be proportional to the product of the surface wind speed and the gradient of the property away from the surface. More detailed aspects of GCMs will be considered in Chapter 5.

2.2.5 The interactive biosphere

The many roles of the biosphere of importance to the climate include: transfer of moisture from the soil into the atmosphere; modification of the albedo, which changes the amount of radiation absorbed by the climate system; responsibility for the exchange of carbon and other chemicals; modification of the surface roughness, which alters the exchange of momentum. Only recently has the interactive nature of the plant life of the planet been included in climate models. The first approach has been to delineate geographic boundaries of biomes (vegetation groups characterized by similar species) by using simple predictors available from the GCM such as temperature, precipitation and possibly sunshine or cloudiness. Currently, attempts are being made to evaluate these methods using palaeo-reconstructions of vegetation cover during past epochs. Some modellers have included simplified succession models into their GCMs, and have been able to make sub-gridscale features of the terrestrial biosphere interactive. These interactive biosphere models are still in their infancy but may provide useful predictions of future responses of the biosphere, including the issue of possible future CO_2 fertilization of the biosphere.

2.3 HISTORY OF CLIMATE MODELLING

As climate models are readily described in terms of a hierarchy (e.g. Figure 2.1), it is often assumed that the simpler models were the first to be developed, with the more complex GCMs being developed most recently. This is not the case. The first atmospheric general circulation climate models were being developed in the early 1960s, concurrently with the first RC models. On the other hand, energy balance climate models, as they are currently recognized, were not described in the literature until 1969 and the first discussion of two-dimensional SD models was in 1970.

The first atmospheric general circulation climate models were derived directly from numerical models of the atmosphere designed for short-term weather forecasting. These had been developed during the 1950s and, around 1960, ideas were being formulated for longer-period integrations of these numerical weather prediction schemes. It is in fact rather difficult to identify the transition point in many modelling groups. For example, Syukuro Manabe joined the National Oceanic and Atmospheric Administration's (NOAA) Geophysical Fluid Dynamics Laboratory (GFDL) in the USA in 1959 to collaborate in the numerical weather prediction efforts, but was to go on to become one of the world leaders in the climate modelling community. Scientists concerned with extending numerical prediction schemes to encompass hemispheric or global domains were also studying the radiative and thermal equilibrium of the Earth–atmosphere system. It was these studies which prompted the design of the RC models, which were once again spearheaded by Manabe, the first of these being published in 1961.

Other workers, such as Julián Adem, also expanded the domain of numerical weather prediction schemes in order to derive global climate models. The low-resolution thermodynamic model first described by Adem in 1965 is an interesting type of climate model, since it lies part way towards the apex of the climate modelling pyramid (Figure 2.1) although the methodology is simpler in nature than that of an atmospheric GCM. Similar in basic composition to an EBM, Adem's model includes, in a highly parameterized way, many dynamic, radiative and surface features and feedback effects, giving it a 'higher' position on the modelling pyramid.

Mikhail Budyko and William Sellers published descriptions of two very similar EBMs within a couple of months of each other in 1969. These models did not depend upon the concepts already established in numerical weather prediction schemes, but attempted to simulate the essentials of the climate system in a simpler way. The EBMs drew upon observational data derived from descriptive climatology, suggesting that major climatic zones are roughly latitudinal. As a consequence of the intrinsically simpler parameterization schemes employed in EBMs, they could be applied to longer time-scale changes than the atmospheric GCMs of the time. It was the work by Budyko and Sellers, in which the possibility of alternative stable climatic states for the Earth was identified, which prompted much of the interest in simulation of geological time-scale climatic change. Concurrently with these developments, RC models, usually globally averaged, were being applied to questions of atmospheric disturbance, including the impact of volcanic eruptions and the possible effects of increasing atmospheric CO_2.

The desire to improve numerical weather forecasting abilities also prompted the fourth type of climate model: the SD model. A primary goal for dynamical climatologists was seen to be the need to account for the observed state of averaged atmospheric motion, temperature and moisture on time-scales shorter than seasonal but longer than those characteristic of mid-latitude cyclones. One group of climate modellers preferred to design relatively simple low-resolution SD models to be used to illuminate the nature of the interaction between forced stationary long waves and travelling weather systems. Much of this work was spearheaded in the early 1970s by John Green. Theoretical study of large-scale atmospheric eddies and their transfer properties, combined with observational work, led to the parameterizations employed in two-dimensional climate models.

By 1980, this diverse range of climate models seemed to be in danger of being overshadowed by one type: the atmospheric GCM. Although single-minded individuals persevered with the development of simpler models, considerable funding and almost all the computational power used by climate modellers was being consumed by atmospheric GCMs. However, by the mid- to late 1980s a series of occurrences of apparently correct results being generated for the wrong reason by these highly non-linear and highly complex models prompted many modelling groups to move backward, in an hierarchical sense, in order to try to isolate the essential processes responsible for the results which are observed from more comprehensive models. When only the most topical (e.g. doubled CO_2) model experiments are considered, the trend has been for GCM experiments to replace simpler modelling efforts. For example, in 1980–81, from a total of 27 estimates of the global temperature change due to CO_2 doubling, only seven were made by GCMs. By 1993–94, GCMs produced 10 of 14 estimates published. The IPCC process has, however, placed great emphasis on the value of results from simple models such as the 'box' models, described in Chapter 3. The strategy of intentionally utilizing an hierarchy of models was originally proposed in the 1980s by scientists such as Stephen Schneider at the US National Center for Atmospheric Research. More recently, the soundness of a hierarchy of climate modelling tools has been championed by Tom Wigley.

In 1969, Kirk Bryan at GFDL developed the ocean model which has become the basis for most current ocean GCMs. The model has been modified and has become widely know as the Bryan–Cox–Semtner model. Albert Semtner and Robert Chervin have constructed a model version which is 'eddy resolving', and as a consequence have pushed the simulations to higher and higher resolution (currently 1/6 degree). Others have chosen to implement the model in non-eddy resolving form and have been able to run the model at 2° resolution for direct coupling with an atmospheric model.

Even though this three-dimensional ocean model dates back to the late 1960s, most global climate models treated the oceans in much simpler ways until the early 1990s. The original GCMs used fixed ocean temperatures based on observed averaged monthly or seasonal values. This 'swamp' model allows the ocean to act only as an unlimited source of moisture. Naturally, it is very difficult in such a model to disturb the climate away from present-day conditions when such large areas of the globe remain unchanged. Following this, in the late 1980s, computation of the heat storage of the mixed layer of the ocean (approximately 70–100 m) was the most common approach. In this model the lower deep ocean layer acts only as an infinite source and sink for water. The mixed layer approach is appropriate for time-scales ≤ 30 years, beyond which the transfer of heat to lower levels becomes significant.

The desire to make climate models more realistic has led to the involvement of many disciplines in the framework of climate modelling, and hence to the realization that no one discipline can assume constancy in the variables prescribed by the others. Joseph Smagorinsky, who pioneered much of the early development in numerical weather prediction and steered the course of one of the flagships of climate modelling, NOAA's Geophysical Fluid Dynamics Laboratory, when commenting on the exponential growth in climate modelling research, noted that at the international conference on numerical weather prediction held in Stockholm in June 1957, which might be considered the first international gathering of climate modellers, the world's expertise comprised about 40 people, all loosely describable as physicists. In 1995, the IPCC Second Scientific Assessment (Working Group I alone) had around 80 lead authors and over 400 contributors. A complete list of all climate modellers would now number many thousands and encompass a wide variety of disciplines. Interdisciplinary ventures have led to both rapid growth in insight and near-catastrophic blunders. Also, increasing complexity in narrowly defined areas, such as land-surface climatology, has forced upon modellers the recognition that other characteristics of their models, such as the diurnal cycle of precipitation, are being poorly predicted. The inclusion of more complex parameterizations of various subsystems, for example sea ice, is of little value if the atmospheric forcing in polar regions is inadequate. The tuning process, which accompanies the addition of new model components, might, in this situation, 'soak up' these errors. Modellers must maintain a holistic view of their model.

2.4 SENSITIVITY OF CLIMATE MODELS

An important stage in the development of climate models is a series of sensitivity tests. Modellers examine the behaviour of their modelled climate system by altering one component and studying the effect of this change on the model's climate.

Equilibrium climatic states

As an example of the change in an internal variable we can consider the variation in the albedo, α, as a function of the mean global temperature in an EBM. Above a certain temperature, T_g, the planet is ice-free and the value of the albedo is independent of temperature. As it becomes colder, we expect the albedo to increase as a direct result of increases in ice and snow cover. Eventually the Earth becomes completely ice-covered, at temperature T_i, and further cooling will produce no further albedo change. This could be expressed in the form

$$\begin{aligned} \alpha(T) &= \alpha_i && \text{for } T \leqslant T_i \\ \alpha(T) &= \alpha_g && \text{for } T \geqslant T_g \\ \alpha(T) &= \alpha_g + b(T_g - T) && \text{for } T_i < T < T_g \end{aligned} \tag{2.2}$$

T_i is usually assumed to be 273 K but may range between 263 and 283 K. If we are concerned with equilibrium conditions (i.e. when the left-hand side of equation (2.1) is zero) we can calculate $R\uparrow$ for a series of temperatures and $R\downarrow$ for a series of albedos and show the results graphically. The points of intersection of the curves occur when emitted and absorbed radiation fluxes balance (i.e. $R\downarrow = R\uparrow$) which represent the equilibrium situations (Figure 2.5). Any slight imbalances between the fraction of the incident solar radiation, S, absorbed, $S(1 - \alpha(T))$, and the top of the atmosphere-emitted longwave flux, approximated by $\varepsilon\sigma T^4$ where ε is the emissivity, lead to a change in the temperature of the system at the rate $\Delta T/\Delta t$, the changes serving to return the temperature to an equilibrium state. However, there are three equilibrium solutions, as shown in Figure 2.5: an ice-free Earth (1), a completely glaciated Earth (3) and an Earth with some ice (2) (e.g. the present situation of the planet). All are possible.

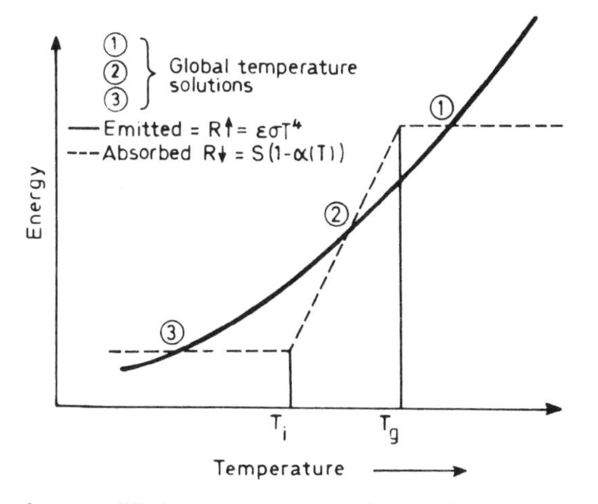

Figure 2.5 The three equilibrium temperature solutions for a zero-dimensional global climate model are shown at the intersection between the curves of emitted infrared radiation $R\uparrow$ and absorbed solar radiation $R\downarrow$. They are: (1) an ice-free Earth; (2) an Earth with some ice; (3) a completely ice-covered Earth (reproduced by permission of Longman from Henderson-Sellers and Robinson, *Contemporary Climatology*, 1986)

Equilibrium conditions and transitivity of climate systems

Such a simple model has some very obvious limitations. However, it not only shows one means of analysing the results of climate models, it also indicates some of the more general problems associated with the solutions, in particular, the question of whether or not all three equilibrium states identified are 'stable' and capable of persisting for long periods of time. Many non-linear systems, even ones which are far simpler than the climate system, have a characteristic behaviour termed almost intransitivity. This behaviour is illustrated in Figure 2.6. If two different initial states of a system evolve to a single resultant state as time passes, the system is termed a transitive system. State A for this transitive system would then be considered the solution or normal state, and all perturbed situations would be expected to evolve to it. At the other extreme, an intransitive system has at least two equally acceptable solution states (A and B), depending on the initial state.

Difficulty arises when a system exhibits behaviour which mimics transitivity for some time, then flips to the alternative state for another (variable) length of time and then flips back again to the initial state and so on. In such an almost intransitive system it is impossible to determine which is the 'normal state', since either of the two states can continue for a long period of time, to be followed by a quite rapid and perhaps unpredictable change to the other. At present, geological and historical data are not detailed enough to determine for certain which of these system types is typical of the Earth's climate. In the case

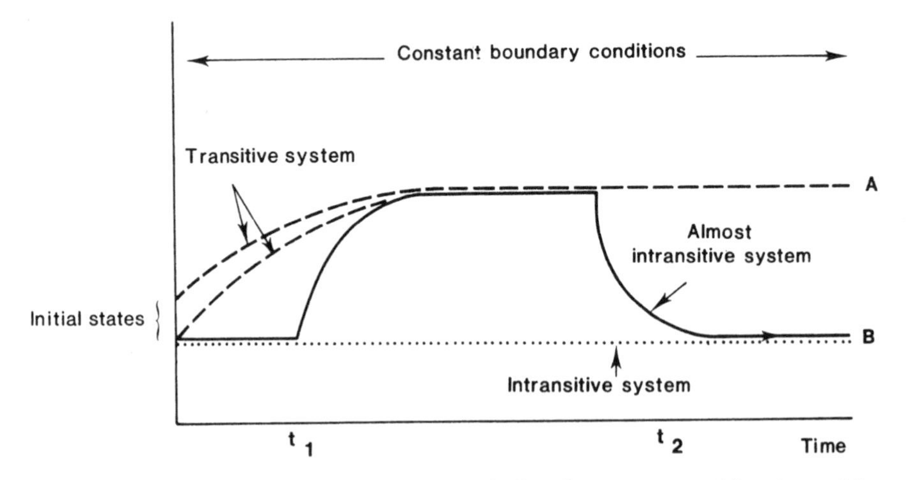

Figure 2.6 The behaviour of the three types of climatic system: transitive, intransitive and almost intransitive with respect to an initial state. In a transitive system two different initial states evolve to the same resultant state, A. An intransitive system exhibits the 'opposite' behaviour with more than one alternative resultant state. The characteristic of an almost intransitive system is that it mimics transitive behaviour for an indeterminate length of time and then 'flips' to an alternative resultant state (reproduced by permission of US National Academy of Sciences, 1975)

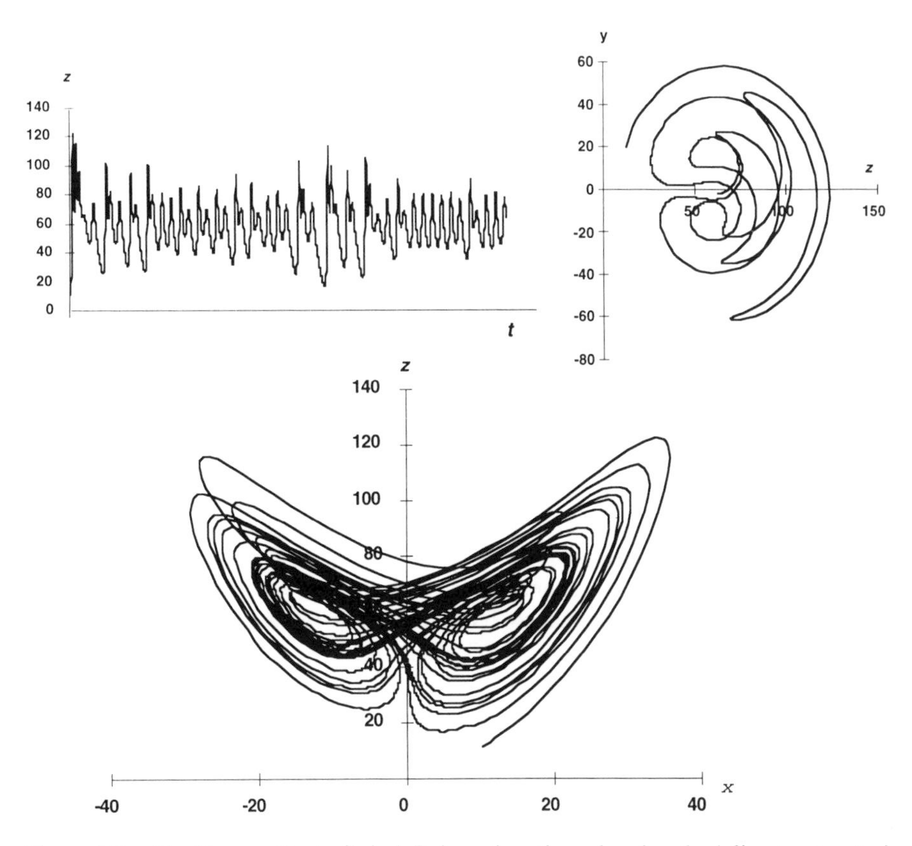

Figure 2.7 The 'Lorenz Butterfly'. A Poincaré section, showing the 'climate attractor' for the simple climate model constructed by Edward Lorenz in the 1960s. The system is characterized by three variables, which pin-point the state of the system in a three-dimensional space. The apparently disordered behaviour of the system indicated in the graph in the top left conceals the structure which is apparent when the system is examined in three dimensions. Since the system never repeats itself exactly, the track never crosses itself

of the Earth, the alternative climate need not be so catastrophic as a complete glaciation or the cessation of all deep ocean circulation. It is easy to see that, should the climate turn out to be almost intransitive, successful climate modelling will be extremely difficult. Current studies of the climate as a chaotic system have focused on determining the characteristics of a climate attractor. The behaviour of the simple model of Edward Lorenz (Figure 2.7) has been used as an example of such an attractor, but no definitive conclusions have been reached on the nature of this attractor (if it exists) and still no clear statements can be made regarding the transitivity of the climate system.

Stability of model results

Great care must be taken in choosing the constants for any parameterization scheme in any model. If values have been determined solely from empirical evidence it may be that they are appropriate only for the present day, with the result that the model is likely to be constrained to predict the present-day situation and is less likely to be able to respond realistically to perturbations. For 'external stability' we can test the response of the model to perturbations in the solar constant, since this is a convenient method of exploring climate model structure. Figure 2.8 shows the way in which $\bar{T}$ changes as the total incident radiation, μS, changes. Reduction of the solar constant to some critical value ($\mu_c S$) means that the number of solutions is reduced from two to one. Below $\mu_c S$, no solution is possible. This point is termed the bifurcation point. For values of incoming radiation, μS, less than $\mu_c S$, temperatures are so low that the albedo, $\alpha(T, \phi)$, becomes very close to or equal to 1 and thus it is impossible to regain energy

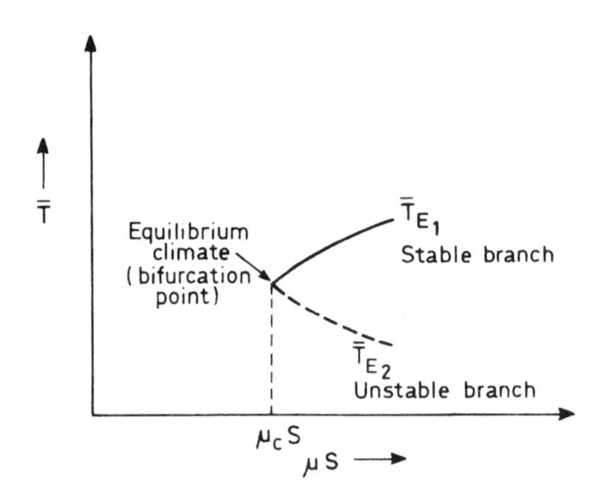

Figure 2.8 The equilibrium climate bifurcation point. For values of the solar luminosity given by μS where μ is a fractional premultiplier of the solar constant S, such that $\mu S > \mu_c S$, there are two solutions, whereas below this critical value no solutions exist. Changes in solar radiation lead to either a stable or an unstable equilibrium climate, illustrated here by the two equilibrium branches

balance. However, if some limit is put on how high the albedo may become, as is usually the case, e.g. $\alpha \leqslant 0.75$, the solution becomes what might be described as an ice-covered Earth. 'Internal stability' concerns the response of each branch in Figure 2.8 to perturbations from equilibrium which are created by internal factors. To determine if temperatures will return to equilibrium after the perturbation, we can use a time-dependent formulation and postulate a new value for which is close to the equilibrium one already calculated at that level of μS. This change can be computed iteratively until it is determined whether the values do regain the original $\bar{T}$ solution. If it is regained, then the solution is said to be 'internally stable'. In the case shown in Figure 2.8, only the top branch is stable because the model preserves $\bar{T}$ as proportional to μS. Using this method it is possible to determine whether the model is transitive or intransitive. The identification of almost intransitivity is not possible in this manner.

2.5 PARAMETERIZATION OF CLIMATIC PROCESSES

The climate system is a physical/chemical/biological system possessing infinite degrees of freedom. Any attempt to model such a highly complex system is fraught with dangers. It is (unfortunately) necessary to represent a distinct part, or more usually many distinct parts, of the complete system by inaccurate or semi-empirical mathematical expressions. Worse still is the need to neglect completely many parts of the complete and highly complex system. This process of neglect/semi-empirical or inaccurate representation is termed parameterization. Parameterization can take many forms. The simplest form is the null parameterization where a process, or a group of processes, is ignored. The decision to neglect these can only be made after a detailed consideration of their importance relative to other processes being modelled. Unnecessary computing time should not be spent on processes that can be adequately represented in some simpler way, or on processes that have relatively little effect on the climate at these scales.

Climatological specification is a form of parameterization which has been widely used in most types of model. In the 1970s, it was not uncommon to specify oceanic temperatures (with a seasonal variation), and in some of these models the clouds were also specified. In climate sensitivity experiments, it is important to recognize all such prescriptions because feedback features of the climate system have been suppressed. Only slightly less hazardous is the procedure by which processes are parameterized by relating them with reference to present-day observations: the constants or functions describing the relationship between variables are 'tuned' to obtain agreement. It is important that physically unrelated processes are not tuned together by this method. For example, the association of gradients in two different variables need not mean that the two are physically related. At best, this procedure presumes that constants and relationships appropriate to today's climate will still be applicable should some aspect of the climate alter.

The most advanced parameterizations have a theoretical justification. For instance, in some two-dimensional zonally averaged dynamical models the fluxes of heat and momentum are parameterized via baroclinic theory (in which the eddy fluxes are related to the latitudinal temperature gradient). The parameterization of radiative transfer in clear skies is another example. All that needs to be known is the vertical variation of temperature and humidity. Unfortunately, these parameterizations can lead to problems of uneven weighting because another process of equal importance cannot be adequately treated. In the case of heat and momentum transport by eddies, the contribution to these fluxes from stationary waves forced primarily by the orography and the land/ocean thermal contrast cannot be so easily considered. In radiation schemes, the parameterization of cloudy sky processes is not as advanced as that for clear skies.

Interactions in the climate system

The interactions between processes in any model of the climate are crucially important. Wiring diagrams which show all these interactions are often used to illustrate the complexity of incorporating them all adequately. A most important concept in climate modelling is that the relative importance of processes and the way that different processes interlink is a strong function of the time-scale being modelled. The whole concept of parameterization is subsumed by this assertion. Establishing whether a system is likely to be sensitive to the parameterization used for a particular process often depends upon the response time of that feature as compared with other 'interactive' features. It is pointless to invoke a highly complex, or exceedingly simplistic, parameterization if it has been constructed for a time-scale different from that of the other processes and linkages in the model. The adage 'choosing horses for courses' is fundamental to the art of climate modelling.

As the climate system depends upon scales of motion and interactions ranging from molecular to planetary dimensions, and from time-scales of nanoseconds to geological eras, parameterizations are a necessary part of the modelling process. In particular, a decision is generally made very early in model construction about the range of space- and time-scales which will be modelled explicitly. Figure 2.9 illustrates the difficulty faced by all climate modellers. The constraints of computer time and costs and data availability restrict the prognostic (or predictive) mode. Outside this range there are 'frozen' boundary conditions and 'random variability'. Thus the two examples shown in Figure 2.9 illustrate the range of prognostic computations for (i) a medium-range weather forecast model, and (ii) an EBM focused upon examining the effect of Milankovitch variations on the climate. In both cases, longer time-scales than those of concern to the modeller are considered as invariant, and shorter time-scales are neglected as being random fluctuations, the details of which are of too short a period to be of interest.

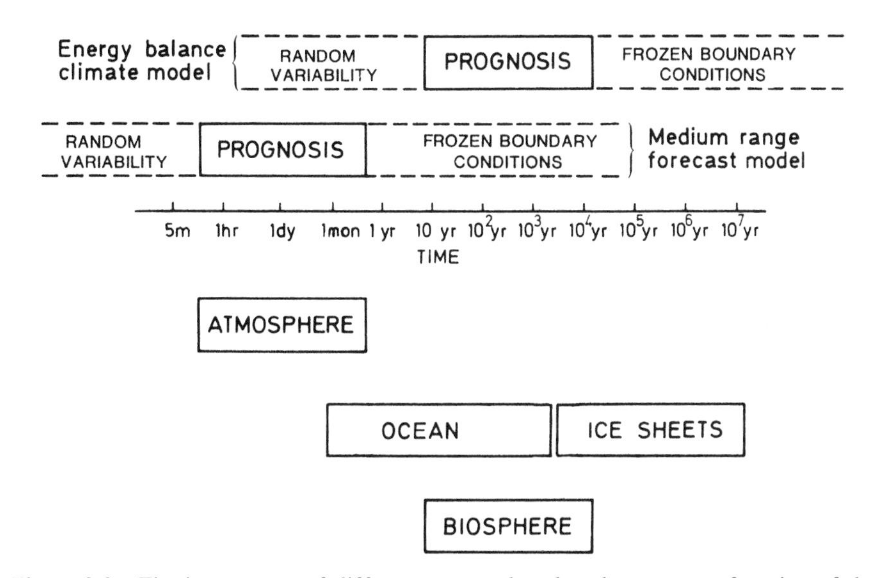

Figure 2.9 The importance of different temporal scales changes as a function of the type of model. The domain in which the model simulates the behaviour of the system is labelled 'prognosis'. It is expected that processes which fluctuate very rapidly compared with the prognostic timescales will contribute only small random variability to the model predictions, while processes which fluctuate very slowly compared with the prognostic timescale can be assumed to be constant. Two types of model are shown: an EBM and a medium-range weather forecast model

Parameterizations must be mutually consistent. For instance, if two processes produce feedback effects of opposite sign, it is important that one process is not considered in the other's absence. An example is the effect that clouds have on the radiative heating of the atmosphere. Longwave radiation causes a comparatively rapid cooling at the cloud top, whereas the absorption of solar radiation results in heating. To consider the effect of clouds on only one of the two radiation fields may be worse than neglecting the effect of clouds entirely.

Figure 2.10 portrays an hierarchical averaging scheme for the climate system. The averaging processes are described in terms of a single variable, which could be as simple a component of the climate system as temperature, but could alternatively be, for example, representative of the carbon budget. There are two averaging subsystems in the lower part of the diagram, the one on the right-hand side being based on an initial averaging of the mean state in the vertical, followed by zonal and/or meridional averaging, while the one on the left-hand side is averaged first around latitude zones.

A traditional view of the averaging diagram in Figure 2.10 would be that the simplest approximations to the climate system (models) lie at the bottom of the diagram (cf. the base of the climate modelling pyramid: Figure 2.1), with

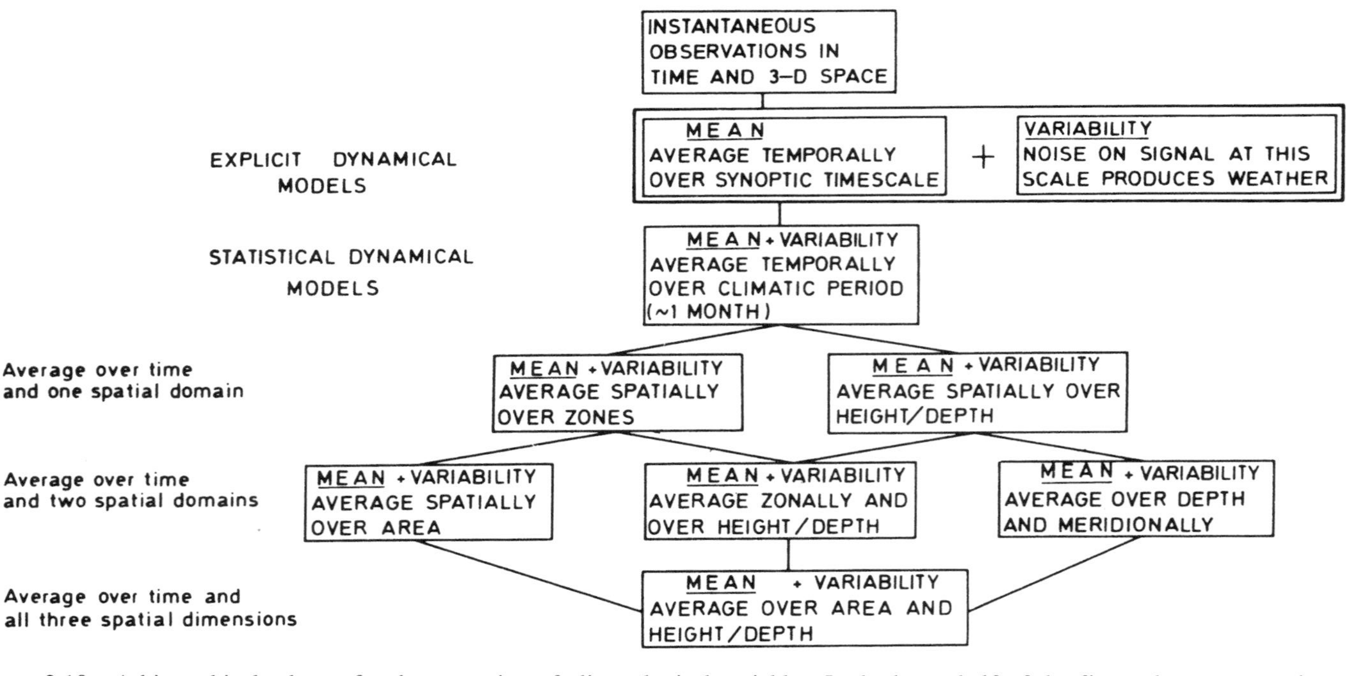

Figure 2.10 A hierarchical scheme for the averaging of climatological variables. In the lower half of the figure the representations of the climate system on the right-hand side involve averaging first over the atmospheric column, whereas the representations on the left-hand side involve zonal averaging first (adapted from Saltzman, 1978)

increasing resolution being synonymous with increasing (and perhaps more desirable) complexity on ascent through the diagram. The apex of this diagram would presumably be that radiative and diffusive processes would be described at the molecular level in GCMs. Clearly such an ultimate goal is absurd, although it sometimes seems to be consistent with the desire for increasing complexity in a few GCM modelling groups. An alternative view might be that some of the more sophisticated lower-resolution SD models might contain the maximum information currently available/verifiable for very long-term integration periods. These would, therefore, be adequate and appropriate models since the climate system over long time-scales would be deemed to be insensitive to higher-resolution features. Thus the key element in any model is the method of parameterization, whereby processes that cannot be treated explicitly are instead related to variables that are considered directly in the model. An example is in EBMs where only the surface temperature is calculated explicitly. Since poleward transport of heat by atmospheric motions is important, this transport has to be parameterized in some way relating to the surface temperature, such as the latitudinal temperature gradient. In GCMs, those processes which operate on scales too small to be resolved by the model (sub-gridscale processes), like convective clouds, can, and do, exert influence on the atmosphere and must be parameterized in terms of available model variables.

The need for observations

All climate models need observed values for part of their input especially in order to specify the boundary conditions, and all require observational data with which to compare their results. Some variables, such as surface pressure, are available worldwide and pose only the problem of evaluating the accuracy of the observed data set. Others, however, are sparse in either time or space. Knowledge of sea-ice extent is largely dependent on satellite observations, so that there is only a short observational record. Thus it is difficult to compare observations with any long-term average values obtained from models. Similarly, global coverage for cloud observations is relatively recent, so that it is difficult even to initialize models with the present-day cloud distribution. Certainly, the improved cloud climatologies which are being developed as more satellite observations become available are needed before more complex models can be evaluated.

2.6 SIMULATION OF THE FULL, INTERACTING CLIMATE SYSTEM: ONE GOAL OF MODELLING

Despite their limitations, coupled OAGCMs (cf. Section 2.2.4) represent the most complete type of climate model currently available. They illustrate the tremendous advances in our understanding of the atmosphere and our ability to model it over the thirty or so years since the first numerical climate models were produced. They do not yet, however, incorporate all aspects of the climate system and are therefore not at the absolute apex of the pyramid in Figure 2.1. They can provide a great deal of information about the present climate and the possible effects of future perturbations. That these predictions are often contradictory is inevitable, given our relatively poor knowledge of present conditions and incomplete understanding of the controlling processes. If a model is developed on sound theoretical principles, incorporates rational, and balanced, parameterization schemes, accounts for the major processes acting in the climate system and has been adequately tested against the present conditions, its results should be treated with respect. The results provide at least an indication of the possible future climate conditions created by a perturbation in the forces which control our present climate. The rest of the book is structured so that the concepts upon which full three-dimensional models are based are introduced sequentially. Chapter 3 underlines the fundamental basis of climate modelling: the energy balance. Chapter 4 describes models which operate in only one or two dimensions and assist in providing insight into the operation of the full climate system.

The overt goal of the text is therefore clear: we are aiming towards Chapters 5 and 6 in which GCMs, including coupled atmosphere–ocean models, are explained and the process of evaluating and using climate model results is described. The other equally valid and important goal is less obvious. Throughout the book we have tried to choose examples to illustrate and enhance understanding of the mechanisms controlling the climate, their complexities, time- and space-scales and interactions. Both goals are worthy of considerable effort.

RECOMMENDED READING

Adem, J. (1965) Experiments aiming at monthly and seasonal numerical weather prediction. *Monthly Weather Review* **93**, 495–503.

Adem, J. (1979) Low resolution thermodynamic grid models. *Dynamics of Atmosphere and Oceans* **3**, 433–451.

Bourke, W., McAvaney, B., Purl, K. and Thurling, R. (1977) Global modelling of atmospheric flow by spectral methods. *In* J. Chang (ed.), *Methods in Computational Physics, Vol. 17*. Academic Press, New York, pp. 267–324.

Bryan, K. (1969) A numerical method for the study of the world ocean. *Journal of Computational Physics* **4**, 347–376.

Budyko, M.I. (1969) The effect of solar radiation variations on the climate of the Earth. *Tellus* **21**, 611–619.

Garcia, R.R., Stordal, F., Solomon, S. and Kiehl, J.T. (1992) A new numerical model of the middle atmosphere. 1. Dynamics and transport of tropospheric source gases. *Journal of Geophysical Research* **97**, 12967–12991.

Gates, W.L. (1979) The effect of the ocean on the atmospheric general circulation. *Dynamics of Atmosphere and Oceans* **3**, 95–109.

Green, J.S.A. (1970) Transfer properties of the large scale eddies and the general circulation of the atmosphere. *Quarterly Journal of the Royal Meteorological Society* **96**, 157–185.

Hansen, J.E., Johnson, D., Lacis, A.A., Lebedeff, S., Lee, P., Rind, D. and Russell, G. (1981) Climate impact of increasing atmospheric CO_2. *Science* **213**, 957–1001.
Hasselmann, K. (1976) Stochastic climate models. Part 1. Theory, *Tellus* **28**, 473–485.
Held, I.M. and Suarez, M.J. (1978) A two level primitive equation model designed for climate sensitivity experiments. *Journal of the Atmospheric Sciences* **35**, 206–229.
MacKay, R.M. and Khalil, M.A.K. (1994) Climate simulations using the GCRC 2-D zonally averaged statistical dynamical climate model. *Chemosphere* **29**, 2651–2683.
Manabe, S. and Bryan, K. (1969) Climate calculations with a combined ocean atmosphere model. *Journal of the Atmospheric Sciences* **26**, 786–789.
Manabe, S. and Möller, F. (1961) On the radiative equilibrium and heat balance of the atmosphere. *Monthly Weather Review* **89**, 503–532.
Manabe, S. and Strickler, R.F. (1964) Thermal equilibrium of the atmosphere with a convective adjustment. *Journal of the Atmospheric Sciences* **21**, 361–385.
Potter, G.L., Ellsaesser, H.W., MacCracken, M.C. and Mitchell, C.S. (1981) Climate change and cloud feedback: The possible radiative effects of latitudinal redistribution. *Journal of the Atmospheric Sciences* **38**, 489–493.
Saltzman, B. (1978) A survey of statistical dynamical models of terrestrial climate. *Advances in Geophysics* **20**, 183–304.
Semtner, A.J. (1995) Modelling ocean circulation. *Science* **269**, 1379–1385.
Smith, N.R. (1993) Ocean modelling in a global observing system. *Reviews of Geophysics* **31**, 281–317.
Shine, K.P. and Henderson-Sellers, A. (1983) Modelling climate and the nature of climate models: A review, *Journal of Climatology* **3**, 81–94.
Smagorinsky, J. (1983) The beginnings of numerical weather prediction and general circulation modeling: Early recollections. *In* B. Saltzman (ed.) *Theory of Climate.* Academic Press, New York, pp. 3–38.
Stone, P.H. (1973) The effects of large scale eddies on climatic change. *Journal of Atmospheric Science* **30**, 521–529.
Thompson, S.L. and Schneider, S.H. (1979) A seasonal zonal energy balance climate model with an interactive lower layer. *Journal of Geophysical Research* **84**, 2401–2414.
Washington, W.M., Semtner, A.J. Jr, Meehl, G.A., Knight, D.J. and Meyer, T.A. (1980) A general circulation experiment with a coupled atmosphere, ocean and sea ice model. *Journal of Physical Oceanography* **10,** 1887–1908.

CHAPTER 3

Energy Balance Models

The more it snows
(Tiddely pom),
The more it goes
(Tiddely pom),
The more it goes
(Tiddely pom),
On snowing ...'

From *The House at Pooh Corner,* by A.A. Milne (1928).
Reproduced by permission of Methuen Children's Books,
McClelland and Stewart, Toronto, and E.P. Dutton, a
division of NAL Penguin Inc.

3.1 BALANCING THE PLANETARY RADIATION BUDGET

There is an excellent book by Abbott which describes a world called 'Flatland' which is inhabited by two-dimensional beings and, finally, visited by a strange three-dimensional object: a sphere. The sphere passes through Flatland and is perceived by the inhabitants as being only a series of discs of changing radius. This glimpse of the three-dimensional 'reality' is impossible for most Flatlanders to comprehend. Climate modellers, on the other hand, are only too painfully aware of the multi-dimensional nature of the climate system. Those who design and work with one- and two-dimensional models are not uncomprehending of the missing dimensions, but have chosen to use a simpler model type. They have two main reasons: (i) these models are simpler and therefore cheaper to integrate on computers and thus can be used for much longer or very many more integrations than full three-dimensional GCMs; (ii) being simpler, the models therefore represent particular features of the climate system more simply because other confusing features are removed. Thus modellers, unlike Flatlanders, recognize complexity and intentionally seek to reduce it. In this chapter we explore some of their reasons and results.

Balancing the planetary radiation budget offers a first, simple approximation to a model of the Earth's climate. The radiation fluxes and the equator-to-pole energy transport are the fundamental processes of the climate system which are incorporated in EBMs. Originally, interest was stimulated by the independent results of Budyko and of Sellers in 1969. While many of the questions which these studies raised have since been answered, these models remain interesting tools for studying climate. This chapter describes how EBMs are constructed and outlines how these models have been used both to study and to illustrate characteristic components of the climate system.

3.2 THE STRUCTURE OF ENERGY BALANCE MODELS

The simplest method of considering the climate system of the Earth, and indeed of any planet, is in terms of its global energy balance. Viewing the Earth from outside, one observes an amount of radiation input which is balanced (in the long term) by an amount of radiation output. Since over 70% of the energy which drives the climate system is first absorbed at the surface, the surface albedo will be predominant in controlling energy input to the climate system. The output of energy will be controlled by the temperature of the Earth but also by the transparency of the atmosphere to this outgoing thermal radiation. An EBM can take two very simple forms. The first form, the zero-dimensional model, considers the Earth as a single point in space having a global mean effective temperature, T_e. The second form of the EBM considers the temperature as being latitudinally resolved. Figure 3.1 illustrates these two approaches.

3.2.1 Zero-dimensional EBMs

In the first case the climate can be simulated by considering the radiation balance. The total energy received from the Sun per unit time is $\pi R^2 S$ where R is the radius of the Earth. The total area of the Earth is, however, $4\pi R^2$. Therefore the time-averaged energy input rate is $S/4$ over the whole Earth. Hence,

$$(1 - \alpha)S/4 = \sigma T_e^4 \tag{3.1}$$

where α is the planetary or system albedo, S is the solar constant (1370 W m^{-2}) and σ is the Stefan–Boltzmann constant. If the atmosphere of the planet contains gases which absorb thermal radiation, then the surface temperature, T_s, will be greater than the effective temperature, T_e. The increment ΔT is known as the *greenhouse increment* and depends upon the efficiency of the infrared absorption. Thus the surface temperature can be calculated if ΔT is known, since

$$T_s = T_e + \Delta T \tag{3.2}$$

For the Earth, the greenhouse increment due to the present atmosphere is about $\Delta T = 33$ K, and hence combining equations (3.1) and (3.2) gives, for $\alpha = 0.3$, $T_e = 288$ K. (Note that the only prognostic variable in an EBM is the temperature, characterized as a surface temperature.)

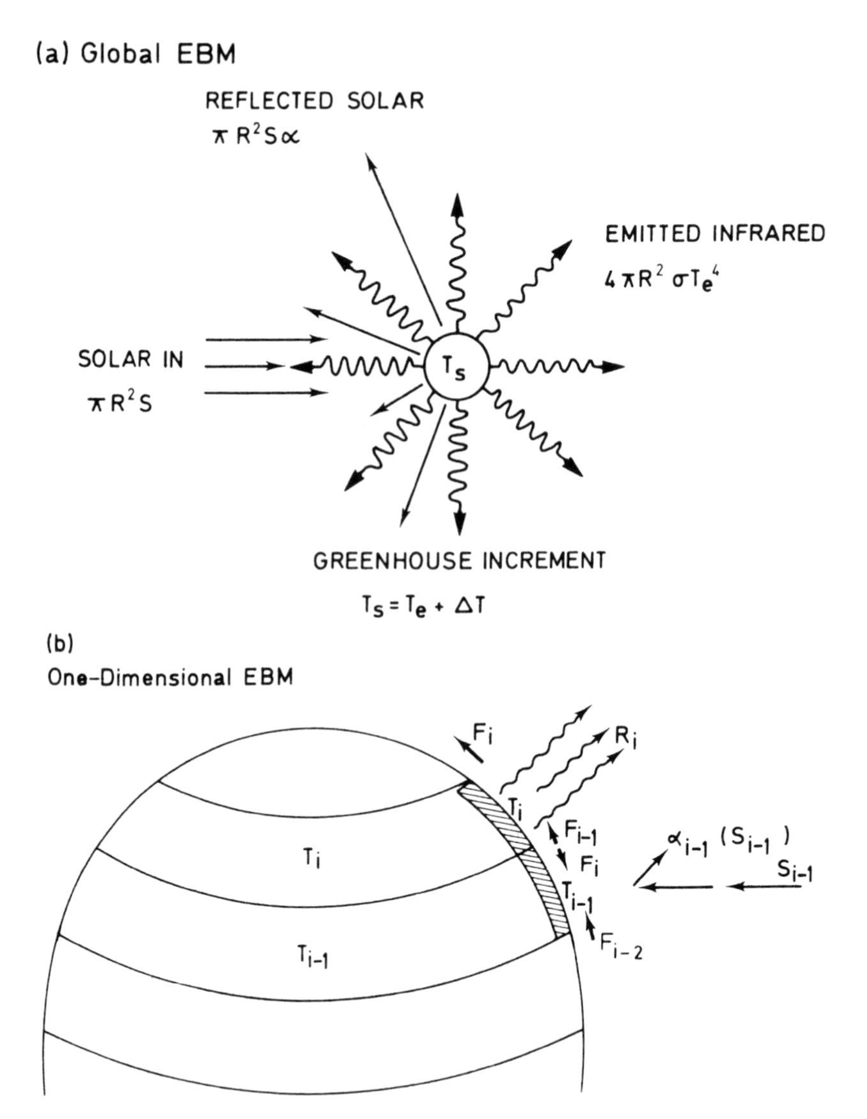

Figure 3.1 Energy transfers in (a) a global EBM, and (b) a full zonal EBM

If the planetary features were different, for example if the solar luminosity were $S = 2619\ \mathrm{W\,m^{-2}}$ and $\alpha = 0.7$, then $T_e = 242$ K. These are the values appropriate to the planet Venus which, while being closer to the Sun and hence enjoying greater incident solar radiation, is almost completely cloud-covered and thus has a very high planetary albedo. The albedo dominates the radiation balance, resulting in an effective temperature which is slightly lower than that of the Earth. However, the atmosphere of Venus is extremely dense and composed almost entirely of carbon dioxide. Hence there is a very much greater greenhouse effect on Venus. The surface temperature of Venus has been found by spacecraft to be ~730 K and, although it is now believed that not the whole of this ΔT is due to greenhouse absorbers, they certainly contribute substantially. The other major contributor to surface heating is adiabatic warming in large regions of descending air (not included in EBMs).

In a simple EBM, the incoming and outgoing energy for the globe are balanced and a single climatic variable (the surface temperature, T_s) is calculated, i.e. T is the dependent variable for which the 'climate equations' are solved. The rate of change of temperature, T, with time, t, is caused by a difference between the top-of-the-atmosphere (or planetary) net incoming, $R\downarrow$, and net outgoing, $R\uparrow$, radiative fluxes (per unit area).

$$mc\,\Delta T/\Delta t = (R\downarrow - R\uparrow)A_E \tag{3.3}$$

where A_E is the area of the Earth, c is the specific heat capacity of the system and m is the mass of the system.

This is a very general equation with a variety of uses. If, for example, the system we wish to model is an outdoor swimming pool, we can calculate the rate of temperature change in timesteps of 1 day from equation (3.3). Suppose the pool has surface dimensions 30×10 m, is well mixed and is 2 m deep. Since 4200 J of energy are needed to raise the temperature of 1 kg of water by 1 K (4200 $\mathrm{J\,kg^{-1}\,K^{-1}}$ is the specific heat of water), and 1 $\mathrm{m^3}$ of water has a mass of 1000 kg, the pool has a total heat capacity equal to $2.52 \times 10^9\ \mathrm{J\,K^{-1}}$. If we assume that the difference between the absorbed radiation and the emitted radiation from the pool ($R\downarrow - R\uparrow$) is 20 $\mathrm{W\,m^{-2}}$ for 24 h, then the difference in energy content of the pool for each 24-h timestep is $20 \times 30 \times 10 \times 24 \times 60 \times 60$ J. Then, from equation (3.3)

$$2.52 \times 10^9\,\Delta T = 20 \times 30 \times 10 \times 24 \times 60 \times 60$$

$$\Delta T(\text{in 1 day}) = \frac{5.184 \times 10^8}{2.52 \times 10^9} \approx 0.2\ \mathrm{K} \tag{3.4}$$

Thus, at this rate, it would take about a month to raise the temperature of the pool water by 6 K.

On the Earth, the value of c is largely determined by the oceans. The specific heat ($\mathrm{J\,kg^{-1}\,K^{-1}}$) for water is around four times that for air, and the mass of the ocean is also much greater than that of the atmosphere. For instance, if we assume that the energy is absorbed in the first 70 m of the ocean (the average global depth of the top or mixed layer) and that approximately 70% of the Earth's surface is covered by oceans, then the value for C (the total heat capacity) comes from

$$C = \rho_w c_w d A_E 0.7 = 1.05 \times 10^{23}\ \mathrm{J\,K^{-1}} \tag{3.5}$$

where ρ_w is the density of water, c_w its specific heat capacity, d is the depth of the mixed layer and A_E is the Earth's surface area.

For our simple EBM of the Earth, the energy emitted, $R\uparrow$, can be estimated using the Stefan–Boltzmann law and the surface temperature, T. This value must be corrected to take into account the infrared transmissivity of the atmosphere τ_a, since $R\uparrow$ is the planetary flux. Therefore we can write

$$R\uparrow \approx \varepsilon\sigma T^4 \tau_a \tag{3.6}$$

The absorbed energy, $R\downarrow$, is a function of the solar flux, S, and the planetary albedo such that $R\downarrow = (1 - \alpha)S/4$. equation (3.3) therefore becomes

$$\frac{\Delta T}{\Delta t} = \frac{1}{C}\left\{\frac{S}{4}(1 - \alpha) - \varepsilon\tau_a\sigma T^4\right\} \tag{3.7}$$

This equation can be used to ascertain the equilibrium climatic state by setting $\Delta T/\Delta t = 0$. This use is complementary to the timestep mode described above. The result represents an 'ultimate' or equilibrium solution of the equation when the change in temperature has ceased. In this case

$$(1 - \alpha)\frac{S}{4} = \varepsilon\tau_a\sigma T^4 \tag{3.8}$$

Using values of $S = 1370\ \mathrm{W\,m^{-2}}$, $\alpha = 0.3$, $\varepsilon\tau_a = 0.62$ and $\sigma = 5.67 \times 10^{-8}\ \mathrm{W\,m^{-2}\,K^{-4}}$ gives rise to a surface temperature of 287 K, which is in good agreement with the globally averaged surface temperature today.

An alternative use of equation (3.7) is similar to the calculation of the swimming pool warming rate made above. Here a timestep calculation of the change in T is made. This could be a response to an 'external' forcing agent, such as a change in solar flux or in the heat capacity of the oceans resulting from changes in their depth or area. Alternatively, the response could be determined by an 'interactive' climate calculation when one of the internal variables (e.g. α) alters.

3.2.2 One-dimensional EBMs

In the case where we consider each latitude zone independently,

$$S_i(1 - \alpha(T_i)) = R\uparrow(T_i) + F(T_i) \tag{3.9}$$

Note that we now have an additional term $F(T_i)$ which refers to the loss of energy by a latitude zone to its colder neighbour or neighbours. So far we have ignored any storage by the system since we have been considering the climate on time-scales where the net loss or gain of stored energy is small. Any stored energy would simply appear as an additional term, $Q(T_i)$ on the right-hand side of equation (3.9).

Since the zero-dimensional model (equation (3.8)) is a simplification of equation (3.9), further discussion will consider the latitudinally resolved model and look in detail at the role of the terms involved.

Each of the terms in equation (3.9) is a function of the predicted variable T_i. The surface albedo is influenced by temperature in that it is increased drastically when ice and snow are able to form. The radiation emitted to space is proportional to T^4, although over the temperature range of interest (~250–300 K), this dependence can be considered linear. The horizontal flux out of the zone is a function of the difference between the zonal temperature and the global mean temperature. The storage term, an attempt to account for the effect of the oceans, depends on the difference between the present temperature and a long-term mean temperature. The albedo is described by a simple step function such that

$$\alpha_i \equiv \alpha(T_i)\begin{cases} =0.6 & T_i \leqslant T_c \\ =0.3 & T_i > T_c \end{cases} \tag{3.10}$$

which represents the albedo increasing at the snowline; T_c the temperature at the snowline, is typically between -10 and 0°C. Because of the relatively small range of temperatures involved, radiation leaving the top of the latitude zone can be approximated by

$$R_i \equiv R\!\uparrow\!(T_i) = A + BT_i \tag{3.11}$$

where A and B are empirically determined constants designed to account for the greenhouse effect of clouds, water vapour and CO_2. The rate of transport of energy can be represented as being proportional to the difference between the zonal temperature and the global mean temperature by

$$F_i \equiv F(T_i) = K(T_i - \overline{T}) \tag{3.12}$$

where K is an empirical constant.

Incorporation of equations (3.11) and (3.12) into equation (3.9) forms an equation which can be rearranged to give the equation

$$T_i = \frac{S_i(1 - \alpha_i) + K\overline{T} - A}{B + K} \tag{3.13}$$

Given a first-guess temperature distribution and by devising an appropriate weighting scheme to distribute the solar radiation over the globe (because of the tilt of the Earth's axis, a simple cosine distribution with latitude does not work in the annual mean), successive applications of this equation will eventually yield an equilibrium solution. An alternative course of action is to calculate the time evolution of the model climate explicitly by including a term representing the thermal capacity of the system. The former method results in computationally faster results, but the latter allows for more experimentation. Such models are relatively simple to construct on a personal computer, either in an elementary programming language, as is illustrated in Section 3.4, or in a spreadsheet, as is demonstrated on the CD.

3.3 PARAMETERIZING THE CLIMATE SYSTEM FOR ENERGY BALANCE MODELS

The model described above illustrates the basic principles of energy balance climate modelling. In this section we shall consider further each of the parameterization schemes and how they are developed.

As mentioned in Chapter 2, the first EBMs were found to be alarmingly sensitive to changes in the solar constant. Small reductions in solar constant appeared to cause catastrophic and irreversible glaciations of the entire planet. Such an effect, although extreme, suggests that such models might be utilized in studying large-scale glaciation cycles. This is indeed the case, but some preparation and background work on the mechanisms in the model need to be undertaken before glaciation cycles can be simulated.

Albedo

The albedo parameterization in EBMs is based simply on the surface albedo being greater when the temperature is low enough to allow snow and ice formation. Two simple parameterizations are that the albedo increases instantaneously to an ice-covered value (equation (3.10)), and a description, which might seem more appropriate, that the albedo increases linearly over a temperature interval within which the Earth can be said to be becoming increasingly snow-covered.

$$\begin{aligned} \alpha(T_i) &= b(\phi) - 0.009T_i \qquad && T_i < 283 \text{ K} \\ \alpha(T_i) &= b(\phi) - 0.009 \times 283 \qquad && T_i \geqslant 283 \text{ K} \end{aligned} \qquad (3.14)$$

Using empirical constants, $b(\phi)$, allows for the inclusion of a latitudinal variation of ice-free albedo which is not affected by temperature. The change in planetary albedo at the poles can then be made to be around half of that at the equator when the ice-free surface is replaced with an ice-covered one. This allows for the higher albedo of the ice-free ocean and enhanced atmospheric scattering which occurs at the low solar elevations near the poles. The sensitivity is reduced by a factor of two, but remains too high to explain a paradox termed the 'faint Sun–warm early Earth paradox'. This conundrum stems from the inference that, although the solar luminosity was only about 70% of its present value during the first aeon of the Earth's history, the surface of the Earth seems not to have been glaciated to the extent which would be suggested by these EBM calculations (i.e. during the period from 3.5 to 4.5 thousand million years ago there is little evidence for widespread glaciation).

The explanation for the apparent gross instability of the Earth atmosphere system to small perturbations in solar constant lies in the close coupling in the parameterizations of the temperature and planetary albedo. This tight relationship is, perhaps, not a good representation of the real system since, although the surface albedo is strongly influenced by temperature, the planetary albedo is affected by the presence of clouds and is also a function of latitude. For example, as latitude increases, the effect on the planetary albedo of adding more snow or ice tends to decrease.

The fundamental flaw in albedo parameterization is the assumption of a very strong connection between the planetary albedo and the surface albedo. Clouds are responsible for the reflection of 70–80% of the radiation which is reflected by the Earth. There is no clear relationship between surface temperature and cloudiness which further reduces the connection between surface temperature and planetary albedo. In our parameterization of the albedo described above, by considering only the effect of ice and snow cover, it would appear at first glance that clouds have been ignored in the formulation of EBMs. This is acceptable because the effect of an increase in cloudiness on the amount of absorbed solar radiation is almost exactly countered by the effect of clouds in retaining a greater proportion of emitted infrared radiation.

Outgoing infrared radiation

The Earth is constantly emitting radiation. Some of the radiation which is emitted is absorbed by the atmosphere and re-emitted back to the ground. Parameterizations will involve some method of accounting for this greenhouse effect. One formulation is to match outgoing longwave radiation to surface temperature and to devise a linear relationship between the two. This was the method which was included in equation (3.11). An alternative formulation is to modify the black body flux by some factor which accounts for the amount re-emitted downwards by the atmosphere, e.g.

$$R_i = \sigma T_i^4[1 - m_i \tanh(19T_i^6 \times 10^{-16})] \tag{3.15}$$

where m_i is the factor representing the atmospheric opacity. This formulation was derived empirically by Sellers. All parameterizations of infrared radiation in EBMs follow one or other of these structures.

Heat transport

The simplest form of heat transport which may be incorporated into an EBM is that of equation (3.12). Here the flux out of a latitude zone is equal to some constant multiplied by the difference between the average temperature of the zone and the global mean temperature. A more complex method is to consider each of the transporting mechanisms separately, with the flux divergence being given by

$$\mathrm{div}(F) = \frac{1}{\cos\phi}\frac{\partial}{\partial y}[\cos\phi(F_{\mathrm{o}} + F_{\mathrm{A}} + F_{\mathrm{q}})] \quad (3.16)$$

the three terms on the right representing transports due to ocean, atmosphere and latent heat, respectively:

$$F_{\mathrm{o}} = -K_{\mathrm{o}}\frac{\partial T}{\partial y}$$

$$F_{\mathrm{A}} = -K_{\mathrm{A}}\frac{\partial T}{\partial y} + \langle v\rangle T$$

$$F_{\mathrm{q}} = -K_{\mathrm{q}}\frac{\partial q(T)}{\partial y} + \langle v\rangle qT \quad (3.17)$$

where K_{o}, K_{A} and K_{q} are all functions of latitude, ϕ, $q(T)$ is the water vapour mixing ratio, $\langle v\rangle$ is the zonally averaged wind speed and y is the distance in the poleward direction.

More realistic parameterizations might be expected to be more complicated. There are the two basically different methods of incorporating the div(F) term: the Newtonian form developed by Budyko (equation (3.12)) and the eddy diffusive mixing formula developed by Sellers (equation (3.16)). The choice is always to weigh the extra detail such as is offered by Sellers against the decreased computational time of Budyko's method.

3.4 A *BASIC* ENERGY BALANCE CLIMATE MODEL

This type of climate model is a useful teaching/learning tool. The program listed in Figure 3.2 was originally written for undergraduate use at the University of Liverpool in the mid-1980s. The program was designed to run on some of the earliest personal computers, and in its present form will be runnable on most modern computers capable of MS-DOS™ emulation.

The formulation of the EBM has been kept as simple as possible. The equations are those described in Section 3.2. The albedo parameterization is a simple 'on–off' step function based on a specified temperature threshold (see equation (3.10)). The emitted longwave radiation is a linear function of the zonal surface temperature (see equation (3.11)) and the transport term is given by equation (3.12). The following sections contain a brief summary of the model presented in Figure 3.2 and suggest some exercises which demonstrate the model's behaviour.

```
5  REM '''''''''''''''''''''''''''''''''''''''''''''''''''''''''''''''''''
10 REM Energy Balance Model 1986 K.McGuffie & A. Henderson-Sellers
12 REM The model is designed to run on an IBMPC and does.
14 REM Other machines may require some alterations. Here CLS=clear screen,
16 REM . is used for formatted output, LPRINT sends output to a
18 REM printer. The code is designed to be as transparent as possible.
19 REM Unclosed quotes in output statements may not work on some machines.
20 FOR I = 1 TO 9
30 READ LATZ$(I)
40 NEXT I
50 DATA "80-90","70-80","60-70","50-60","40-50","30-40","20-30","10-20"," 0-10"
60 E$=CHR$(27):CLS
70 C=3.8
80 Q=342.5
90 A=204
100 B=2.17
110 IN= 3.14159/18!
120 PIBY2=3.14159/2!
130 DEF FNR(X)=INT(100*X)/100
140 FOR LAT = 1 TO 9
150 READ S(LAT)
160 NEXT LAT
170 DATA 0.5,0.531,0.624,0.77,0.892
180 DATA 1.021,1.12,1.189,1.219
190 PRINT
200 PRINT
210 PRINT
220 PRINT "*******************************************************************
230 PRINT "                              ENERGY
240 PRINT "                              BALANCE
250 PRINT "                              MODEL
260 PRINT "*******************************************************************
270 PRINT
280 PRINT "          Copyright      A. Henderson-Sellers
290 PRINT "                          & K. McGuffie
300 PRINT "                                             (1986)
310 PRINT
320 PRINT "     This model is similar to those of Budyko and Sellers.
330 PRINT "     You will be offered the opportunity to alter the
340 PRINT "     values of the parameters which control the model climate.
350 PRINT "
360 PRINT
370 PRINT "
380 PRINT "               Press the space bar to continue
390 GOSUB 3120
400 CLS
410 PRINT
420 PRINT "          There are various possibilities for changing
430 GOSUB 3160
440 PRINT "          the model climate.     You can then test the
450 GOSUB 3160
460 PRINT "          sensitivity of this climate to changes in the
470 GOSUB 3160
480 PRINT "          solar constant.    That you should observe the
490 GOSUB 3160
500 PRINT "          changes due to your changing of the model
510 GOSUB 3160
520 PRINT "          parameters is also of importance in understanding
530 GOSUB 3160
540 PRINT "          the nature of this model.
550 PRINT
560 PRINT
570 PRINT "          Press space to continue "
580 GOSUB 3120
590 FOR I = 1 TO 9
600 READ ALAND(I)
610 NEXT I
620 DATA 0.5,0.5,0.452,0.407,0.357,0.309,0.272,0.248,0.254
630 AICE=.62
640 TCRIT=-10
650 C=3.8
660 Q=342.5
```

Figure 3.2 BASIC computer code for the energy balance climate model. The code was designed for an IBM PC-type computer

```
670 A=204
680 B=2.17
690 CLS
700 PRINT "                    * * * * * * * * * * * * * * * * * * * *
710 PRINT "                              M A I N   M E N U
720 PRINT "                    * * * * * * * * * * * * * * * * * * * *
730 PRINT
740 PRINT
750 PRINT "         There are 3 main parameterization schemes within the
760 PRINT "         model.  You may make alterations to any or all of them
770 PRINT "         at any one time.  Any which you choose not to alter
780 PRINT "         will be filled by default values.
790 PRINT
800 PRINT
810 PRINT "   Choice                         Parameterization
820 PRINT "-----------------------------------------------------------------
830 PRINT "    (1)                           ALBEDO
840 PRINT "    (2)                           LATITUDINAL TRANSPORT
850 PRINT "    (3)                           LONGWAVE RADIATION TO SPACE
860 PRINT
870 PRINT "    (4)                           RUN
880 PRINT
890 PRINT "   Enter the number of your choice
900 N$=INKEY$
910 IF N$="1" THEN GOTO 960
920 IF N$="2" THEN GOTO 1600
930 IF N$="3" THEN GOTO 1800
940 IF N$="4" THEN GOTO 2020
950 GOTO 900
960 CLS
970 PRINT "* * * * * * * * * * * * * * * * * * * * * * * * * * * * * * * * *
980 PRINT "                   P A R A M E T E R I Z A T I O N   O F
990 PRINT "                               A L B E D O
1000 PRINT "* * * * * * * * * * * * * * * * * * * * * * * * * * * * * * * *
1010 PRINT
1020 PRINT
1030 PRINT "         There are three things which you may alter
1040 PRINT
1050 PRINT "            1.    The temperature at which the surface "
1060 PRINT "                  becomes ice covered."
1070 PRINT
1080 PRINT "            2.    The albedo of this ice covered surface
1090 PRINT
1100 PRINT "            3.    The albedo of the underlying ground.
1110 PRINT
1120 PRINT "            4.    Return to main menu
1130 PRINT
1140 PRINT " Choose which one you want"
1150 AL$=INKEY$
1160 IF AL$="1" THEN GOTO 1210
1170 IF AL$="2" THEN GOTO 1280
1180 IF AL$="3" THEN GOTO 1350
1190 IF AL$="4" THEN GOTO 690
1200 GOTO 1150
1210 CLS
1220 PRINT
1230 PRINT
1240 PRINT "   The default value is -10 deg.C"
1250 PRINT
1260 INPUT "   What is the new value you want for TCRIT ? ",TCRIT
1270 GOTO 960
1280 CLS
1290 PRINT
1300 PRINT "   the albedo of the ice is currently 0.62."
1310 PRINT
1320 INPUT "   What is your new value for this albedo ? ",AICE
1330 IF AICE > .99 OR AICE < 0! THEN 1280
1340 GOTO 960
1350 CLS
1360 PRINT
1370 PRINT "         The albedos look like this from north to equator "
1380 PRINT
1390 PRINT "         (1)          80-90   ";ALAND(1)
```

Figure 3.2 *Continued*

```
1400 PRINT "          (2)          70-80   ";ALAND(2)
1410 PRINT "          (3)          60-70   ";ALAND(3)
1420 PRINT "          (4)          50-60   ";ALAND(4)
1430 PRINT "          (5)          40-50   ";ALAND(5)
1440 PRINT "          (6)          30-40   ";ALAND(6)
1450 PRINT "          (7)          20-30   ";ALAND(7)
1460 PRINT "          (8)          10-20   ";ALAND(8)
1470 PRINT "          (9)           0-10   ";ALAND(9)
1480 PRINT
1490 PRINT "  Which one do you want to alter ( zero for none of them )"
1500 PRINT
1510 INPUT "   Enter the number ",I
1520 IF I = 0 THEN GOTO 960
1530 PRINT
1540 PRINT "   The old value in band ",I," is ",ALAND(I)"."
1550 INPUT "   What is your new value ? ",ALAND(I)
1560 IF ALAND(I) >0! AND ALAND(I) < 1! GOTO 1350
1570 GOSUB 3200
1580 GOTO 1550
1590 PRINT
1600 CLS
1610 PRINT
1620 PRINT "* * * * * * * * * * * * * * * * * * * * * * * * * * * * * * * * * *
1630 PRINT "                        T R A N S P O R T
1640 PRINT "* * * * * * * * * * * * * * * * * * * * * * * * * * * * * * * * * *
1650 PRINT
1660 PRINT "      In this case you can alter the rate at which heat is
1670 PRINT "      transported around the model by varying the value of C
1680 PRINT "      in the following equation.
1690 PRINT
1700 PRINT "             Heat Flux = C x ( T(mean) - T(zone) )"
1710 PRINT
1720 PRINT "                  The current value is ",C
1730 PRINT
1740 INPUT "  What is the value you want to use ? ", C
1750 IF C >0 AND C<50 GOTO 690
1760 GOSUB 3200
1770 GOTO 1740
1780 GOTO 690
1790 PRINT
1800 CLS
1810 PRINT "* * * * * * * * * * * * * * * * * * * * * * * * * * * * * * * * * *
1820 PRINT "               L O N G W A V E  L O S S  T O  S P A C E
1830 PRINT "* * * * * * * * * * * * * * * * * * * * * * * * * * * * * * * * * *
1840 PRINT
1850 PRINT "         The longwave loss to space is determined by the
1860 PRINT "         following equation.
1870 PRINT
1880 PRINT "             R = A + B x T(zone) "
1890 PRINT
1900 PRINT "       Currently        A=";A
1910 PRINT "                        B=";B
1920 PRINT
1930 PRINT "  Enter  1 to change them 0 to keep them the same"
1940 R$=INKEY$
1950 IF R$="0" THEN GOTO 690
1960 IF R$="1" THEN GOTO 1980
1970 GOTO 1940
1980 PRINT
1990 INPUT " Enter new value for A";A
2000 INPUT " Enter new value for B";B
2010 GOTO 690
2020 CLS
2030 PRINT
2040 PRINT
2050 PRINT "    What fraction of the solar constant would you like ?"
2060 INPUT "    Your choice >",SX
2070 REM start of routine to calculate temperatures
2080 FOR LAT = 1 TO 9
2090 READ TSTART(LAT)
2100 NEXT LAT
2110 DATA -16.9,-12.3,-5.1,2.2,8.8,16.2,22.9,26.1,26.4
2120 F=1
```

Figure 3.2 *Continued*

```
2130 FOR LAT= 1 TO 9
2140 TEMP(LAT)=TSTART(LAT)
2150 NEXT LAT
2160 FOR H = 1 TO 50
2170 SOLCON=SX*1370!/4!
2180 REM Calculate albedo of zones
2190 LATICE=0
2200 FOR LAT = 1 TO 9
2210 NL=0
2220 ALBEDO(LAT)=ALAND(LAT)
2230 IF TEMP (LAT) > TCRIT THEN GOTO 2440
2240 ALBEDO(LAT) = AICE
2250 IF LAT = 9 THEN GOTO 2430
2260 IF TEMP(LAT+1)<=TCRIT THEN GOTO 2440
2270 DP=(-TCRIT+TEMP(LAT+1))*.1745/(TEMP(LAT+1)-TEMP(LAT))
2280 WORK2=(PIBY2-(LAT+.5)*IN)
2290 LATICE =WORK2+DP
2300 IF DP > .0872564 THEN GOTO 2370
2310 A3=PIBY2-(LAT+1)*IN
2320 A4=A3-IN
2330 A5=(SIN(A4)-SIN(LATICE))/(SIN(A4)-SIN(A3))
2340 NC=ALBEDO(LAT+1)*(1!-A5)+AICE*A5
2350 NL=LAT+1
2360 GOTO 2440
2370 A3=PIBY2-LAT*IN
2380 A4=PIBY2-(LAT-1)*IN
2390 A5= (SIN(LATICE)-SIN(A3))/(SIN(A4)-SIN(A3))
2400 NC=AICE-(AICE-ALBEDO(LAT))*A5
2410 NL=LAT
2420 GOTO 2440
2430 NL=0
2440 NEXT LAT
2450 IF ALBEDO(1) = ALAND(1) THEN NI = 90!/57.296
2460 REM CALCULATE MEAN TEMPERATURE
2470 SM=0
2480 FOR LAT = 1 TO 9
2490 WORK1=PIBY2-(LAT-1)*IN
2500 WORK2=WORK1-IN
2510 AC=ALBEDO(LAT)
2520 IF LAT=NL THEN AC=NC
2530 SM=SM+(SIN(WORK1)-SIN(WORK2))*AC*S(LAT)
2540 NEXT LAT
2550 TX=(SOLCON*(1-SM)-A)/B
2560 REM END OF MEAN TEMP ROUTINE
2570 FOR LAT = 1 TO 9
2580 TM(LAT)=TEMP(LAT)
2590 TEMP(LAT)=(SOLCON*S(LAT)*(1-ALBEDO(LAT))-A+C*TX)
2600 TEMP(LAT)=FNR(TEMP(LAT)/(C+B))
2610 NEXT LAT
2620 REM NOW TEST FOR CONVERGENCE
2630 AM=0
2640 IC=0
2650 FOR LAT= 1 TO 9
2660 MA=ABS(TEMP(LAT)-TM(LAT))
2670 IF MA > AM THEN AM =MA
2680 NEXT LAT
2690 IF AM < .01 THEN IC = 1
2700 IF IC = 1 THEN GOTO 2740
2710 NEXT H
2720 PRINT "  Model has failed to converge - - Think about your input  "
2730 END
2740 REM RESULTS
2750 CLS
2760 PRINT " ------------------------------------------------------------------
2770 PRINT "                                 R E S U L T S
2780 PRINT " ------------------------------------------------------------------
2790 PRINT "           Zone                  Temperature              Albedo
2800 PRINT "        =============            ==========               ======
2810 FOR LAT = 1 TO 9
2820 PRINT "              ";
2830 PRINT LATZ$(LAT);
2840 PRINT "           ";
2850 PRINT USING "£££.£" ;TEMP(LAT);
```

Figure 3.2 *Continued*

```
2860 PRINT USING "                           £££.££";ALBEDO(LAT)
2870 NEXT LAT
2880 LATICE= FNR(LATICE*57.296)
2890 PRINT "            The ice edge is at " LATICE " degrees north."
2900 PRINT
2910 PRINT "            Input parameters "
2920 PRINT "  Fraction of solar constant";FNR(SX)
2930 PRINT "            A=";A;"        B=";B
2940 PRINT "            C=";C
2950 PRINT " Ice albedo= ";AICE;"  Changes at ";TCRIT;" Deg C"
2960 PRINT
2970 PRINT "   Press space bar to continue"
2980 GOSUB 3120
2990 CLS
3000 PRINT
3010 PRINT "                 Do you want to try again ?"
3020 PRINT
3030 PRINT "              (1)      Reset all parameters
3040 PRINT "              (2)      Try again with same parameters
3050 PRINT
3060 RESTORE 600
3070 CH$=INKEY$
3080 IF CH$="1" THEN GOTO 590
3090 IF CH$="2" THEN GOTO 690
3100 GOTO 3070
3110 END
3120 SP$=INKEY$
3130 IF SP$=" " THEN GOTO 3150
3140 GOTO 3120
3150 RETURN
3160 FOR I=1 TO 700
3170 NEXT I
3180 PRINT
3190 RETURN
3200 PRINT "   Illegal response try again"
3210 RETURN
```

Figure 3.2 *Continued*

Description of the EBM

The model is governed by the equation originally devised by both Sellers and Budyko in 1969

$$(\text{Shortwave in}) = (\text{Transport out}) + (\text{Longwave out}) \tag{3.18}$$

which is formulated as

$$S(\phi)\{1-\alpha(\phi)\} = K\{T(\phi)-\bar{T}\} + \{A + BT(\phi)\} \tag{3.19}$$

where

K = the transport coefficient (here set equal to 3.80 W m^{-2} °C^{-1}),
$T(\phi)$ = the surface temperature at latitude ϕ,
$\bar{T}$ = the mean global surface temperature,
A and B are constants governing the longwave radiation loss (here taking values $A = 204.0$ W m^{-2} and $B = 2.17$ W m^{-2} °C^{-1}),
$S(\phi)$ = the mean annual radiation incident at latitude ϕ,
$\alpha(\phi)$ = the albedo at latitude ϕ.

Note that if the surface temperature at ϕ is less than −10 °C the albedo is set to 0.62. The solar constant in the model is taken as 1370 W m^{-2}.

The EBM is designed to be used to examine the sensitivity of the predicted equilibrium climate to changes in the solar constant. If the default values for the variables A, B, K and the albedo formulation are selected, an equilibrium climate which is quite close to the present-day situation is predicted for a fraction = 1 of the solar constant. This equilibrium climate is given in Table 3.1.

Once this equilibrium value for an unchanged solar constant has been seen, the user can modify the fraction of the solar constant prescribed and note the changes in the predicted climate. More importantly, the EBM permits the user to alter the albedo formulation, the latitudinal transport and the parameters in the infrared radiation term and examine the sensitivity of the modified model. The EBM is presented here in a hemispheric form.

EBM model code

In the program listed in Figure 3.2, an equilibrium solution is achieved by iterating the calculation of each zonal T_i of equation (3.13). A maximum of 50 iterations is allowed. The snow-free albedo of the planet has been coded as latitude-dependent. The following exercises are useful examples of the types of climate simulation experiments that can be undertaken.

Energy balance model exercises

Exercise 1.

(a) Using the default values of albedo, K, A and B, determine what decrease in the solar constant is required just to glaciate the Earth completely (ice edge at 0°N).

Table 3.1 EBM simulation display showing input parameters and resultant equilibrium climate

A	B	K
204 W m^{-2}	2.17 W m^{-2} °C^{-1}	3.81 W m^{-2} °C^{-1}
$TC = -10$	$AC = 0.62$	
Fract. of solar const. = 1		
Latitude	**Temp. (°C)**	**Albedo**
85	−13.5	0.62
75	−12.9	0.62
65	−4.8	0.45
55	1.8	0.40
45	8.5	0.36
35	16.0	0.31
25	22.3	0.27
15	26.9	0.25
5	27.7	0.25

(b) Select some other values of A, B, K and the albedo formulation and repeat Exercise 1 (a).

Exercise 2.

(a) Various authors have suggested different values for the transport coefficient, K. For instance, Budyko (1969) originally used $K = 3.81\ \mathrm{W\,m^{-2}\,{}^{\circ}C^{-1}}$, and Warren and Schneider (1979) used $K = 3.74\ \mathrm{W\,m^{-2}\,{}^{\circ}C^{-1}}$. How sensitive is the model's climate to the particular value of K?

(b) Investigate the climate that results when using very small or very large values of K. How sensitive are these different climates to changes in the solar constant? Try and 'predict' how you think the model will behave before you perform the experiment.

Exercise 3.

(a) Observations show that land will be totally snow-covered during winter for an annual mean surface temperature of 0 °C, and oceans totally ice-covered all year for a temperature of about −13 °C. The model specifies a change from land/sea to snow/ice at −10 °C. Alter this 'critical' temperature and investigate the change in the climate and the climatic sensitivity to changing the solar constant.

(b) The albedo over snow-covered areas can vary within the limits of 0.5–0.8 depending on vegetation type, cloud cover and snow/ice condition. Investigate the sensitivity of the simulated climate to changing the snow/ice albedo.

Exercise 4.

(a) There have been many suggestions for the values of the constants A and B determining the longwave emission from the planet – some have been dependent on cloud amount. Budyko (1969) originally used $A = 202\ \mathrm{W\,m^{-2}}$ and $B = 1.45\ \mathrm{W\,m^{-2}\,{}^{\circ}C^{-1}}$. Cess (1976) suggested $A = 212\ \mathrm{W\,m^{-2}}$ and $B = 1.6\ \mathrm{W\,m^{-2}\,{}^{\circ}C^{-1}}$. How do these different constants influence the climate and its sensitivity?

(b) Holding A constant, just vary B and investigate the effect on the climate. What does a variation of B correspond to physically?

Exercise 5.

Repeat Exercise 1 with the values of A, B, K and the albedo formulation which you believe are 'best' (i.e. most physically realistic for the present-day climate). Once the Earth is just fully glaciated, begin to increase the fractional solar constant. Determine how much of an increase in the solar constant is required before the ice retreats from the equator. Do you understand the value?

Once the simple exercises described above have been completed and the model *understood*, it can be modified to run a sequence of timesteps or to produce plots of time-varying values. The next section describes some other types of experiments which can be conducted with EBMs similar to this.

3.5 ENERGY BALANCE MODELS AND GLACIAL CYCLES

So far we have looked at the components and the results of EBMs. In this section, the results of some EBM experiments will be examined. In previous sections, we have ignored seasonality and have also neglected the effect of the oceans as a heat source and sink. In this section we will examine how EBMs have been used in climate simulation experiments. EBMs have been used extensively in the study of palaeoclimates. One common experiment is to introduce the effect of orbital (Milankovitch) variations and changed continental configurations on an EBM.

Geochemical data suggest a positive correlation between CO_2 and temperature over the last 540 million years. A notable exception to this is the Late Ordovician glaciation (around 440 million years ago), which occurred at a time when the atmospheric CO_2 content is believed to have been around 15 times as high as it is today. Reduced solar luminosity compensated in part for this, but experiments with EBMs have shown that the configuration of the continents was such that the ice sheets could coexist with high CO_2 levels. With the benefit of the insight gained from such EBM studies it has been possible to go on to perform more detailed calculations with a GCM, which have confirmed the hypothesis based on the EBMs. The advantage of EBMs in this kind of problem is the ease with which many different experiments can be performed. Since information on boundary conditions for model simulations is poor, the simple model offers the chance to test a range of situations before embarking on expensive calculations with a GCM.

We have already mentioned the rapid glaciation of the modelled Earth as a result of a decreased solar constant. Energy balance models incorporate the cryosphere, which is the frozen water of the Earth, as if it were a thin, high-albedo covering of the Earth's surface. The solution of the governing equation of an EBM for various values of solar constant is shown in Figure 3.3. This is an illustration of the solution of a simple, zero-dimensional model. It shows the fundamental characteristic of non-linear systems. A slow decrease in the solar constant from initial conditions for the present day means a gradual decrease in temperature until the point is reached (point A) where a runaway feedback loop causes total glaciation and a rapid drop in temperature (solid line to point B). When the solar constant is then increased the process is not immediately reversed; the temperature follows a different route until, at a value of the solar constant greater than that of the present day (point C), temperatures rise again (dashed line). The modelled climate exhibits hysteresis.

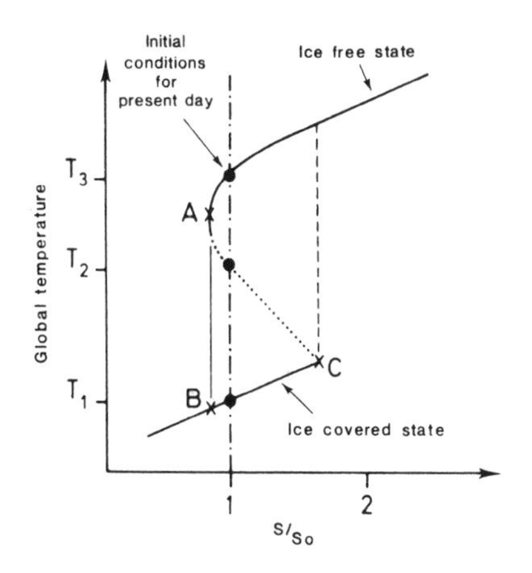

Figure 3.3 Characteristic solution of an EBM, plotted here as global mean temperature as a function of fraction of present-day solar constant. The dotted line represents a branch of the solution which, while being mathematically correct, is physically unrealistic. This lack of physical realism is due to the fact that on this branch increasing energy input results in a decreased temperature. More complex parameterizations within EBMs induce more complex shaped curves

The formulation of an EBM in 'time-dependent' form changes the nature of the interpretation of the 'physical' branch in Figure 3.3. This branch now represents the presence of a small, unstable ice cap. The notion that ice caps which are smaller than some characteristic length scale are unstable is referred to as the small ice cap instability (SICI), or sometimes as the thin ice cap instability (TICI). The phenomenon has been proposed as a mechanism for the initiation and growth of the Greenland and Antarctic ice sheets.

Much of the response of ice sheets to climate fluctuations depends on their thermal inertia. To make effective models of ice sheets, it is necessary to consider the ice sheet as more than a simple, thin covering of ice or snow. Some modellers have developed ice sheet models which extend the simple thin ice sheet model of the EBMs to be more realistic. In contrast, most GCMs do not deal with the growth and decay of ice sheets as, even now, the time-scales over which the ice sheets change is much longer than typical GCM integrations. In current GCMs, ice sheets continually collect snow, but the important loss mechanism, iceberg formation, is not included in the model. A more fundamental problem with modelling ice sheets is that we still know very little about the properties of the ice sheets and the way in which they change in response to climate forcing.

Figure 3.4 shows schematically two types of ice sheet. Figure 3.4(a) characterizes the major northern hemisphere ice sheets and Figure 3.4(b) characterizes the type of ice sheet which forms when a land mass exists at a pole, as is the case currently in the southern hemisphere. Provided the snowline is below the level of the ice or bedrock topography, then an ice sheet can exist. Once the ice sheet has acquired height, then it can be sustained even if the

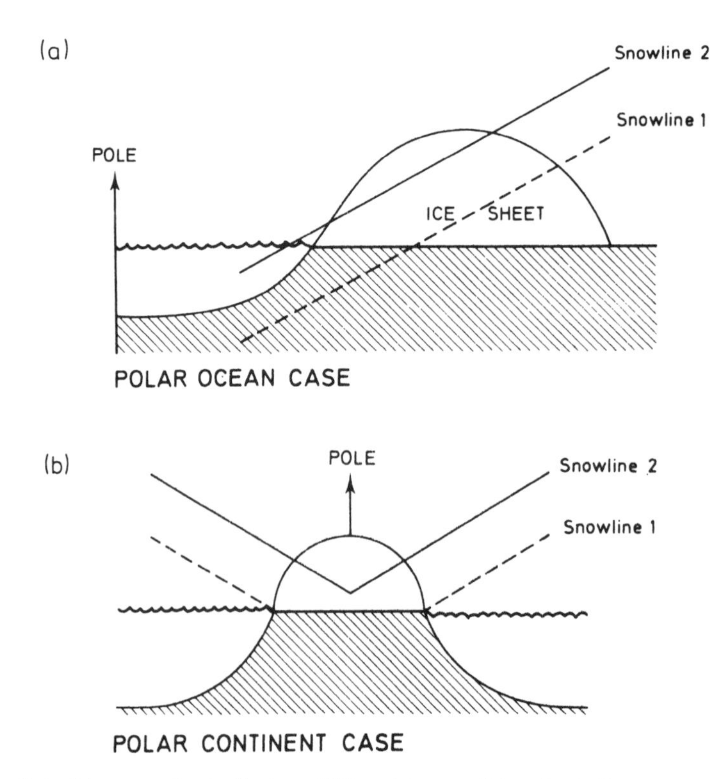

Figure 3.4 There are basically two different types of ice sheet: those occurring when there is a polar ocean and those which occur on a polar continent. In both cases it is possible for the ice sheet to persist even when the snowline is above ground level

snowline moves above the bottom of the ice sheet (i.e. snowline 1 changes to snowline 2). Such a situation is found in Greenland at the present time. The extent of an ice sheet is governed by the balance between accretion and ablation. Accretion is by precipitation, while ablation is the result of melting or calving (i.e. disintegration into the ocean). As the ice sheet grows, two effects compete.The extra height means a greater area of snow-fall because the higher areas are colder (a positive feedback). However, as the growth continues, the amount of moisture reaching the top of the ice sheet becomes less and the amount of snow falling on the top of the ice sheet also decreases, providing a natural halt on ice-sheet growth (a negative feedback). Since the ice sheet behaves like a plastic material, some continental configurations allow the ice to spread laterally. As the ice spreads, then the height is reduced and the spread continues. When there is a continent at the pole, such spreading is not possible beyond the limit of the coast. An ice sheet model can be coupled to representations of the response of the lithosphere to ice load and to an EBM such as that described earlier in this chapter to make a combined model. Some models of this sort have been shown to exhibit internal variability. The components interact to form a constantly varying climate even without external forcing such as the Milankovitch variations (cf. the 'climate' atttractor in Figure 2.7).

Continental ice sheets and permafrost extent typically vary on time-scales of approximately 1000–10 000 years (see Table 1.1), although shorter time-scale effects have been suggested. The results from detailed cryospheric EBMs as early as 1980 showed that the influence of an ice sheet on the radiation balance was small if sea-ice and snow cover were already incorporated. On the other hand, the inclusion of the ice sheet height–accumulation feedback loop discussed above substantially increased climate sensitivity. This means that once the ice sheet has formed over a portion of the continent, additional accumulation forces the edge of the ice sheet further towards the equator.

In modelling the response of ice sheets to Milankovitch variations, a range of sensitivity experiments has shown that the final outcome is highly dependent on the values of the input parameters. By combining an ice sheet model similar to that shown in Figure 3.4(a) with a two-dimensional EBM, it is possible to simulate the glacial/interglacial cycles over the past 240 000 years. Although the ice sheet model simulates growth well, it is found that the observed rapid dissipation of ice sheets can only be simulated by a parameterization of the calving and subsequent melting of icebergs. In the real world, this would raise sea-level and, by undercutting the ice sheets, form further icebergs. In the model of an ice sheet many different factors must be incorporated, the complexity of the formulation being related to the projected use of the model.

3.6 BOX MODELS – ANOTHER FORM OF ENERGY BALANCE MODEL

The concept of computing the energy budget of an area or subsystem of the climate system can be extended and modified to produce other forms of energy balance models. These models are not strictly EBMs and are often termed box models. A very elementary box model was considered early in this chapter (Section 3.2) in the example of the solar-heated swimming pool. That model had two boxes: one 'box' being the water and the other the air overlying the pool. A more complex consideration involves a more realistic parameterization of the energy transfer between the air and the pool, and interactive variation of other elements such as the radiative forcing. Following the same formulation, a simple column EBM can be used to consider the likely effect upon global temperatures of rising levels of atmospheric CO_2.

3.6.1 A simple box model of the ocean–atmosphere

The column EBM, used as an example here, represents the ocean–atmosphere system by only four 'compartments' or 'boxes': two atmospheric (one over land, one over ocean), an oceanic mixed layer and a deeper diffusive ocean (Figure 3.5(a)). The heating rate of the mixed layer is calculated by assuming a constant

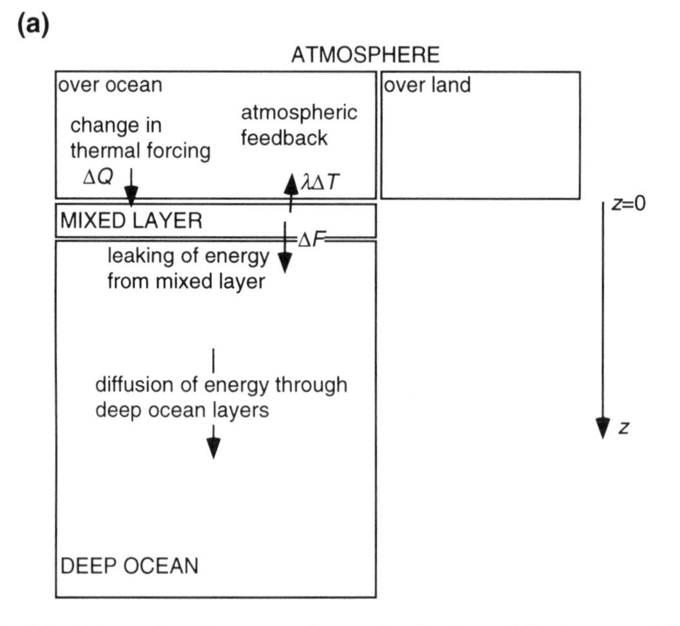

Figure 3.5 (a) Schematic diagram of a simple box-diffusion model of the atmosphere–ocean system. (b) Isolines of temperature change to 1980 (CO_2 level of 338 ppmv) as a function of the CO_2-doubling temperature change and the 1850 initial CO_2 level for two pairs of ocean diffusivity and mixed layer depth: upper diagram, $K = 10^{-4}$ m^2 s^{-1}, $h = 70$ m; lower diagram, $K = 3 \times 10^{-4}$ m^2 s^{-1}, $h = 110$ m. Results are based on a full numerical solution of the equations described in Wigley and Schlesinger (1985). (reproduced by permission from *Nature*, **315**, 649–652. Copyright © 1985 Macmillan Journals Ltd)

depth in which the temperature difference, ΔT, due to some perturbation, changes in response to: (i) the change in the surface thermal forcing, ΔQ; (ii) the atmospheric feedback, expressed in terms of a climate feedback parameter, λ; (iii) the leakage of energy permitted into the underlying waters. This energy, ΔF, acts as a surface boundary condition to the deep ocean (i.e. below the mixed layer) in which the turbulent diffusion coefficient, K, is assumed to be a constant. The equations describing the rates of heating in the two 'layers' are thus:

(i) for the mixed layer (total heat capacity C_m)

$$C_m \frac{d\Delta T}{dt} = \Delta Q - \lambda \Delta T - \Delta F \tag{3.20}$$

(ii) for the deeper waters

$$\frac{\partial \Delta T_0}{\partial t} = K \frac{\partial^2 \Delta T_0}{\partial z^2} \tag{3.21}$$

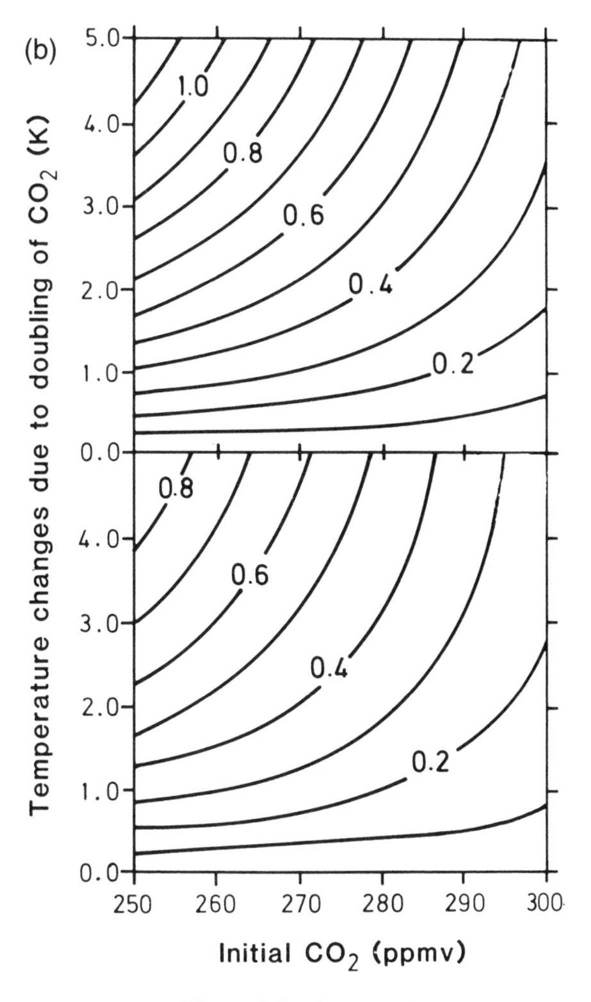

Figure 3.5 *Continued*

This latter equation may be evaluated at any depth, z (measured vertically downwards from zero at the interface), or calculated numerically using a vertical grid. In either case, the heat source at the top surface of the deep water is the energy 'leaking' out of the mixed layer, ΔF, which thus acts as a surface boundary condition to the lower-level differential equation. However, a simpler parameterization can be utilized by assuming that at the interface there is continuity between mixed layer temperature change ΔT and deeper layer temperature change evaluated at the interfacial level, $\Delta T_0(0, t)$, i.e.

$$\Delta T_0(0, t) = \Delta T(t) \tag{3.22}$$

With this formulation, the value of ΔF can be calculated from

$$\Delta F = -\gamma \rho c K \left\{ \frac{\partial \Delta T_0}{\partial z} \right\} \Bigg|_{z=0} \tag{3.23}$$

and used in equation (3.20). In this last equation, γ is the parameter utilized to average over land and ocean and has a value between 0.72 and 0.75, ρ is the density of water and c its specific heat capacity.

The model described by equations (3.20) and (3.21) can be used to evaluate different atmospheric forcings, related to possible impacts of increasing atmospheric carbon dioxide. There are two possible forms for the change, ΔQ: either an instantaneous 'jump'

$$\Delta Q = a \tag{3.24}$$

or a gradual increase

$$\Delta Q = bt \exp(\omega t) \tag{3.25}$$

Using both these possible forms for ΔQ, it is possible to compare a full numerical solution of the model with an approximation, which is gained by considering an infinitely deep ocean for which ΔF can be given by the expression

$$\Delta F = \gamma \mu \rho c h \frac{\Delta T}{(\tau_{\mathrm{d}} t)^{1/2}} \tag{3.26}$$

where μ is a tuning coefficient evaluated by comparison with the numerical solution, h is the mixed layer depth and τ_{d} ($= \pi h^2/K$) is a characteristic time for exchange between the mixed layer and the deep ocean. Substituting equation (3.26) into equation (3.20) results in an ordinary differential equation

$$\gamma \frac{\mathrm{d}\Delta T}{\mathrm{d}t} + \Delta T \left\{ \frac{1}{\tau_{\mathrm{f}}} + \frac{\mu \gamma}{(\tau_{\mathrm{d}} t)^{1/2}} \right\} = \frac{\Delta Q}{\rho c h} \tag{3.27}$$

where $\tau_{\mathrm{f}} = \rho c h / \lambda$, which can then be solved analytically using a prescribed functional form for ΔQ. For the two expressions, given here as equations (3.24) and (3.25), values for the temperature increment over a period of 130 years (1850–1980) can be deduced (Figure 3.5(b)) for chosen values of K and h. Here two sets of parameter values are shown. Using the CO_2 values observed for 1958 (315 ppmv) and 1980 (338 ppmv), the coefficients b and ω are easily evaluated from the equation for the increase of CO_2, which, corresponding to equation (3.25), is

$$C(t) = C_0 \exp(Bt \exp(\omega t)) \tag{3.28}$$

The two coefficients are determined by choice of initial (1850) CO_2 concentrations (horizontal axis in Figure 3.5(b)), from which the coefficient b in equation (3.25) can be calculated as

$$b = \frac{B\Delta Q_{2x}}{\ln 2} \tag{3.29}$$

where the atmospheric forcing resulting from a doubling of CO_2, ΔQ_{2x}, is related to the chosen values for the climate feedback parameter, λ (where λ here is the same as λ_{TOTAL} defined in Section 1.4.4), and the assumed value for the CO_2 doubling temperature change, ΔT_{2x}, (vertical axis in Figure 3.5(b)) is

$$\Delta Q_{2x} = \lambda \Delta T_{2x} \tag{3.30}$$

From these diagrams it is apparent that for reasonable estimates of initial (viz. 1850 baseline) carbon dioxide concentration (270 ppmv), the expected 1850–1980 temperature increment of the mixed layer for a wide range (0–5 K) of expected temperature increments due to a doubling of CO_2 is well in accord with observations. (Note that the observed air temperature increments must be assumed to be equal to the mixed layer temperature increases over the same period by assuming long-term quasi-equilibrium.) A numerical implementation of this simple box model is available on the CD (Appendix B).

3.6.2 A coupled atmosphere, land and ocean energy balance box model

It is possible to increase the level of complexity incorporated into a box model, such as that described in the previous section, so that other features can be resolved. Figure 3.6 illustrates the components of an energy balance box model which includes separate subsystems for northern and southern hemisphere land, ocean mixed layer, ocean intermediate layer and deep oceans. This model separates the atmospheric response over land and ocean and incorporates polar sinking of oceanic water into the deep ocean (the formation of deep water). Despite these features, the model is essentially a box advection–diffusion model, although it includes seasonally varying mixed layer depth and is forced with a seasonally varying insolation.

As with all relatively simple models, some features are prescribed. For example, hemispherically averaged cloud fraction is prescribed as a seasonally varying feature. As the land is hemispherically averaged, there is no opportunity to incorporate a temperature–surface albedo feedback in this sort of model. Despite these constraints, this simple box model can be used to investigate sensitivity to features which have not yet been effectively incorporated into the coupled ocean–atmosphere GCMs such as those discussed in Chapter 5. For example, the response of atmospheric and mixed layer temperatures to feedback processes involving changes in vertical diffusivity and changes in vertical velocities can be computed explicitly. Figure 3.7 shows the response of the atmospheric temperatures over the land and over the ocean, and of the oceanic mixed layer temperature of both hemispheres following a transient

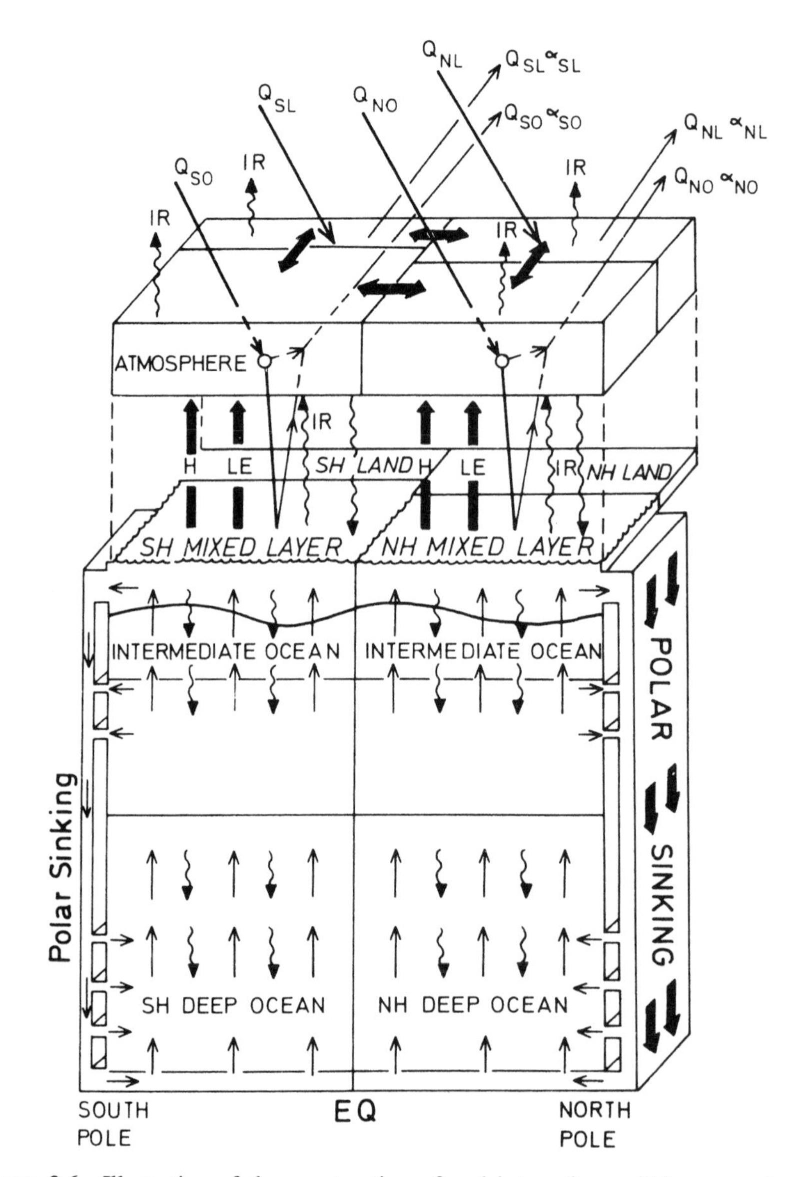

Figure 3.6 Illustration of the construction of and interactions within a complex box model of the Earth's climate system which includes hemispheric and land/ocean resolution and oceanic deep water formation (reproduced by permission from Harvey and Schneider (1985) *J. Geophys. Res.*, **90**, 2207–2222, copyright by the American Geophysical Union)

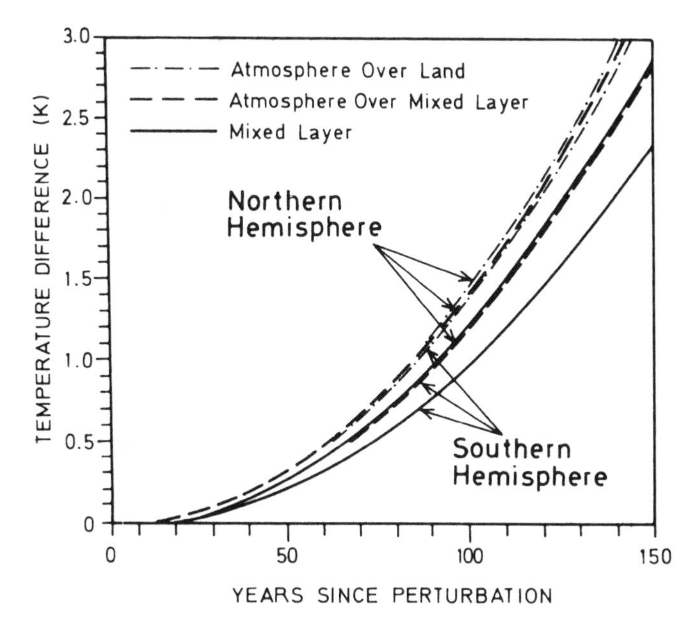

Figure 3.7 Effect of increasing CO_2 on the climate of the sophisticated box model of the climate system shown in Figure 3.6 (reproduced by permission from Harvey and Schneider (1985) *J. Geophys. Res.*, **90**, 2207–2222, copyright by the American Geophysical Union)

CO_2 perturbation simulated by a change in the parameterization of the infrared (IR) emission to space, where

$$\text{Emitted IR} = A' + B'T + (\text{cloud term}) \tag{3.31}$$

and the transient increase in atmospheric CO_2 causes a change in A' given by

$$\Delta A'(t) = -2.88 \times 10^{-4}\, t^2 \tag{3.32}$$

where t is the time, in years, since 1925.

In this model, oceanic vertical velocities can change in perturbed climatic states. The results in Figure 3.7 follow from the velocity increase in the northern hemisphere and the decrease in the southern hemisphere. There is a faster mixed layer warming, which reduces the lag of the mixed layer warming behind the atmospheric warming in the northern hemisphere as compared with the response in the southern hemisphere. These results suggest that more detailed analysis of oceanic feedback effects is required than can apparently be accomplished at present by three-dimensional coupled ocean–atmosphere models. These box models often rely on GCMs to calibrate transport and diffusion coefficients and are thus only as representative as these GCMs. In the IPCC Second Scientific Assessment, models like this have been used to examine the likely thermal expansion of the oceans, considering a wider range

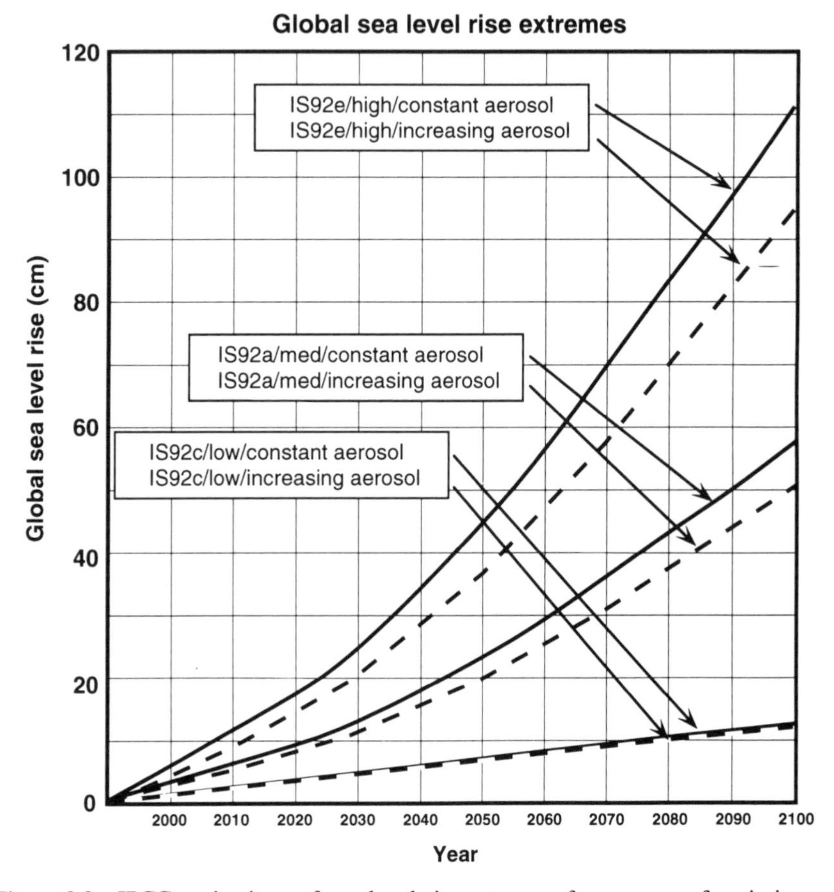

Figure 3.8 IPCC projections of sea-level rise extremes for a range of emissions and aerosol scenarios (after Houghton *et al.*, 1996)

of fossil fuel usage futures than is possible with (expensive) GCMs. Figure 3.8 shows the sea-level rise predicted for a range of futures including changing levels of tropospheric aerosols.

3.7 ENERGY BALANCE MODELS: DECEPTIVELY SIMPLE MODELS

Although they are of very simple construction, EBMs are extremely valuable tools in our study of the climate system. By forcing an EBM with heat flux anomalies which are random, it is possible to investigate the relationship between this 'weather' and variability on longer time-scales. Even simple EBMs can generate useful information on decadal and longer-term variability. They cannot tell us about the variability due to changes in ocean circulation, for example, but they do offer information on the 'passive' aspects of variability. In this chapter we have intentionally emphasized the simple basis of EBMs – the energy fluxes into and out of the planet as a whole (and of a region or climate subsystem) must balance unless there is cooling or heating. This concept is fundamental to climate modelling. It will recur in Chapter 4, where the heating rates of atmospheric layers are computed for the energy balance, and in Chapter 5, where each of the components of global climate models are seen to be driven by their energy balance.

The other topic which has been stressed in this chapter is computing. We wanted to underline that the basis of practically all climate modelling is (relatively) simple mathematical formulations and parameterizations represented in, and executed by, very fast computers. We have taken space to list the full code of one EBM in Figure 3.2. The code of an atmospheric GCM written in a similar high-level language (most of them are currently written in FORTRAN, which is fairly similar to BASIC) would be as thick as a substantial dictionary. More sophisticated coupled GCMs have codes whose page listings are thicker than a stack of encyclopaedias but, despite this apparent complexity, the exercises posed in this chapter could usefully be considered with reference to more complex models. Indeed, EBM-type analyses are commonly performed on the output of GCMs. It is therefore helpful to keep in mind the fundamental concepts developed in this chapter and to return, often, to the deceptively simple basis of the models described. The principle of energy balance is fundamental to the construction of physically based climate models, and the concept of using models to reveal and interpret the nature of the climate system and its behaviour is to be found throughout the remainder of this book.

RECOMMENDED READING

Abbott, E.A. (1963) *Flatland: A Romance of Many Dimensions. 5th edn.* Barnes and Noble, New York, 108 pp.

Budyko, M.I. (1969) The effect of solar radiation variations on the climate of the Earth. *Tellus* **21**, 611–619.

Cess, R.D. (1976) Climatic change, a reappraisal of atmospheric feedback mechanisms employing zonal climatology. *Journal of the Atmospheric Sciences* **33**, 1831–1843.

Hansen, J., Russell, G., Lacis, A., Fung, I., Rind, D. and Stone, P. (1985) Climatic response times: Dependence on climate sensitivity and ocean mixing. *Science* **229**, 857–859.

Harvey, L.D.D. and Schneider, S.H. (1985) Transient climate response to external forcing on $100–10^4$ year time-scales. 2. Sensitivity experiments with a seasonal, hemispherically averaged, coupled atmosphere, land, and ocean energy balance model. *Journal of Geophysical Research* **90**, 2207–2222.

Lee, W.-H. and North, G.R. (1995) Small ice cap instability in the presence of fluctuations. *Climate Dynamics* **11**, 242–246.

Murphy, J.M. (1995) Transient response of the Hadley Centre coupled ocean–atmosphere model to increasing carbon dioxide. Part III. Analysis of global mean response using simple models. *Journal of Climate* **8**, 496–514.

North, G.R., Cahalan, R.F. and Coakley, J.A. (1981) Energy balance climate models. *Reviews of Geophysics and Space Physics* **19**, 91–121.

Oerlemans, J. and van der Veen, C.J. (1984) *Ice Sheets and Climate*. Reidel, Dordrecht, 217 pp.

Sellers, W.D. (1969) A global climatic model based on the energy balance of the Earth–atmosphere system. *Journal of Applied Meteorology* **8**, 392–400.

Warren, S.G. and Schneider, S.H. (1979) Seasonal simulation as a test for uncertainties in the parameterization of a Budyko–Sellers zonal climate model, *Journal of the Atmospheric Sciences* **36**, 1377–1391.

Wigley, T.M.L. and Schlesinger, M.E. (1985) Analytical solution for the effect of increasing CO_2 on global mean temperature. *Nature* **315**, 649–652.

Wigley T.M.L. and Raper, S. (1992) Implications for climate and sea level of revised IPCC emissions scenarios. *Nature* **357**, 293–300.

CHAPTER 4

Computationally Efficient Models

It is not clear whether the authors [of a previous paper] intended to question the necessity of meridional circulations for the maintenance of the kinetic energy of the atmosphere. Such an intention would mean a complete change of the whole foundation of dynamic meteorology, and I doubt strongly that the genius of the authors, recognized by all meteorologists, will be sufficient for that goal.

E. Palmen (1949) (*Journal of Meteorology* **6**, 429–430)

4.1 WHY LOWER COMPLEXITY?

Although models which include only one or two dimensions can explicitly trace their origin in the need to perform calculations with limited computing resources, they remain a vital tool for climate modellers today. Modellers now look to these models as a means of examining a particular aspect or aspects of the climate system in as efficient a manner as possible or as a means of developing and testing new parameterizations. In this chapter we will examine how these models are constructed and how they are put to use and note that, as innovative modelling techniques are developed, the distinction between one-, two- and three-dimensional models is becoming less clear.

4.2 ONE-DIMENSIONAL RADIATIVE–CONVECTIVE MODELS

In earlier chapters the importance of the greenhouse effect was noted. This effect is due to the absorption of the upwelling thermal infrared radiation which has been emitted by the surface of the Earth. If none of the gases in the Earth's atmosphere possessed absorption features in the wavelength region in which the Earth emits radiation, there would be no greenhouse addition and the surface temperature would be equal to the planetary effective radiative temperature (see Figure 3.1). The greenhouse absorbers not only affect the surface temperature, they also modify the atmospheric temperature by their absorption and emission of radiation.

Radiative–convective climate models are one-dimensional models like the EBMs described in Chapter 3. In this case, however, the dimension is the vertical. These models resolve many layers in the atmosphere and seek to compute the atmospheric and surface temperatures. They can be used for sensitivity tests and, importantly, offer the opportunity to incorporate more complex radiation treatments than can be afforded in GCMs.

RC models derive a temperature profile for the atmosphere by dividing it into a number of layers. Suppose we divide the Earth's atmosphere into two layers so that each layer just absorbs the infrared radiation incident on it and, for this very simple example, let each layer be described as having an infrared optical thickness of $\tau = 1$. The principal absorber in the Earth's atmosphere is water vapour, which is contained within the lowest few kilometres. Thus our two layers can be taken to be centred at heights of 3 km (layer 1) and 0.5 km (layer 2). The infrared energy fluxes are shown in Figure 4.1(a). Both layers radiate as black bodies upwards and downwards, and the ground radiates up. Since the planet is emitting at its effective temperature, T_e, and because all radiation from below is absorbed by the top layer, σT_e^4 must equal σT_1^4 and thus $T_1 = T_e$. The energy balance of the first atmospheric layer is (emitted = absorbed) or $\sigma T_2^4 = 2\sigma T_1^4 = 2\sigma T_e^4$. In general the temperature of layer n can be

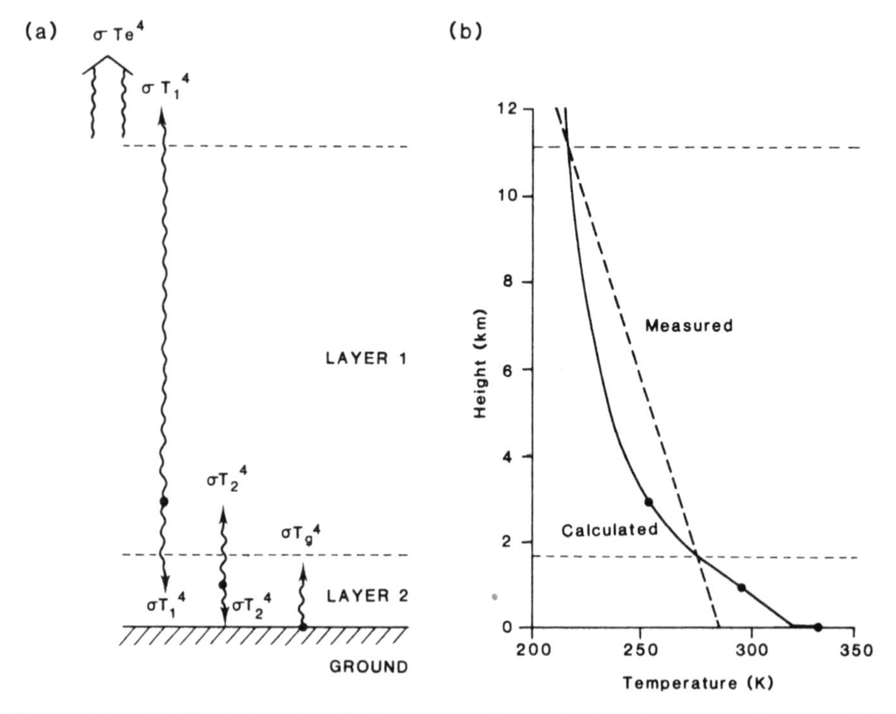

Figure 4.1 (a) The infrared radiative energy fluxes in the simple, two-layer model described. (b) The radiative equilibrium temperature profile calculated using the very simple model described in the text compared with a lapse rate of 6.5 K km^{-1}. This lapse rate is achieved by convection since the radiative equilibrium profile is unstable. (Richard M. Goody and James C. G. Walker, *Atmospheres*, © 1972, p. 56. Reproduced by permission from Prentice-Hall, Inc., Englewood Cliffs, New Jersey)

shown to be related to the effective temperature by

$$T^4(n) = \tau_{TOTAL}(n) T_e^4 \tag{4.1}$$

where $\tau_{TOTAL}(n)$ is the total infrared optical thickness from the top of the atmosphere to layer n. In our simple case, because each layer has $\tau = 1$, then $\tau_{TOTAL}(n) = n$. From equation (4.1) we can now calculate that the temperature of the top layer is equal to the effective temperature (255 K) and that the lower-layer temperature is 303 K. The surface temperature can be obtained by considering the radiative budget of the second atmospheric layer. Here $2\sigma T_2^4 = \sigma T_g^4 + \sigma T_1^4$ so that $T_g = (3T_e^4)^{0.25}$, which is 335 K.

These approximations are drastic, but the resultant temperature/height profile (Figure 4.1(b)) is quite close to the radiative equilibrium profiles which were first produced by RC models. Note that the surface air temperature is lower than the ground temperature, and the upper layer temperature is lower than the lower one. Compared with this standard (i.e. observed) lapse rate (the decrease of temperature with height), which is about $6.5\ \text{K km}^{-1}$, our calculated radiative temperature profile is unstable. Thus if a small parcel of air were disturbed from a location close to the surface it would tend to rise because it would be warmer than the surrounding air. Its temperature would decrease at (roughly) the observed lapse rate, so that at some arbitrary height (H) its temperature would be greater than that of the atmosphere. It would therefore continue to rise. Such a rising parcel of air would carry energy upwards and the resulting convection currents would mix the atmosphere, in this example, throughout the whole depth of the troposphere. The convective mixing would alter the temperature profile until the atmosphere was dynamically stable.

This convective adjustment of a radiatively produced temperature profile is the essence of RC models. In the above, highly simplified, discussion we have made several gross assumptions, but the basic concept of a radiatively computed temperature profile being adjusted to stability by parameterized convection is sound.

The structure of global radiative–convective models

The RC model is a single column containing the atmosphere and bounded beneath by the surface. This bounded column usually represents the globally averaged conditions in the Earth–atmosphere system. As the name of the model type indicates, radiation and convection are treated explicitly. The radiation scheme is detailed and occupies the vast majority of the total computation time, while the 'convection' is accomplished by a numerical adjustment of the temperature profile at the end of each timestep. In addition, some RC models also include cloud prediction schemes. The atmosphere is divided into a number of layers not necessarily of equal thickness. The layering can be defined with respect to height or pressure, but it is more common to introduce the non-dimensional vertical coordinate, σ (sigma), which is a function of pressure

$$\sigma = \frac{p - p_{\mathrm{T}}}{p_{\mathrm{s}} - p_{\mathrm{T}}} \tag{4.2}$$

where p is the pressure, p_{T} the (constant) 'top-of-the-atmosphere' (usually a low stratosphere location) pressure and p_{s} the (variable) pressure at the Earth's surface. The top of the atmosphere has $\sigma = 0$ and the surface always has $\sigma = 1$. The σ values at selected layer boundaries in an 18-layer atmosphere are given in Figure 4.2. In this example model, which is included on the CD, the layers extend from the surface to around 42 km, where the pressure is assumed to be zero. The sigma coordinate system, when used in GCMs, avoids the complications associated with model levels intersecting mountain ranges. The choice of sigma level values is arbitrary. In the RC model on the CD, the levels are evenly spaced in the sigma coordinate system.

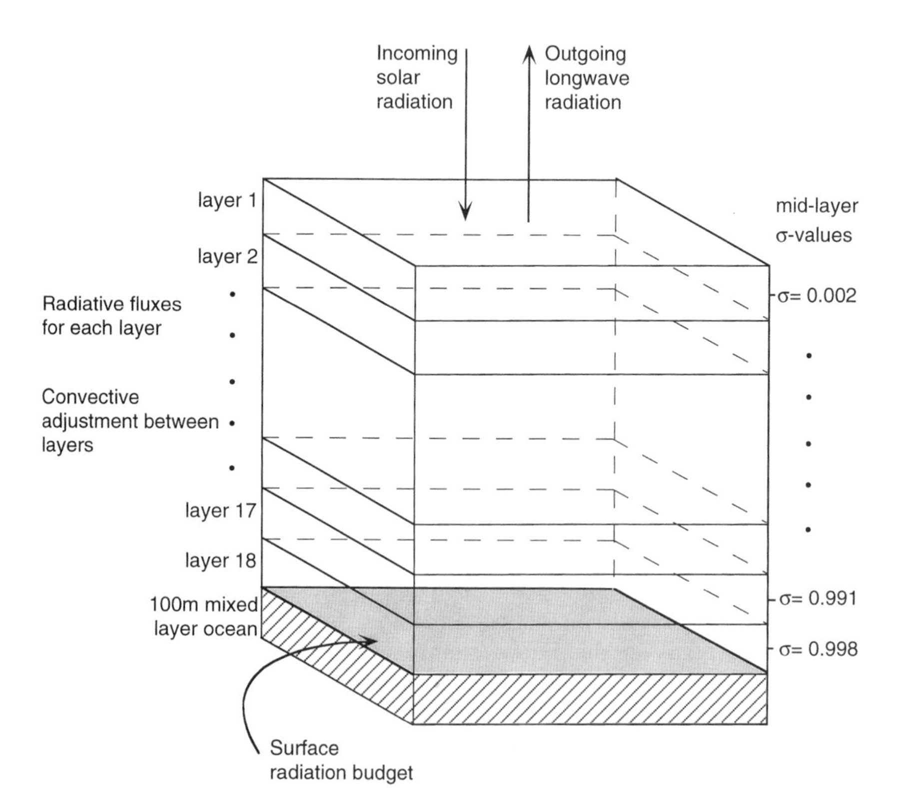

Figure 4.2 The vertical structure of an RC model. This particular model (which is included on the CD) has 18 layers and includes a 100 m ocean slab to represent the thermal inertia of the ocean mixed layer

4.3 RADIATION: THE DRIVER OF CLIMATE

Radiation is fundamental to the climate. Solar radiation is absorbed and infrared radiation emitted, with these two terms balancing over the globe when averaged over a few years. This simple energy balance was the basis of the EBMs described in Chapter 3, but in EBMs the way in which radiation is absorbed, transferred and re-emitted by the atmosphere was ignored. In this chapter we stress the radiative transfer processes: the heating of the surface by absorption of shortwave energy, and the heating and cooling of the atmosphere by absorption and emission of infrared radiation.

The principles involved in radiative computations in climate models are most readily illustrated by considering a very simple global model in which a single cloud or aerosol layer is spread homogeneously over the surface (Figure 4.3). The incident solar radiation of S W m^{-2} can be traced as it interacts with the cloud. A part is reflected, a part absorbed and a part transmitted; the cloud albedo, α_c, and shortwave absorptivity, a_c, control these interactions. The remaining solar flux interacts with the surface. Here only reflection and absorption occur and the reflected ray can be followed through the cloud and out to space. The infrared radiation emitted from the surface is partially absorbed by the cloud. Here it is assumed that the infrared absorption of the cloud is equal to the infrared emissivity of the cloud, ε. The downwelling radiation from the cloud adds to the surface heating, thus contributing to the greenhouse effect.

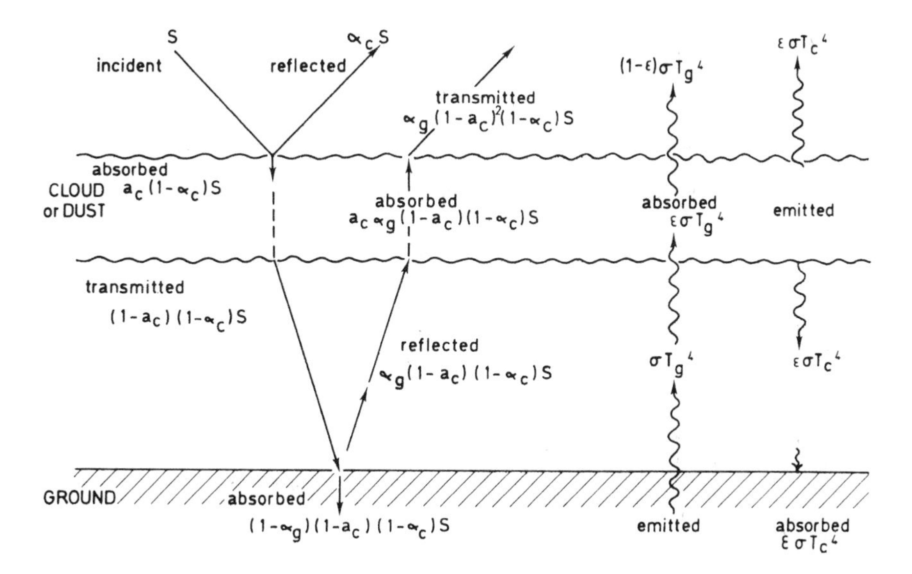

Figure 4.3 Idealized radiative interaction between a cloud or dust layer and the surface. There is assumed to be no reflection from the cloud base. The solar radiation suffers absorption on each passage through the layer. The infrared absorption is equal to the emissivity

Three main assumptions have been made in this simple model: (i) there is no reflection of the upward-travelling shortwave radiation by the cloud; (ii) the surface emissivity has been set equal to unity; (iii) the cloud/dust absorption in the infrared wavelength region is equal to ε. With all these assumptions, three energy balance equations can be written if it is also assumed that the model has reached equilibrium. Absorbed and emitted or reflected radiation at each level are equated, so that

$$S = \alpha_c S + \alpha_g(1 - a_c)^2(1 - \alpha_c)S + \varepsilon\sigma T_c^4 + (1 - \varepsilon)\sigma T_g^4 \quad (4.3)$$

$$a_c(1 - \alpha_c)S + a_c\alpha_g(1 - a_c)(1 - \alpha_c)S + \varepsilon\sigma T_g^4 = 2\varepsilon\sigma T_c^4 \quad (4.4)$$

$$(1 - \alpha_g)(1 - a_c)(1 - \alpha_c)S + \varepsilon\sigma T_c^4 = \sigma T_g^4 \quad (4.5)$$

Equations (4.3)–(4.5) represent the energy balances at the top of the atmosphere, the cloud level and the surface, respectively. These equations can be solved directly by giving values for the dust/cloud shortwave absorption, a_c, albedo, α_c, its infrared emissivity, ε and the surface albedo, α_g. Alternatively, the surface albedo term and the cloud temperature term can be eliminated from the equations leaving an expression for T_g.

$$\sigma T_g^4 = \frac{S(1 - \alpha_c)}{(2 - \varepsilon)}(2 - a_c) \quad (4.6)$$

Taking the value of S as a quarter of the solar flux at the planet, i.e. $S = 343\ \mathrm{W\,m^{-2}}$, and considering first the cloudless atmosphere case, if $\alpha_c = 0.08$, appropriate to scattering by atmospheric molecules alone, $a_c = 0.15$, representing absorption by the atmosphere, and $\varepsilon = 0.4$, then the surface temperature is 283 K. This is quite close to the global average surface temperature of ~288 K.

We can now consider the addition of 'clouds' to this simple model: first a volcanic aerosol cloud and second a water droplet cloud. Inserting a volcanic aerosol into an otherwise cloudless atmosphere will increase the albedo slightly, say $\alpha_c = 0.12$, and will increase the solar absorption and the infrared emissivity, say to $a_c = 0.18$ and $\varepsilon = 0.43$. Thus this volcanic aerosol 'cloud' gives rise to a cooling, with the global mean temperature dropping to 280 K. (Note that Figure 1.13 shows a similar cooling caused by the introduction of a volcanic aerosol cloud into a more complex RC model.)

Alternatively, we could consider the introduction of a water droplet cloud into the original cloudless atmosphere. If we assume that the cloud cover is to be approximately the same as that of the present day, then α_c increases, say to 0.30, a_c increases slightly to 0.20 and ε increases to 0.90. With these values for partly cloudy skies, the globally averaged surface temperature is 288 K. The introduction of this water droplet cloud has increased the calculated surface temperature because, with the radiative characteristics we have chosen, the greenhouse effect of the cloud is greater than the albedo effect. If, however, we alter the selected value of α_c very slightly so that $\alpha_c = 0.40$ and allow the other values to remain the same then the surface temperature becomes 277 K. The cloud albedo effect has 'beaten' the greenhouse effect this time. The sensitivity of this simple model's climate (as represented by the computed surface temperature) illustrates the interconnected role of various parameters in radiative transfer calculations and, in particular, the importance of the relative impact on the absorption of incoming solar radiation and the emission of infrared radiation.

The following treatment of atmospheric radiation is more complex than this stylized model but still somewhat simplified. The principles are the same in most models, the major complicating features being detailed consideration of the dependence of scattering and absorption on wavelength and other atmospheric variables.

The temperature profile of the atmospheric column is computed by calculating the net radiative heating in each layer. Calculations are made in terms of the layer potential temperature θ_i given by

$$\theta_i = T_i(p_0/p)^{\kappa} \tag{4.7}$$

where $p_0 = 1000$ hPa is a reference pressure and $\kappa = R/c_p = 0.286$ (R is the gas constant for air and c_p is its specific heat at constant pressure).

The simplified atmosphere we will consider is shown in Figure 4.4. The radiation is assumed to be absorbed by layers A_1 (between model levels 0 and 2) and A_3 (between model levels 2 and 4). The shortwave and longwave radiation are treated separately. The shortwave radiation includes all the solar radiation, the attenuation of this radiation by Rayleigh scattering, its reflection from the Earth's surface and from clouds, and its absorption in the atmosphere and in clouds. The longwave radiation includes all emissions by the atmosphere, clouds and the Earth's surface. The ground temperature, T_g, needed in order to evaluate the evaporation, the sensible heat flux from the surface and the net longwave surface radiation, is determined from the heat balance at the Earth's surface.

4.3.1 Shortwave radiation

In this simple case we divide the incoming solar radiation into two parts by wavelength, the division being somewhere between 0.7 and 0.9 μm. The two wavelength regions can then either be treated identically, or the absorption and scattering can be partitioned by wavelength so that the two parts of the radiation are designated R_s (the shortwave part which is roughly 65% of the total and is subject to Rayleigh scattering) and R_a (the near infrared wavelength part which is roughly 35% of the total and is subject to atmospheric absorption). These can be approximated as

$$R_s = 0.65S \cos \mu \tag{4.8}$$

$$R_a = 0.35S \cos \mu \tag{4.9}$$

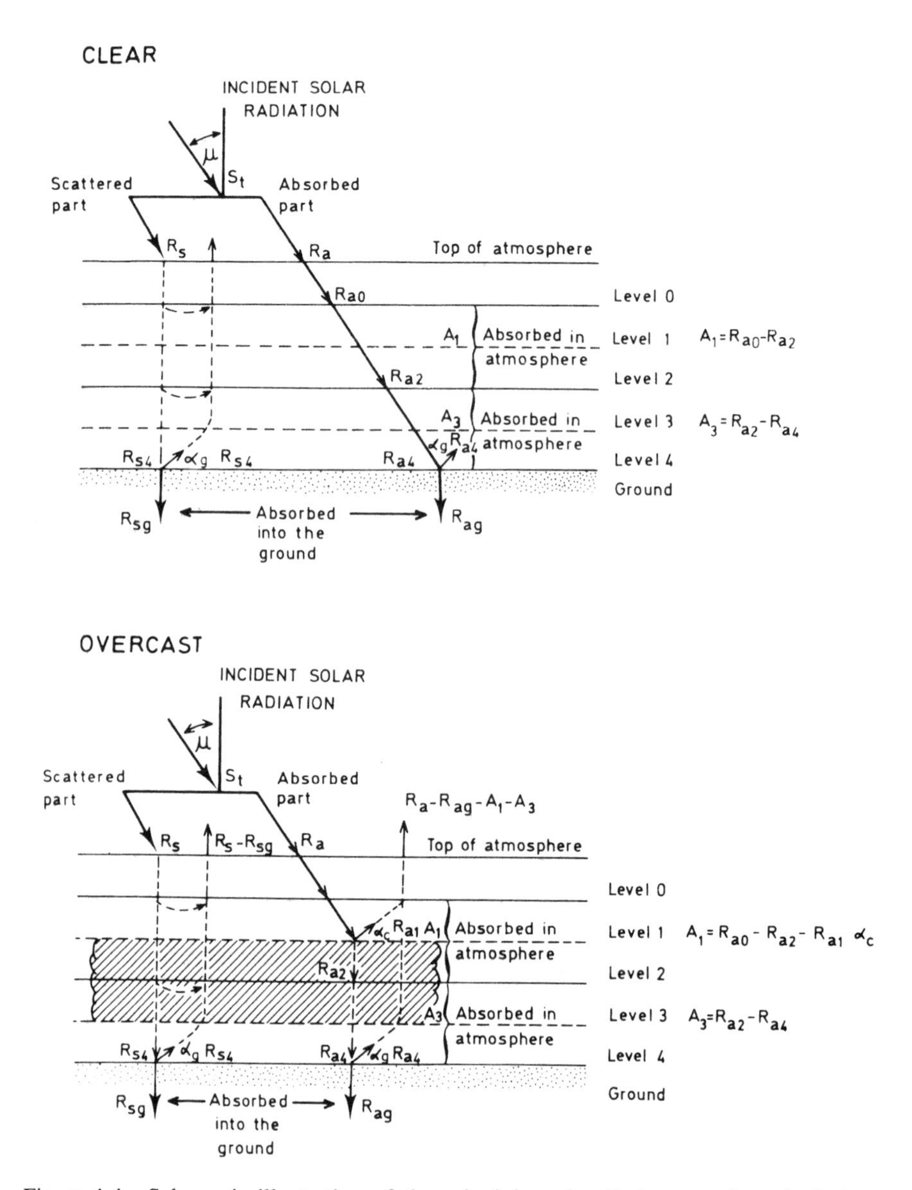

Figure 4.4 Schematic illustration of the principles of radiative transfer calculations. In this example there are only two atmospheric layers. More complex techniques involve many layers and many spectral intervals. All schemes consider only clear and cloudy cases. Intermediate cloud amounts are simulated by weighting these results

where S is the solar constant (adjusted for the Earth–Sun distance, which varies throughout the year) and μ is the zenith angle of the Sun. Some models calculate a mean value of μ, which is used during the hours of daylight. A summary of the disposition of these components of the shortwave radiation for both clear and cloudy skies is given in Figure 4.4 and is described in detail in the following paragraphs.

Albedo

The albedo of the clear atmosphere for the portion of the radiation which is assumed to be subject to Rayleigh scattering is given by

$$\alpha_0 = \min[1, 0.085 - 0.247 \log_{10}\{(p_0/p_s)\cos\mu\}] \tag{4.10}$$

For an overcast atmosphere, the albedo for the scattered part of the radiation is composed of the contributions of Rayleigh scattering (by atmospheric molecules) and of Mie scattering (by cloud droplets). The simplest useful formulation is

$$\alpha_{ac} = 1 - (1 - \alpha_0)(1 - \alpha_c) \tag{4.11}$$

where α_c is the cloud albedo for both R_a and R_s.

Albedos are a function of the surface or cloud type. A reasonable global average surface albedo for the entire solar spectrum is $\alpha_g = 0.10$. For specific surfaces it is often considered advantageous to introduce a spectral dependence since vegetation albedos increase rather sharply at around 0.7 μm while snow and ice albedos begin to decrease at about the same wavelength (Figure 4.5).

Shortwave radiation subject to scattering (R_s)

The part of the solar radiation which is assumed to be scattered does not interact with the atmosphere, except to be partly scattered back to space. Thus the only part with which we are concerned is that amount which reaches, and is absorbed by, the Earth's surface. This is given by the expressions

$$\begin{aligned} R'_{sg} &= R_s \frac{(1-\alpha_g)(1-\alpha_0)}{(1-\alpha_0\alpha_g)} \quad \text{for clear sky} \\ R''_{sg} &= R_s \frac{(1-\alpha_g)(1-\alpha_{ac})}{(1-\alpha_{ac}\alpha_g)} \quad \text{for overcast sky} \end{aligned} \tag{4.12}$$

Multiple reflections between sky and ground or between cloud base and ground are accounted for by the terms in the denominators. For partly cloudy conditions (neither clear nor overcast) the scattered radiation absorbed at the Earth's surface is

$$R_{sg} = NR''_{sg} + (1 - N)R'_{sg} \tag{4.13}$$

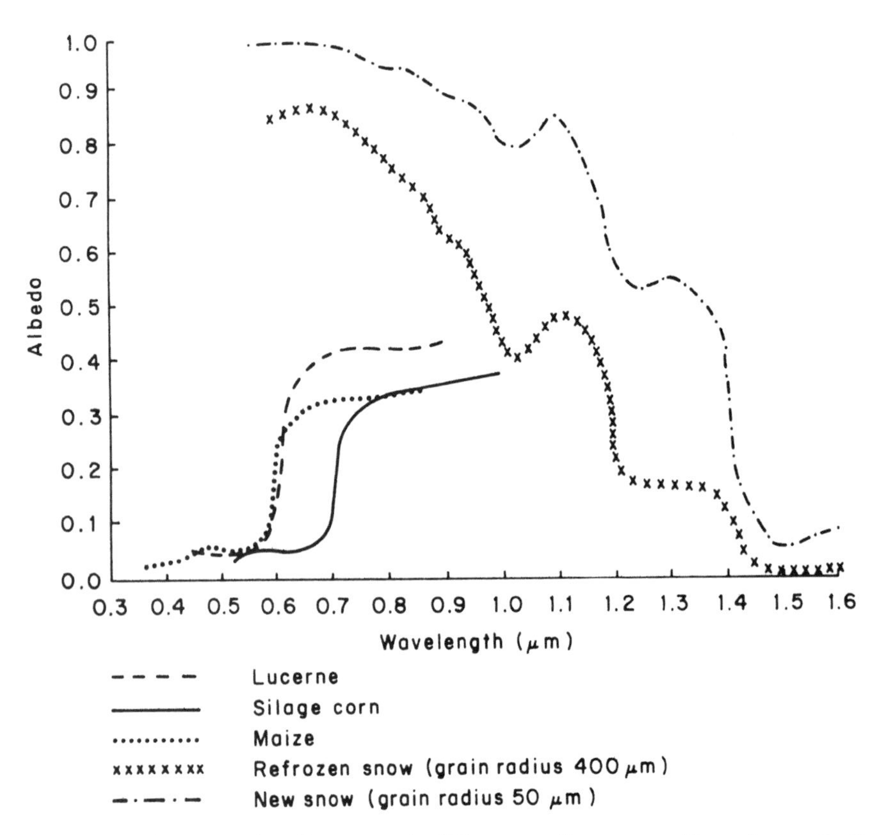

Figure 4.5 The spectral reflectance of different surface types. Note the markedly different spectral characteristics of snow surfaces and vegetated surfaces (reproduced by permission from Henderson-Sellers and Wilson (1983) *Rev. Geophys. Space Phys.*, **21**, 1743–1778, American Geophysical Union)

where N is the fractional cloudiness of the sky. Cloud albedo is a function of both the character and thickness of the cloud, as is illustrated by the experimental results shown in Figure 4.6. The absorption of this radiation by the ground affects the ground temperature and subsequently affects the longwave emission from the ground and the ground-level heat balance.

Shortwave radiation subject to absorption (R_a)

The solar radiation subject to absorption is distributed as heat to the various layers in the atmosphere and to the Earth's surface. The absorption depends upon the effective water vapour content as well as the amounts of ozone. Usually these absorptivities are computed semi-empirically using formulae appropriate to wide spectral intervals. For a cloudy sky the absorption in a cloud is generally prescribed as a function of cloud type only (e.g. Table 4.1).

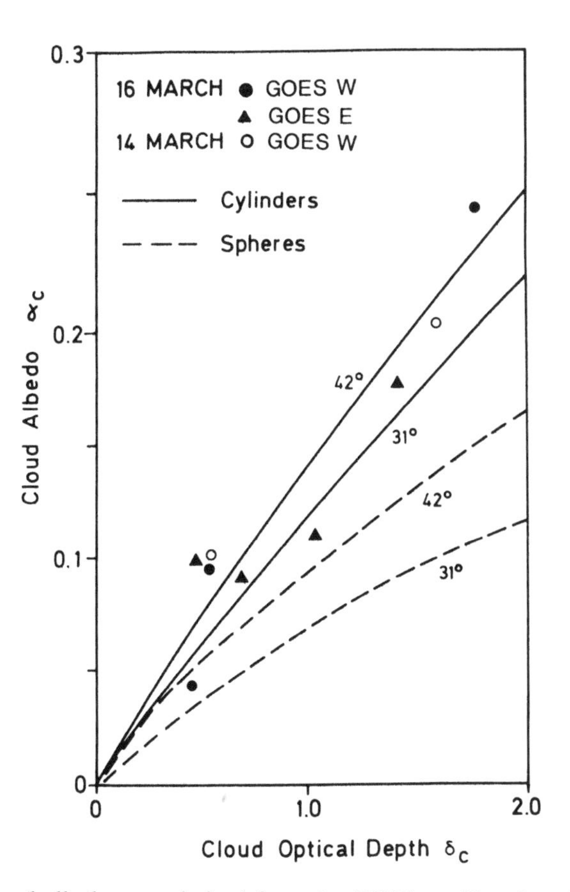

Figure 4.6 Cloud albedos, α_c, derived from the GOES satellite plotted as a function of lidar-derived cloud optical depths, δ_c. Also shown are theoretically predicted albedos, for two solar zenith angles, computed using ice spheres and ice cylinders (the latter being much more like the real cloud ice crystals) as the scattering objects (after Platt *et al.*, 1980)

Table 4.1 Typical albedos and absorptivities of clouds

Cloud level	Albedo	Transmissivity	Absorptivity
High	0.25	0.75	0.005
Middle	0.60	0.38	0.02
Low	0.70	0.36	0.035

The incoming beam becomes diffuse within any cloud, and its path is assumed to be 1.66 times the vertical thickness of the cloud. The factor 1.66 is often termed the Elsasser factor and is derived by assuming that the diffuse radiation is isotropic. Below the cloud the beam is still diffuse, and the factor 1.66 for path length is retained.

When the sky is partly cloudy, the total flux at level i is given by a weighted average of the clear and overcast fluxes:

$$R_{ai} = NR''_{ai} + (1 - N)R'_{ai} \tag{4.14}$$

That part of the flux subject to absorption which is actually absorbed by the ground is given (from Figure 4.4) by

$$R'_{ag} = (1 - \alpha_g)R'_{a4} \tag{4.15}$$

for clear sky, and by

$$R''_{ag} = \frac{(1 - \alpha_g)R''_{a4}}{(1 - \alpha_c\alpha_g)} \tag{4.16}$$

for completely cloudy (overcast) sky, where the factor $1/(1 - \alpha_c\alpha_g)$ again accounts for multiple reflections between the ground and cloud base. For partly cloudy skies, the radiation absorbed by the ground is therefore given by

$$R_{ag} = NR''_{ag} + (1 - N)R'_{ag} \tag{4.17}$$

The total solar radiation absorbed by the ground will be the sum of that part of the solar radiation spectrum which is subject to (atmospheric) absorption, that is absorbed by the ground, and that part of the spectrum subject to atmospheric scattering, that is absorbed by the ground, thus giving

$$R_g = R_{ag} + R_{sg} \tag{4.18}$$

4.3.2 Longwave radiation

The calculation of the longwave radiation, like that of the shortwave radiation, is based on an empirical transmission function depending primarily upon the amount of water vapour. The net longwave radiation at any level can be expressed as the sum of two terms (Figure 4.7)

$$F_{(\text{net})} = F\downarrow - F\uparrow \tag{4.19}$$

The upward flux at $z = h$ for radiation at some wavelength λ consists of the sum of two terms:

$$F_\lambda\uparrow(h) = B_\lambda[T(0)]\tau_\lambda(h, 0) + \int_0^h B_\lambda[T(z)](\mathrm{d}/\mathrm{d}z)\tau_\lambda(h, z)\,\mathrm{d}z \tag{4.20}$$

The first term in equation (4.20) is the infrared flux arriving at $z = h$ from the surface. It is given by the surface flux $B_\lambda[T(0)]$ multiplied by the infrared

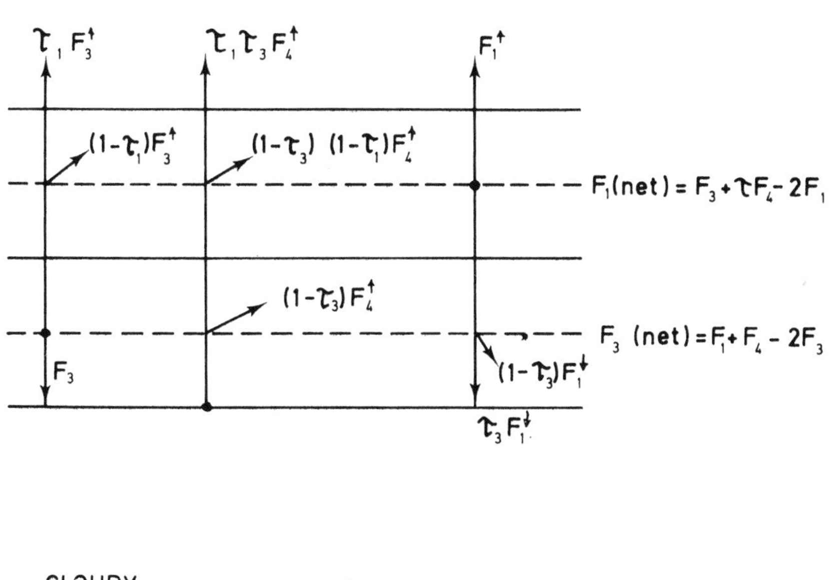

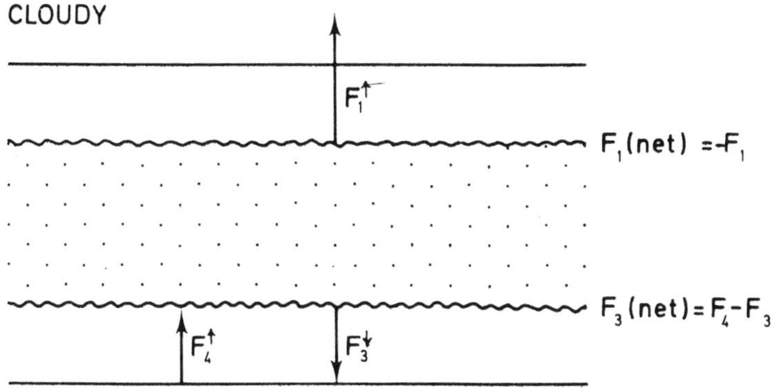

Figure 4.7 Schematic illustration of the transfer of terrestrial (longwave) radiation. There are fewer processes involved than for the case of shortwave radiation, but greater uncertainties in physical mechanisms make this portion just as difficult to deal with. (N.B. Both the shortwave transfer (Figure 4.4) and the longwave interactions shown here are considerably simplified by the assumption of horizontally infinite clouds. No climate models yet incorporate finite clouds and cloud edge effects)

transmittance of the atmosphere, τ_λ. The second term in equation (4.20) is the contribution to the total upward flux from emission of infrared radiation by atmospheric gases below the level $z = h$. Unlike the surface emission, which is nearly ideal black body radiation, the atmospheric emission as incorporated into the z derivative of τ_λ is highly wavelength-dependent. This is a result of the selective absorption by CO_2 or H_2O in certain spectral regions. The downward infrared flux is composed solely of atmospheric emission, since the incoming planetary infrared radiation from space is essentially zero. The downwelling radiation at a layer at $z = h$ is given by

$$F_{\lambda}\downarrow(h) = \int_{h}^{\infty} B_{\lambda}[T(z)](\mathrm{d}/\mathrm{d}z)\tau_{\lambda}(h, z)\mathrm{d}z \tag{4.21}$$

The transmittance, τ_{λ}, of the atmosphere for infrared radiation in the wavelength interval $(\lambda, \lambda + \Delta\lambda)$ and between atmospheric altitudes $z = h_1$ and $z = h_2$ can be expressed in terms of the optical thickness of the atmosphere in that wavelength region. The degree of detail used in the wavelength integration depends upon the computational power available and also on the type of application. The availability of fast computer routines for performing certain mathematical calculations often makes it desirable to rewrite a problem such as this in an alternative form. Two alternative methods of computing τ_{λ} use either a representation formulated in terms of exponential integrals or one involving proportionality to exponentials. Although the latter form is simpler to interpret, there are fast numerical routines available to calculate the exponential integrals as rapidly as exponentials, and the former are more accurate. Such techniques are common in climate modelling and the development of fast, efficient algorithms is an important contribution to this subject.

For some sensitivity experiments where overlap between absorption spectra of different atmospheric constituents can affect the sensitivity of the model, detailed integration through at least some wavelength regions will be required. In other types of experiments, such as those concerned with cloud–radiation interactions for example, where the radiative fluxes are not strongly affected by the perturbed variable, band-averaging can be satisfactory. Band-averaging uses a wavelength (or more usually, frequency) averaged value of the Planck function, $B_{\lambda}(T)$, in equation (4.20) together with an averaged value of transmissivity. In this case the layer-averaged transmissivity, τ_i, can be calculated from the layer temperature and pressure and the amount of gaseous absorber.

Generally, other uncertainties in a model, especially those generated by the somewhat arbitrary nature of the convective adjustment, will be greater than those associated with the difference between band-averaging or wavelength integration in the infrared calculations.

4.3.3 Heat balance at the ground

The ground temperature, T_g, is obtained from the heat balance at the ground. The treatment of the heating of the ground usually depends upon the assumed character of the ground or underlying surface. The albedo is wavelength-dependent and can also be dependent upon the surface moisture and vegetation. The ground can either be considered to be a perfect insulator with zero heat capacity, or a heat capacity can be specified. The total flux of heat across the air/ground interface is given by

$$R_g + F - \varepsilon\sigma T^4 - H_L - H_S = \text{stored energy} \tag{4.22}$$

where H_S is the sensible heat flux from the surface, H_L is the flux of latent heat due to evaporation from the surface, R_g is the solar radiation absorbed by the ground and F is the downwelling longwave radiation at the surface.

4.4 CONVECTIVE ADJUSTMENT

The computational scheme described so far defines a radiative temperature profile, $T(z)$, determined solely from the vertical divergence of the net radiative fluxes. Computation of globally averaged vertical radiative temperature profiles for clear sky conditions and with either a fixed distribution of

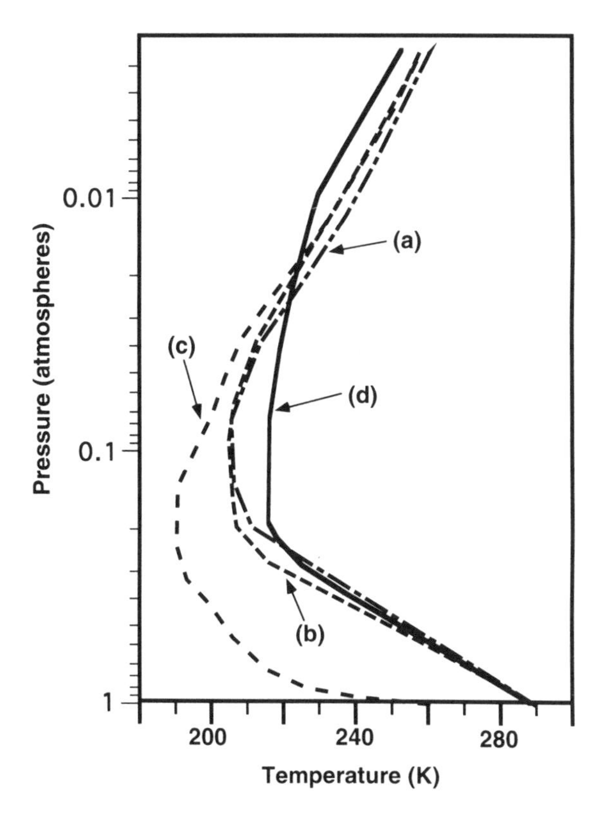

Figure 4.8 A comparison of vertical temperature profiles obtained using different values for the critical lapse rate in an RC model. Profiles obtained using (a) 6.5 K km^{-1}, (b) using the moist adiabatic lapse rate, (c) no convective adjustment (radiative effects only) are compared with (d) the 1976 US standard atmosphere (after MacKay and Khalil, 1991)

relative humidity or a fixed distribution of absolute humidity yields very high surface temperatures, and a temperature profile which decreases extremely rapidly with altitude. (In other words, despite the assumption made about the humidity distribution, the computed radiative temperature profiles $T(z)$ have large lapse rates in the lower troposphere, considerably in excess of the mean value given by a widely used standard atmosphere which has a lapse rate $\gamma = 6.5\ \mathrm{K\,km^{-1}}$, e.g. Figure 4.1.) This observed (critical) lapse rate, γ_c, is less than the computed $-\partial T/\partial z$ in the lower troposphere because the radiative equilibrium profiles are modified by free and forced vertical (moist and dry) convection and the vertical heat transport due to large-scale eddies. Radiative

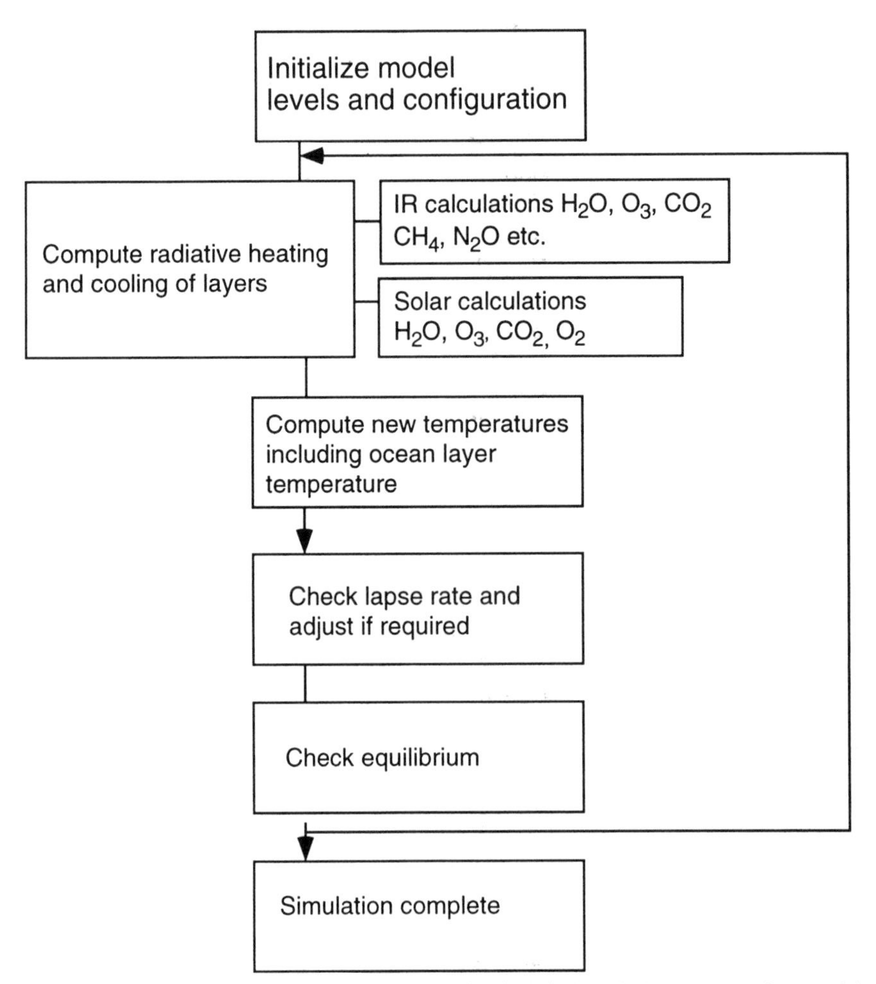

Figure 4.9 Schematic of the code structure for a typical radiative–convective model

equilibrium profiles are unstable to vertical (moist and dry) convection. Thus the tendency of radiative heating alone to produce large lapse rates adjacent to the Earth's surface, as shown in Figure 4.8, is offset by a rapid vertical transfer of heat.

By the mid-1960s, it was realized that if column models were to produce meaningful values of surface and vertical temperatures, it was necessary that the computed unstable profiles be modified. This modification was termed the 'convective adjustment'. It must be noted that it is not a computation of convection, but rather a numerical re-evaluation which is applied whenever the critical lapse rate γ_c is exceeded in the time evolution of the numerical calculation. The temperature difference between vertical layers is adjusted to the critical lapse rate, γ_c, by changing the temperature with time according to the integrated rate of heat addition. It can be shown that if continuity of temperature across the radiation/convection interface is satisfied at one time in the course of a time-dependent calculation, it will be satisfied for all later times. Although the term 'convective adjustment' is still used, modern parameterizations of convection are more physically based.

The complete structure of a basic RC model is shown in Figure 4.9. The flow continues until the atmospheric temperature profile converges to some final, equilibrium state. This convergence was well illustrated in one of the

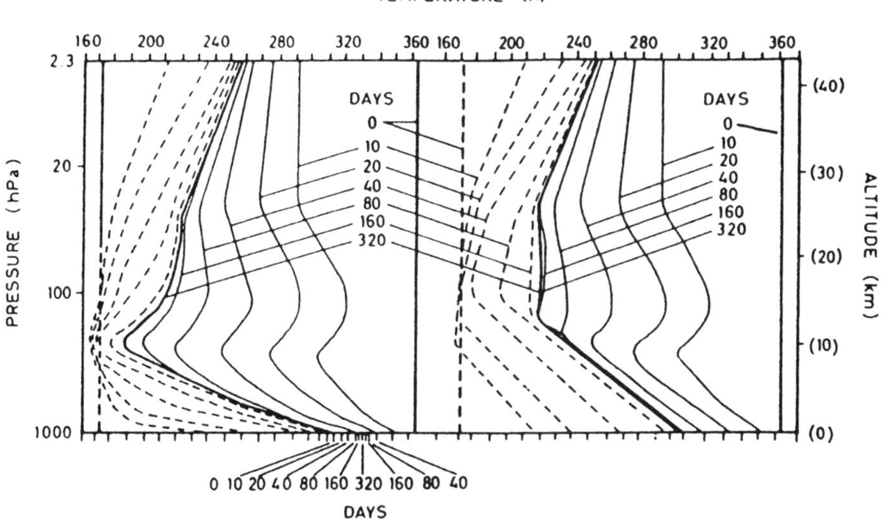

Figure 4.10 The left- and right-hand sides of the figure, respectively, show the approach to states of pure radiative and RC equilibrium. The solid and dashed lines show the approach from a warm and cold isothermal atmosphere, respectively (reproduced by permission from the American Meteorological Society from Manabe and Strickler (1964) *Journal of the Atmospheric Sciences*, **21**, 361–385)

earliest works describing one-dimensional vertically resolved models (Figure 4.10). In this case two versions of the model are shown: convergence to purely radiative equilibrium and convergence to 'thermal' equilibrium. The latter is similar to the now more usual RC model except that in this early (1964) model the absolute humidity, rather than the relative humidity, was prescribed. The latter condition has since been shown to be more appropriate as it varies little with atmospheric temperature. Thus the flow diagram in Figure 4.9 shows the calculation of the water vapour mixing ratio in atmospheric layers using the assumed, constant vertical profile of relative humidity.

4.5 SENSITIVITY EXPERIMENTS WITH RADIATIVE–CONVECTIVE MODELS

Radiative–convective models are particularly useful for studying the probable effects of perturbations to the radiative characteristics of the atmosphere. Their disadvantage is that they are usually formulated in terms of a global averaged value. On the other hand, this makes them computationally very efficient, and more time can therefore be taken in making spectrally detailed calculations.

The RC model can be summarized by saying that the vertical temperature profile of the atmosphere plus surface system, expressed as a vertical temperature set, T_i, is calculated in a timestepping procedure, such that

$$T_i(z, t + \Delta t) = T_i(z, t) + \frac{\Delta t}{c_\mathrm{p}\rho}\left[\frac{\mathrm{d}F_\mathrm{r}}{\mathrm{d}z} + \frac{\mathrm{d}F_\mathrm{c}}{\mathrm{d}z}\right] \tag{4.23}$$

Here the temperature, T_i, of a given layer, i, with height z and at time $t + \Delta t$ is a function of the temperature of that layer at the previous time t and the combined effects of the net radiative and 'convective' energy fluxes deposited at height z. In equation (4.23), c_p is the heat capacity at constant pressure, ρ is the atmospheric density and $\mathrm{d}F_\mathrm{r}/\mathrm{d}z$ and $\mathrm{d}F_\mathrm{c}/\mathrm{d}z$ are the net radiative and convective flux divergences.

There are two common methods of using RC models: either to gain an equilibrium solution after a perturbation, or to follow the time evolution of the radiative fluxes immediately following a perturbation. In the first case, the timestepping continues until there is a balance between the top-of-the-atmosphere shortwave and longwave fluxes. The radiative–convective model on the CD includes a simple mixed layer ocean model (e.g. Figure 4.2). This introduces some thermal inertia into the system so that time-dependent simulations can be undertaken. The latter method was used to compute the impact of the eruption of Mount Agung on surface temperatures. Simulations similar to this can be conducted using the 1DRC model included on the CD.

Sensitivity to humidity

Table 4.2 compares the predictions of ΔT (the increase in surface temperature) for differently formulated one-dimensional RC models for a perturbation in the form of a doubling of the atmospheric carbon dioxide from 300 to 600 ppmv. Model 1 has fixed absolute humidity. Hence the amount of water vapour in the atmosphere does not change and, in response to the external perturbation of CO_2, temperatures increase and relative humidity decreases. The resulting temperature increase of ~1.2 K can be thought of as a basic RC result since this model does not incorporate any feedback effects (i.e. neither atmospheric water vapour nor clouds have changed in response to the temperature change, and λ_{TOTAL}, as defined in Section 1.4.4, is equal to λ_B). Model 2, by contrast, has fixed relative humidity. This means that as the temperature increases, the saturation vapour pressure increases and thus, because the relative humidity is the ratio of actual vapour pressure to saturated vapour pressure, the actual vapour pressure must also increase. This extra water vapour must be the result of surface evaporation. It introduces a positive feedback of $\lambda = -1.41\ \mathrm{W\,m^{-2}\,K^{-1}}$ so that λ_{TOTAL} decreases and the surface temperature increase predicted is 1.94 K. The difference between the results from models 1 and 2 illustrates the effect of evaporation on radiative exchanges and its importance for any climate prediction model. Model 3 uses a convective adjustment to the moist adiabatic lapse rate rather than the value of $6.5\ \mathrm{K\,km^{-1}}$ that is used in models 1 and 2. It produces a slightly lower predicted temperature increase since the lapse rate decreases as additional water vapour is added to the atmosphere, i.e. two feedbacks of opposite sign

Table 4.2 Equilibrium surface temperature increase due to doubled CO_2 (300–600 ppmv): results from a suite of one-dimensional model sensitivity experiments (modified by permission from Hansen *et al.*, 1981)

Model*	Description	ΔT (K)	Feedback factors†	
			f	λ_{TOTAL}
1	Fixed absolute humidity, $6.5\ \mathrm{K\,km^{-1}}$, fixed cloud altitude	1.22	1	3.75
2	Fixed relative humidity, $6.5\ \mathrm{K\,km^{-1}}$, fixed cloud altitude	1.94	1.6	2.34
3	Same as 2, except moist adiabatic lapse rate replaces $6.5\ \mathrm{K\,km^{-1}}$	1.37	0.7	5.36
4	Same as 2, except fixed cloud temperature replaces fixed cloud altitude	2.78	1.4	2.68

* Model 1 has no feedbacks affecting the atmosphere's radiative properties.
† The feedback factors f (dimensionless) and λ_{TOTAL} ($\mathrm{W\,m^{-2}\,K^{-1}}$) are those defined in Section 1.4.4 and are two commonly used methods of representing the effect of each added process on model sensitivity to doubled CO_2.

combine. The difficulty of selecting an 'appropriate' global lapse rate to which convective adjustment should be made is considerable.

Comparison of models 4 and 2 illustrates the importance of cloud temperature and height effects. In model 4, the clouds, which are set at a constant, empirically determined amount, are at fixed temperatures rather than at the fixed heights used previously. The clouds therefore move to higher altitudes as the CO_2 perturbation increases temperature. Hence the computed surface temperature must be raised further so that the planetary (top-of-the-atmosphere) energy balance is maintained, resulting in a predicted surface temperature increase, $\Delta T = 2.8$ K, which is considerably larger than the ~1.9 K for fixed cloud altitude.

Sensitivity to clouds

Absorption of solar energy by the entire Earth–atmosphere system is most simply specified in terms of the albedo of that system, usually using separate albedos for the cloudless part of the Earth–atmosphere system and the cloud-covered part. Using an RC model, it was shown in 1972 (Figure 4.11) that increasing the cloud amount by about 8% (retaining fixed cloud top height and albedo) would lower the global mean surface temperature by 2 K, whereas raising the level of the cloud top height by about 0.5 km (at constant cloud amount and albedo) would produce exactly the opposite effect on surface temperature. Thus it is not possible to generalize globally averaged results to yield information on the local (in latitude and time) effect of variations in cloudiness, since the effect of changes in cloudiness on surface net heating depends upon the local values of the cloud amounts, heights and albedos, the albedo of the surface, the average solar zenith angle and the local vertical distribution of temperature and radiatively active constituents. The radiative calculations do suggest that even the direction of possible cloud feedback on surface temperature is far from obvious.

Sensitivity to lapse rate selected for convective adjustment

The generally accepted value of the critical lapse rate of 6.5 K km^{-1}, to which 'convective' adjustment is made, may not be appropriate for many experiments. Observational data give a globally averaged value for the tropospheric lapse rate closer to 5.5 K km^{-1}. Experiments have also been undertaken in which the moist adiabatic lapse rate was used rather than a fixed value. In one such model for a doubled CO_2 experiment the resulting change in surface temperature was 0.79 K, as opposed to 1.94 K when the 6.5 K km^{-1} value was used. These differences are similar to those found in another, different, model in which a 2.36 K increase in surface temperature from a doubling of atmospheric CO_2 concentration was computed using a band-averaged calculation. A temperature increase of only 1.9 K was found when using a

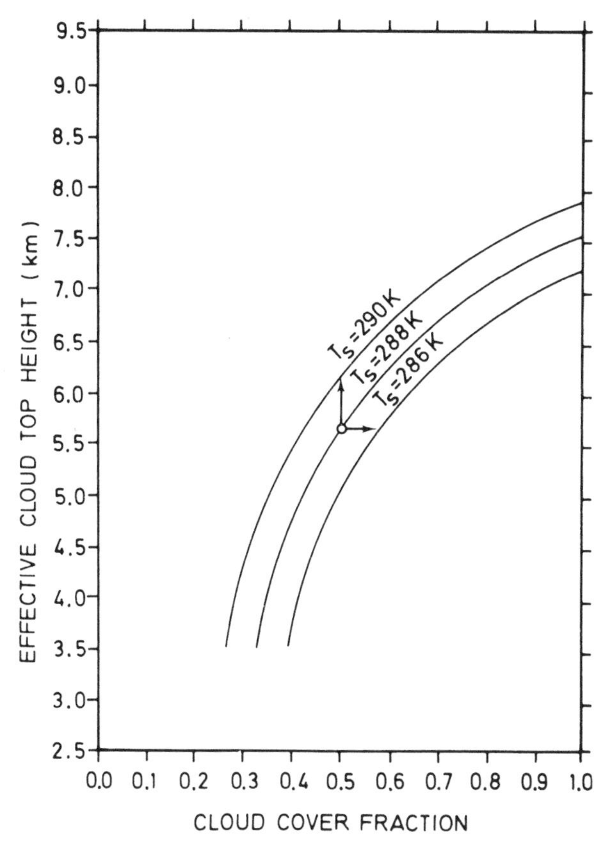

Figure 4.11 Effective cloud-top height and cloud-cover fraction giving the indicated equilibrium surface temperatures. From a present-day of 0.5 for cloud cover and an effective cloud height of 5.5 km, increasing the cloud amount for constant cloud height decreases the surface temperature, whereas retaining the fractional cover at 0.5 and increasing the cloud height causes a surface temperature increase (reproduced by permission from Schneider, 1972)

model which was identical except for the subdivision into many limited spectral intervals and where the CO_2–water vapour overlap region around 15 μm was divided into four sub-intervals. This lapse rate dependent difference is larger than that indicated in Table 4.2, where the temperature increase computed for doubled CO_2 using the saturated adiabatic lapse rate was 1.37 K.

4.6 DEVELOPMENT OF RADIATIVE–CONVECTIVE MODELS

Inclusion of an interactive cloud prediction scheme into a one-dimensional RC model would imply the incorporation of another facet of the climate pyramid (Figure 2.1), viz. dynamics. In this example, the model used is a one-dimensional, globally averaged RC model in which the atmosphere is divided into 17 layers. The tropospheric lapse rate is set at the standard atmospheric lapse rate of $6.5\,\mathrm{K\,km^{-1}}$ and a fixed relative humidity is maintained.

Cloud prediction

The cloud cover is predicted by being calculated as a proportion of the water mixing ratio, W, in each layer. This mixing ratio will be affected in turn by the latent heating, an indication of the amount of water added, and by precipitation, the water removed. The latent heat flux, H_L, will be determined from the model's convective adjustment and a calculated effective 'atmospheric Bowen ratio', B. It is necessary to make an assumption about the variation of B with altitude. The simplest assumption is that B is constant, i.e. independent of altitude. There are data which seem to substantiate this assumption, but it remains a somewhat dubious claim as the data do not take into account small-scale turbulence.

The net latent heat flux into each layer is found from the convective adjustment which the model calculates at each timestep. The net convective flux into a layer, H_C, is then derived. This net total convective flux can be divided by $(1 + B)$ to give the net latent heat flux

$$\frac{H_C}{1+B} = \frac{H_C}{1+H_S/H_L} = \frac{H_C H_L}{H_L+H_S} = H_C\frac{H_L}{H_C} = H_L \tag{4.24}$$

since H_C, the total convective flux, equals the sum of the latent flux and the sensible flux. These latent fluxes in each layer are then multiplied by the gravitational acceleration divided by the pressure change across the layer to give the flux per unit mass for unit area.

The cloud cover of the previous timestep multiplied by a constant gives the old water mixing ratio, W_o. This constant (here 5.5×10^{-4}) represents a typical mixing ratio for precipitating cloud systems. Thus, greater cloud is linearly equated with higher mixing ratios and thus greater precipitation. The precipitation rate is calculated by multiplying the old mixing ratio, W_o, by another empirical constant ($1.25 \times 10^{-4}\,\mathrm{s^{-1}}$), representing the inverse of the conversion time from cloud droplets into rain. Thus the new water mixing ratio is the sum of three terms

$$W = W_o + \frac{H_C}{(1+B)L} - 1.25 \times 10^{-4} W_o \tag{4.25}$$

the first term being the old ratio, the second the amount of water released through condensation and the third that precipitated out, where L is the latent heat of evaporation.

The new fractional cloud cover, C_i, for each of i layers is then determined from

$$C_i = W/(5.5 \times 10^{-4}) \tag{4.26}$$

Finally, an adjustment is made if the total cloud coverage is greater than unity, i.e. if $\sum C_i > 1$. In this case a proportional amount of cloud is removed from each layer by converting it to rain.

Model sensitivity

This model has been used to investigate the role that changing cloud cover and height may have played in the evolution of the Earth's atmosphere. Figure 4.12

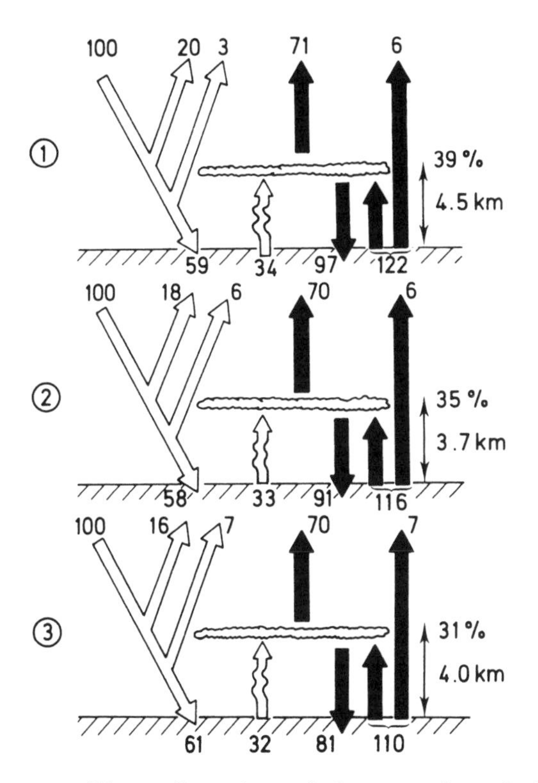

Figure 4.12 Three possible configurations of the atmosphere–hydrosphere of the early Earth. The calculated fluxes have been normalized to the incident solar radiation of the standard case 1 (i.e. 100 units = $(1/4) \times 1100\ \mathrm{W\,m^{-2}}$). Solar fluxes (open arrows) are on the left of each sketch, upward arrows represent atmospheric (left) and surface (right) reflection. The wavy arrow represents convective heat flux from the surface, and infrared fluxes are on the right (solid arrows). Note the variations in mean cloud cover (as a percentage) and cloud height (in kilometres) on the right (reproduced by permission from Cogley and Henderson-Sellers (1984) *Rev. Geophys. Space Phys.*, **22**, 161, copyright by the American Geophysical Union)

shows three possible evolutionary states of the climate system. Case 1 has a lowered solar luminosity appropriate to the early Pre-Cambrian (~4.0×10^9 years ago) of 80% of present-day solar flux, higher atmospheric CO_2 (1650 ppmv) and a surface albedo appropriate to a near global ocean (0.05). Despite the low value of incident solar flux, the computed surface temperature, 277 K, is well above freezing. Other situations can also be examined. If a greater emergence of land is postulated (Case 2), then the surface albedo would be expected to be higher (say 0.10) and the computed surface temperature lower, 274 K. If, additionally, considerable silicate weathering is believed to have occurred, reducing the atmospheric CO_2 to below present-day values (say 330 ppmv), then the mean global surface temperature drops still further to 270 K.

The inclusion of a cloud prediction scheme has caused this RC model to predict surface temperatures for these postulated early Earth situations which are significantly higher than would have been the case if clouds had been prescribed. In particular, the decreasing cloud amount as temperatures fall and the relative increase in cirrus cloud, with its associated greenhouse effect, gives rise to surface temperatures at which sea-ice just melts in the present-day ocean. Thus, results from this RC model, which includes prediction of cloud amounts and heights (used to determine properties), offer a possible solution to the 'reduced solar luminosity-enhanced early surface temperature' paradox described in Section 3.3.

Regional and local applications

The convective adjustment which occurs in the atmosphere means we introduce vertical motion into our climate model and can use this to parameterize movement of atmospheric water and the formation of clouds. The approximations we have made are acceptable for global averages, but once we become interested in a specific region we must consider horizontal transfer as well as vertical motion and radiation. In the next section we introduce the concept of large-scale horizontal motion. It is also worth noting that 'single column' models have been introduced for sensitivity testing of the processes represented in the 'columns' of GCMs. These column models apply, therefore, to the area of a single grid point in a GCM. The horizontal (advective) fluxes are specified, usually from full GCM runs. These column models are further examples of the increased fuzziness between types of climate models.

4.7 TWO-DIMENSIONAL STATISTICAL DYNAMICAL CLIMATE MODELS

Secondary school Geography classes would appeal to the Flatlanders mentioned in Chapter 3 because at this level the general circulation of the atmosphere is often introduced as being composed of cellular circulations. These Hadley,

Ferrel and polar cells are meridional features, i.e. they consist solely of latitudinally averaged movement between zones. Of course this is a gross simplification which ignores the major circulation features in mid-latitudes: the Rossby waves (see Figure 2.2 and Figure A.1 in the glossary in Appendix A).

4.7.1 Parameterizations for two-dimensional modelling

Most two-dimensional SD climate models are constructed to simulate the meridional motions only. The two dimensions they represent explicitly are height in the atmosphere and latitude. Variations around latitude zones (i.e. longitudinal variations) are neither resolved nor described. These models solve numerically the basic equations listed in Table 2.1. In general these models are able to produce realistic simulations of the large-scale two-dimensional flow (Figure 4.13). The fundamental difference between these models and full atmospheric GCMs is that all the variables of interest are zonally averaged values. This zonal averaging is identified below by angle brackets. The equations to be solved are:

Zonal momentum

$$\frac{\partial\langle u\rangle}{\partial t} - f\langle v\rangle + \frac{\partial\langle u'v'\rangle}{\partial y} = F \tag{4.27}$$

Meridional momentum (geostrophic balance)

$$f\langle u\rangle + R\langle T\rangle \frac{\partial}{\partial y}(\ln\langle p\rangle) = 0 \tag{4.28}$$

Hydrostatic balance (vertical component)

$$\frac{\partial(\ln\langle p\rangle)}{\partial z} = \frac{-g}{R\langle T\rangle} \tag{4.29}$$

Thermodynamic balance

$$\frac{\partial\langle T\rangle}{\partial t} + \frac{\partial\langle v'T'\rangle}{\partial y} + \frac{\partial\langle w'T'\rangle}{\partial z} + \langle w\rangle\left\{\frac{g}{\langle\rho\rangle c_{\mathrm{p}}} + \frac{\partial\langle T\rangle}{\partial z}\right\} = \frac{Q}{\langle\rho\rangle c_{\mathrm{p}}} \tag{4.30}$$

Continuity

$$\frac{\partial\langle\rho\rangle\langle v\rangle}{\partial y} + \frac{\partial\langle\rho\rangle\langle w\rangle}{\partial z} = 0 \tag{4.31}$$

where u, v and w are the velocities in the eastward (x), northward (y) and vertical (z) directions, T is the temperature, Q the zonal diabatic heating, F a friction term, R the universal gas constant, c_{p} the specific heat at constant

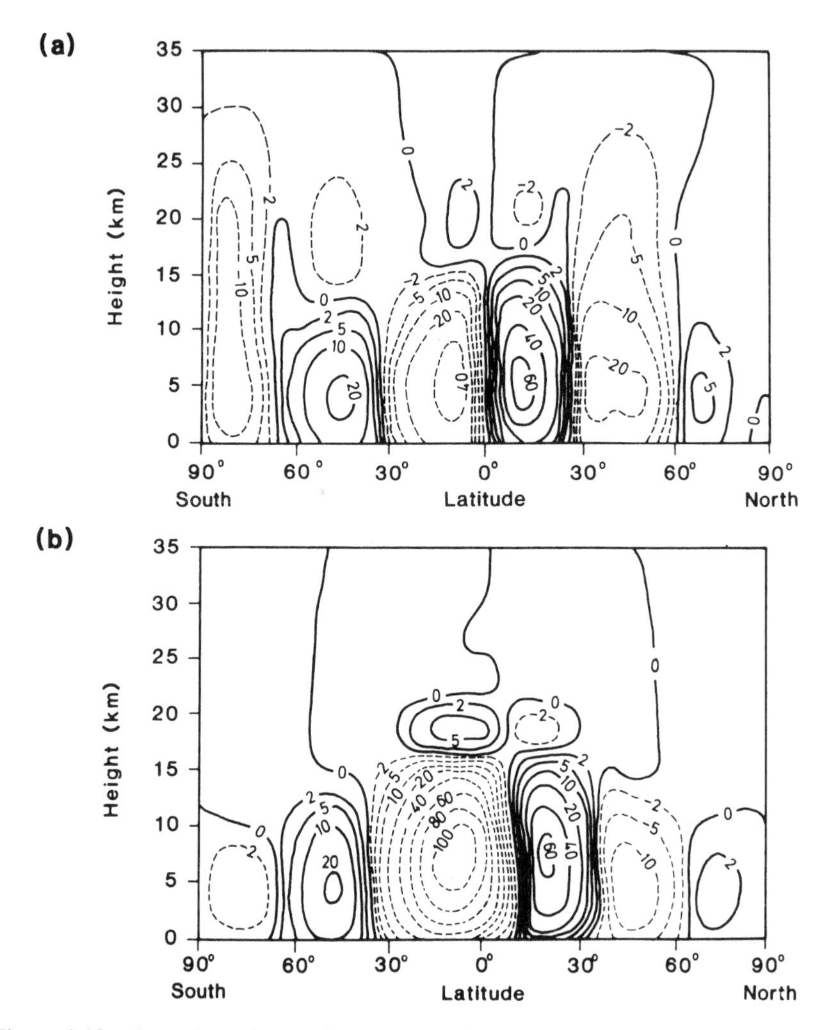

Figure 4.13 Annual and zonal mean meridional mass flux stream function (10^9 kg s^{-1}): (a) as observed (Oort and Peixóto, 1983); (b) as simulated by the Lawrence Livermore two-dimensional SD model (reproduced by permission from MacCracken and Ghan, 1987)

pressure, g the acceleration due to gravity, f the Coriolis parameter and $\langle \rho \rangle = \langle p \rangle / (R \langle T \rangle)$. The primed notation denotes a deviation from the zonal average of these variables.

As these equations are essentially those which are solved in AGCMs, although here they are written in a simpler form, they are worth considering in some detail. The momentum equations are themselves expressed for unit mass, so that the density terms cancel and the momentum (mass × velocity) is represented simply by the velocity component in the direction of interest. Zonal momentum changes (with time) are thus represented by the first term on the left-hand side of equation (4.27). These temporal changes are balanced by the Coriolis term, $f\langle v\rangle$, and the rate of change in the poleward direction of the correlation term, $\langle u'v'\rangle$, i.e. the eddy transport of momentum in the poleward direction. Finally there is an additional frictional dissipation term, F, to be taken into consideration.

The meridional momentum equation is similarly constructed but the small temporal changes are neglected; if they were not, the model might accumulate errors and predict non-zero momentum fluxes at the poles. Consequently, the balance equation is simply between the Coriolis force and the pressure gradient force in the poleward direction, friction being neglected.

The hydrostatic equation is the third component of the conservation of momentum in the atmosphere (Table 2.1). In this case, resolution in the vertical direction yields no Coriolis component, but changes occur in the pressure field (with height). This equation (equation (4.29)) can be rewritten as

$$\frac{1}{\langle p\rangle}\frac{\partial\langle p\rangle}{\partial z} = -\frac{g}{R\langle T\rangle} \tag{4.32}$$

whence, from the ideal gas law, is derived the more common expression

$$\frac{\partial\langle p\rangle}{\partial z} = -\frac{g\langle\rho\rangle R\langle T\rangle}{R\langle T\rangle} = -g\langle\rho\rangle \tag{4.33}$$

The thermodynamic balance exists between the temporal rate of change of zonally averaged temperature and the rate at which temperature is transported both into and out of each latitude zone. This is accomplished by eddies in the lateral (northward) and vertical directions and represented by the two eddy correlation terms (second and third terms on the left-hand side of equation (4.30)). A further term represents vertical transport, taking into account adiabatic heating and cooling due to the compressibility of the atmosphere. The balance is completed by the inclusion of the diabatic heating term on the right-hand side.

The continuity equation says simply that mass can neither be created nor destroyed, i.e. the rate of change of mass in all three dimensions overall is zero. However, since zonal averages are under discussion, the change in the x direction has been averaged out, as expressed by the use of angle brackets, and only two components remain. The sum of these two is zero. Thus in regions where there is net divergence or convergence, there must of necessity be a vertical motion also.

In writing and discussing these 'prototype' equations we have neglected the need for a spherical geometry for a complete treatment. Despite this, the representation is useful. The horizontal component of the eddy momentum flux $\langle u'v'\rangle$ is not only responsible for transferring zonal momentum but also drives the meridional circulations. Figure 4.14 shows mean meridional cross-sections

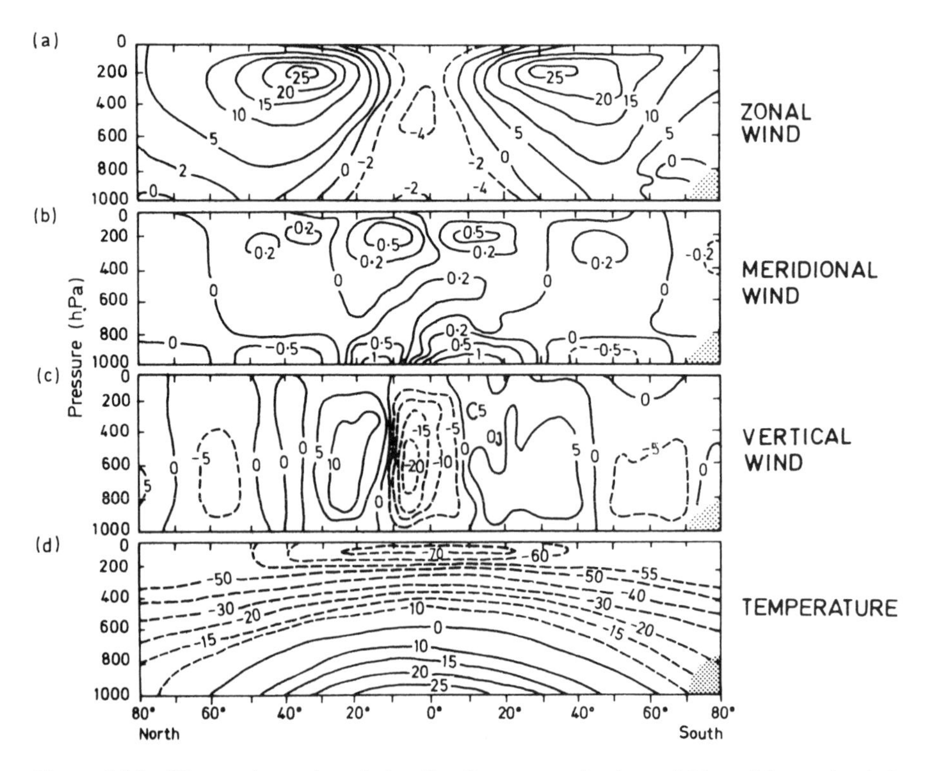

Figure 4.14 Observed zonal statistics for four atmospheric variables: (a) zonal wind (m s^{-1}); (b) meridional wind (m s^{-1}); (c) vertical wind ($\times 10^{-3}$ m s^{-1}); (d) temperature (°C) (reproduced by permission of Academic Press from Oort and Peixóto, 1983)

of (a) zonal wind, (b) meridional wind, (c) vertical motion and (d) temperature. These can be calculated by specifying a vertically and latitudinally varying distribution of eddy momentum fluxes similar to those observed. Note that the magnitude of the three induced velocities differs considerably: the zonal wind can be as high as 30 m s^{-1}, whilst the meridional wind is typically 0.25 m s^{-1} and the induced vertical motion is 0.005 m s^{-1}.

For two-dimensional climate models it is necessary to find representations for the eddy fluxes in equations (4.27)–(4.29) so that this system of equations can be solved numerically. In these models the atmosphere is represented on a latitude versus pressure (height) grid with approximately 10 layers and 10–20 grid points between the poles. Often, considerable effort goes into representing atmospheric radiative processes and surface features, although the main problem remains the characterization of eddy transports.

Eddy transport is of critical importance for determining the equator to pole temperature gradient and the vertically distributed zonal wind field, especially the strength of the jet-stream winds. Early parameterizations of eddy flux simply related eddy transports to gradients of zonal mean variables using empirically determined diffusion coefficients. This representation is similar to the parameterization used in some EBMs for the meridional energy transport (e.g. equation (3.12) in Section 3.2). This parameterization was based on the argument that, since baroclinic waves are driven by the meridional temperature gradient, their eddy transports might also be simply parameterized as being proportional to this gradient. Thus the eddy heat flux is given by

$$\langle v'T' \rangle = -K_{\mathrm{T}} \frac{\partial \langle T \rangle}{\partial y} \tag{4.34}$$

and the eddy momentum flux by

$$\langle u'v' \rangle = -K_{\mathrm{M}} \frac{\partial \langle u \rangle}{\partial y} \tag{4.35}$$

where K_{T} and K_{M} are empirically derived coefficients for temperature and momentum. More detailed study has shown that, while the diffusive representation is fairly reasonable for eddy heat transport, it is a completely inadequate representation of eddy momentum flow, since momentum can be transported up as well as down the meridional gradient of momentum. Consequently, later parameterizations reformulated the transport equations in terms of the potential vorticity gradient.

Originally, the parameterizations described for two-dimensional models were empirically based. However, subsequent theoretical analysis by a number of authors in the early and mid-1970s demonstrated that the diffusion coefficients, as well as the eddy transport itself, may be proportional to the meridional temperature gradient. It was found that the equator to pole temperature gradient was considerably different when computed with EBMs which used a value for the diffusion coefficient dependent on the temperature gradient as opposed to a constant eddy diffusion coefficient (Figure 4.15). This finding suggests that one of the reasons why the energy balance climate models (EBMs) showed considerable sensitivity to a small decrease in solar constant was that the diffusion coefficient in these models was constant, rather than being a function of the temperature gradient. A much greater decrease in the solar constant is necessary to initiate an ice age in a model which includes a temperature gradient dependency of the diffusion coefficient. This is because the temperature gradient remains high at low values of solar input. This recognition may offer another partial solution to the cool Sun–enhanced early surface temperature paradox described earlier (see especially Sections 3.3 and 4.6).

The basis of the parameterization problem is the simplification which is generally made in the solution of the zonal flow equation in baroclinic wave theory. The usual simplifications are to assume that the zonal wind, $\langle u(y, z) \rangle$, is a function of y only (the barotropic solution) or z only (the baroclinic solution). Instability to small disturbances of a zonal wind which varies only in

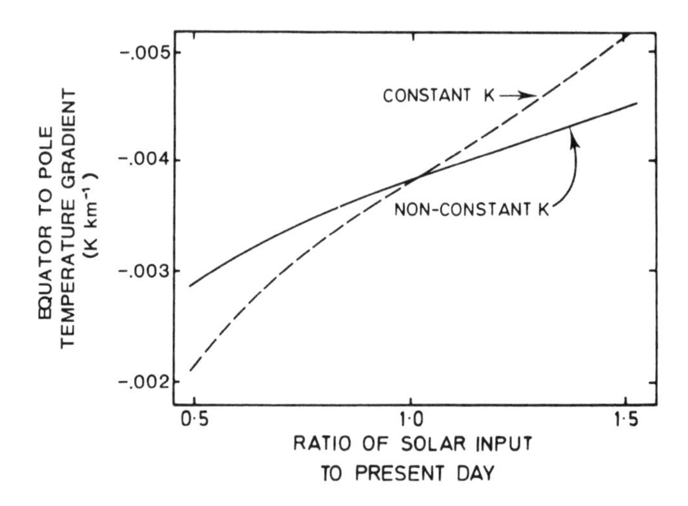

Figure 4.15 The effect of the inclusion of a variable (i.e. itself a function of the temperature gradient) diffusion coefficient, K, on the equator-to-pole temperature gradient in an EBM (reproduced by permission of the American Meteorological Society from Stone (1973) *Journal of the Atmospheric Sciences*, **30**, 521–529)

the vertical (z) direction (a baroclinic instability) converts to eddy energy the energy that is stored in the current latitudinal variation of zonal temperature, $\langle T \rangle$. This energy is released by the eddy flux of heat $\langle v'T' \rangle$. On the other hand, instability of a zonal wind to horizontal (y component) perturbations (a barotropic instability) converts kinetic energy of the zonal wind to eddy energy through the flux of horizontal eddy momentum $\langle u'v' \rangle$.

The parameterization of momentum fluxes is considerably more complex when theoretically based than the parameterization of heat fluxes. It has been shown that potential vorticity is more suitable for the treatment of $\langle u'v' \rangle$, as the eddy momentum flux can be obtained once eddy potential vorticity and eddy heat fluxes have been derived. A further problem with eddy flux parameterizations is the existence in the atmosphere of large stationary eddies forced both by topography and by land/ocean temperature contrasts. Since the parameterization schemes described above represent only transient baroclinic eddies, it is possible that there will be an underestimation of the total eddy transport produced due to the neglect of these stationary eddies. However, observational data suggest that a compensatory mechanism may exist, since total eddy flux seems to be correlated better with observed temperature gradients than is the transient eddy flux alone.

Full formulations of two-dimensional SD models often also include vertical and horizontal eddy transports of water vapour as well as those of heat and momentum described above. Since two-dimensional models attempt to parameterize only the eddy transport, while the mean meridional transport terms are computed explicitly, it is hoped that inadequacies in the eddy transport parameterizations are compensated for in the explicit calculation of meridional transport.

4.7.2 'Column' processes in two-dimensional statistical dynamical models

Vertical motion must be parameterized rather carefully in zonally averaged models. Around any one latitudinal band, there would normally be a range of different surface types and atmospheric states possessing different degrees of instability. The extent of these areas and of the instability would be expected to vary with season and time of day. Unfortunately, zonally averaged conditions can be quite stable even though within a latitude zone there might be many locations where convection would occur. Thus if cloud cover and precipitation are to be computed realistically, an 'on–off' switch in a zonally averaged model would result in zones being either cloudy with precipitation or having clear skies and no rain.

A more complex, and usually empirically derived, formulation is required in 2DSDs to determine the onset and extent of convection, and often of cloud formation processes, instead of the simple convective adjustment used in one-dimensional RC models (Section 4.2). Generally less than 100% cloud cover is predicted within individual layers in an attempt, once again, to avoid 'on–off switching'. Similar sub-gridscale descriptions of vertical convection and cloud formation processes have been derived to limit the extent of precipitation and/or cloud within a grid cell of a GCM (see Chapter 5). Convective precipitation is likely to be the major contributor to rainfall in zonally averaged models. However, large-scale precipitation can also occur if the mixing ratio of an air parcel exceeds the saturation mixing ratio. Since it is most unlikely that zonally averaged relative humidities would ever reach saturation in anything other than the near surface layer, thresholds are often set somewhat lower, say for instance a zonally averaged relative humidity of 80%.

Surface to atmosphere fluxes of momentum and sensible and latent heats are computed using standard bulk aerodynamical formulae:

$$H_{\mathrm{D}} = \rho c_{\mathrm{D}} u^2 \tag{4.36}$$

$$H_{\mathrm{S}} = \rho c_{\mathrm{p}} c_{\mathrm{S}} u (T_{\mathrm{s}} - T_{\mathrm{a}}) \tag{4.37}$$

$$H_{\mathrm{L}} = \rho L c_{\mathrm{L}} u (q_{\mathrm{s}} - q_{\mathrm{a}}) \tag{4.38}$$

where c_{D}, c_{s} and c_{L} are the aerodynamic drag coefficients, u the wind speed evaluated at the same standard height as the drag coefficient, T the temperature, q the water vapour mixing ratio and L the latent heat of vaporization. This type of parameterization of surface to atmosphere fluxes is the same as is used in GCMs (see Chapter 5).

It is possible to incorporate some characteristics of ocean surfaces in two-dimensional models. If oceanic heat fluxes are specified, part of the meridional transport of energy can be subsumed into this oceanic transport term. These models can also be used to perform more detailed studies of ice-albedo feedback, since sea-ice and land snow-cover can be computed independently within a latitude zone, and surface temperatures associated with the different surface types can be computed and stored independently of one another.

The most useful area of application of two-dimensional SD models is probably predicting the effects of small perturbations that are fairly zonally homogeneous, in atmospheric chemistry and palaeoclimate studies. Good examples include the increase in Arctic aerosol loading and transient increases in stratospheric and tropospheric aerosol and stratospheric ozone depletion. Two-dimensional SD models are especially useful in such studies for two reasons: (i) because detecting the signal produced by such small changes in a full three-dimensional model with eddy-related inherent variability would take very many years of climate simulation; (ii) because the slight disturbances in the climatic state do not negate the assumption inherent in the formulation of two-dimensional models that the eddy fluxes can be satisfactorily parameterized. A full two-dimensional SD model should be approximately 100–1000 times faster in execution than a GCM of roughly equivalent resolution and physical detail (see Figure 1.5).

4.8 OTHER TYPES OF COMPUTATIONALLY EFFICIENT MODELS

It is possible to select two domains from the climate system other than the 'height in the atmosphere' and 'latitude zone' which characterize two-dimensional SD models. Indeed a number of modelling groups have 'added' a second dimension to existing one-dimensional models or stripped one dimension out of a 3-D model. For example, the ocean-plus-land 'box' models described in Section 3.6 could be thought of as two-dimensional models.

4.8.1 An upgraded energy balance model

Another variant of an EBM was developed by William Sellers in the 1970s. He worked with the thermodynamical energy equation for the Earth–atmosphere system applied separately to the land- and water-covered portions of each 10° latitude belt of an Earth with a single large continent extending from pole to pole. In other words, this two-dimensional model is a cleverly partitioned EBM. The model was formulated in terms of an idealized continent/ocean system. Except for latitudes between 40° and 70° S, the fraction of each 10° latitude belt occupied by land was set equal to the present-day value, and the land masses were offset in relation to one another in order to give a meridional transport across each 10° latitude circle from water to land, water to water, land to land and land to water similar to that currently observed. The energy equation in each area is a function of the surface, e.g. the equation for the vertical temperature profile, $T(p)$ (p is pressure), is written in terms of the surface temperature, T, and the vertical temperature gradient, $\partial T/\partial p$:

$$T(p) = T_s - (p_s - p)\partial T/\partial p \tag{4.39}$$

where $\partial T/\partial p$ is specified as 0.12 K hPa^{-1}. This two-dimensional model also differs from the basic EBM of Sellers described in Chapter 3 in that the eddy diffusivities are used to parameterize the poleward transport of heat by ocean currents, atmospheric standing waves and transient eddies. The values of the atmospheric eddy coefficients for sensible heat, K_H, and water vapour, K_v, (assumed to be equal) are proportional to the first power of the north–south temperature gradient

$$K_H = K_v = 0.25\,|\Delta T| \times 10^6 \tag{4.40}$$

where ΔT is the temperature gradient between successive 10° latitude belts. In addition, an eddy diffusivity, K_w, for heat transfer by ocean currents was specified as

$$K_w = 1.7 \times 10^5 \times (1 - A_I)A_L \tag{4.41}$$

where A_I is the fractional area of the oceans covered by ice. The factor A_L, being the fraction of a given latitude belt occupied by land, allows for the effect of the continents in channelling the north–south ocean currents. The average value of K_w is about 5×10^4 m^2 s^{-1}, which agrees well with values found in the vicinity of the Gulf Stream, which is about one order of magnitude larger than values considered typical for the more quiescent parts of the oceans. This discrepancy was intentional and was an attempt to account partially for the neglected heat transport by vertical circulations (cf. Section 3.6).

The albedo–temperature feedback is still an important feature of this model, but now cloudiness is treated explicitly by computing surface net radiation for the separate cases of clear and cloudy skies. This model is very much more comprehensive than an EBM, but is still highly tuned as compared with the general circulation climate models which will be described in Chapter 5. Despite some discrepancies between modelled and observed climate, especially in high latitudes, this relatively simple model is able to reproduce seasonal signals in various climatic variables fairly successfully. Although this result might have been anticipated, since much of the climatic response is controlled by thermal inertia, it is still worthy of note because the seasonal range in most, if not all, variables is considerably greater than many anticipated climatic perturbations.

4.8.2 Multi-column RC models

It is possible to set up one-dimensional RC models for a number of latitude zones as characterized by an EBM. Such a combination would constitute a two-dimensional model. An example of such a two-dimensional model is shown in

Figure 4.16. This particular example is based partly on the 1D RC discussed earlier in this chapter and included on the CD. In addition, the surface heat balance equation for land is given by

$$C_L D \frac{\partial T_L}{\partial t} = R_g - I - H_S - H_L \tag{4.42}$$

where C_L, D and T_L are the heat capacity, effective depth and temperature of the land surface, R_g is the solar radiation absorbed and I, H_S and H_L are the infrared, sensible and latent heat fluxes, respectively. The atmospheric balance can be written as

$$\frac{\partial T}{\partial t} = Q_S + Q_I + Q_L + A \tag{4.43}$$

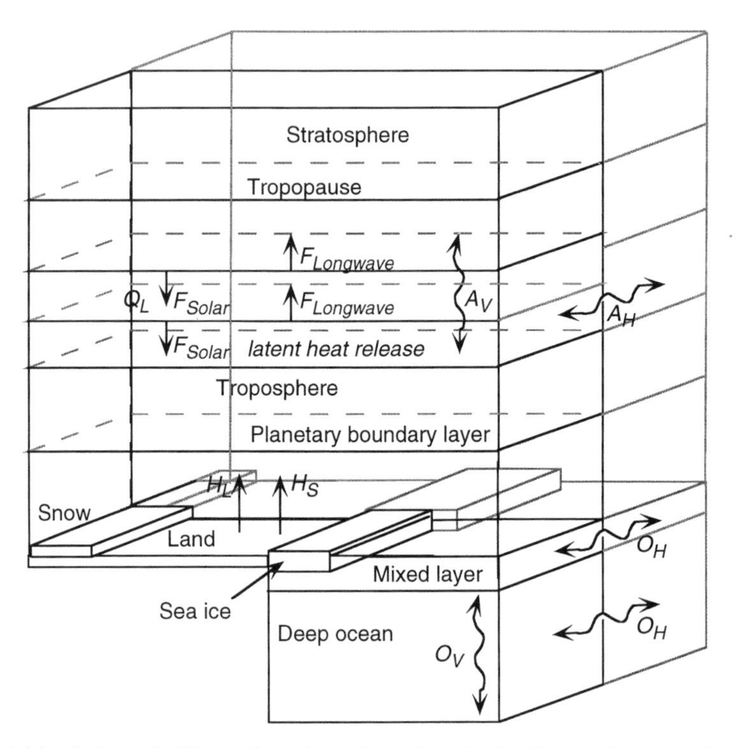

Figure 4.16 Schematic illustration of one latitude belt of a 2D zonal model. F_S and F_L denote shortwave and longwave radiation, respectively. A_V and A_H denote vertical and horizontal heat transport in the atmosphere respectively; Q_L is latent heat release; H_L and H_S are latent and sensible heat fluxes, respectively, for land (with and without snow), sea-ice and ocean; O_V and O_H denote vertical and horizontal heat transport, respectively, in the ocean (after De Haan *et al.*, 1994)

where T is the atmospheric temperature and Q_S, Q_I and Q_L are the heating rates due to solar, longwave and latent heating. A is the heating rate due to dynamical redistribution of heat, which is formulated empirically. The model includes four different surface types, as illustrated in Figure 4.16.

The ocean model is composed of two components: an ocean climate model and an ocean biosphere and chemistry model. The ocean climate model includes heat exchanges with the atmosphere (but not with sea-ice) and prescribed advective flows derived from observational data, as shown in Figure 4.17. The surface heat fluxes are simulated for the Atlantic and Indo-Pacific basins separately and the heat thereby transported from the equator to the poles. The ocean parameterization also includes the role of the ocean biomass in climate, through its uptake of carbon (so reducing atmospheric CO_2 concentrations). The model includes downward transport of substances by phytoplankton and the subsequent settling of marine grazer faeces, each parameterized in terms of fluctuating environmental conditions. The simplified food web includes only phytoplankton and detritus. The phytoplankton are governed by

$$\frac{\mathrm{d}B}{\mathrm{d}t} = B\left(P_{\max} f(I) \frac{N}{N+k} - r - m\right) \tag{4.44}$$

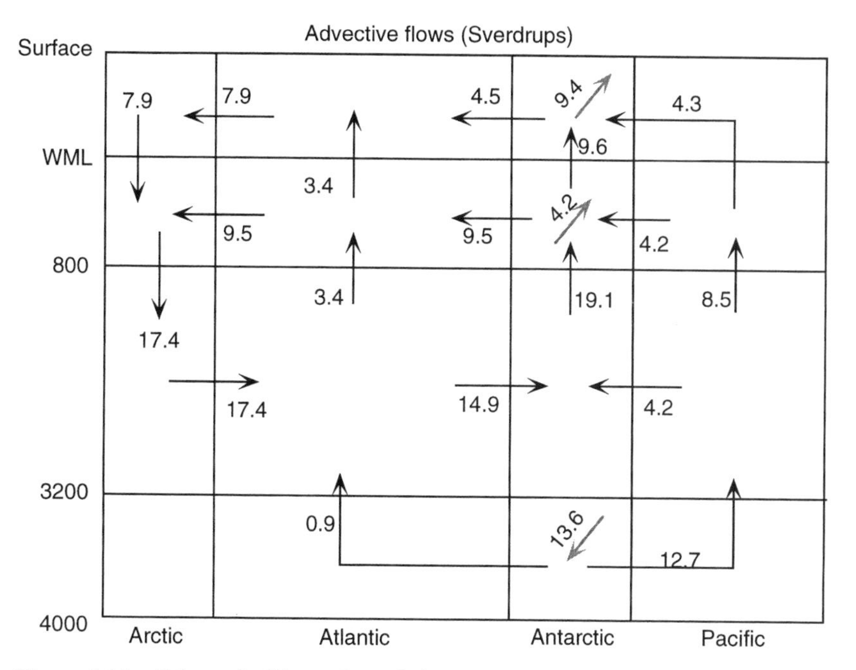

Figure 4.17 Schematic illustration of the ocean circulation as used in the 2D zonal model shown in Figure 4.16. Transports are derived from observational data (after De Haan *et al.*, 1994)

where B is the phytoplankton biomass, N the organic nitrogen, P_{max} the maximum production rate, $f(I)$ a light limitation function, k the half-saturation fraction for N, m the mortality rate and r the respiration rate. The detritus is governed by

$$\frac{\mathrm{d}D}{\mathrm{d}t} = mB - sD \tag{4.45}$$

where D is the detritus concentration and s is the settling rate for detritus.

This model provides for simulation of the main features of the ocean–atmosphere system on a latitudinally averaged basis. The climate sensitivity of the model was found to be within the IPCC range, and the carbon uptake by the ocean biomass within the range of current state-of-the-art carbon cycle models (see Chapter 6). The model is designed as one component of a large integrated assessment model; hence the need for the inclusion of the ocean biomass, which is a major sink for atmospheric carbon (CO_2) and therefore an important factor in long-term climate simulation. Such integrated assessment models are discussed more fully in Chapter 6.

4.8.3 A severely truncated spectral general circulation climate model

An alternative method of achieving a similar position on the climate-modelling pyramid (Figure 2.1), as an embellished zonal model, is to restrict a more complex model. One example is severely to truncate a global spectral model. Figure 4.18, which complements Figure 1.5, illustrates the relative performance (measured qualitatively as a percentage of the dynamical soundness of a GCM) against the computational effort expended. As the difference between the computational domains of spectral and cartesian grid AGCMs will not be explained until Chapter 5, here it is sufficient to say that severe truncation of a spectral model is roughly equivalent to greatly coarsening the spatial grid of a cartesian grid model. In many senses, however, the spectral truncation is less detrimental than the equivalent grid coarsening, since when the resolution is decreased in a cartesian grid model, baroclinically unstable waves are more and more poorly resolved, reducing the dynamical soundness of the model simulation. On the other hand, spectral truncation can be very great before the fundamental dynamics are badly misrepresented. This type of highly truncated GCM can be 'better' than a two-dimensional SD model since the large-scale eddy fluxes of momentum, heat and moisture are calculated explicitly rather than being parameterized. Despite this explicit computation, the computational effort is approximately two orders of magnitude smaller than that for a fully resolved GCM. The loss of spectral resolution means that many features are either poorly predicted or must be improved by better sub-gridscale parameterization. However, in order to improve climate representation, the distortions to the energy cascade, which are severe, must be compensated for by parameterization

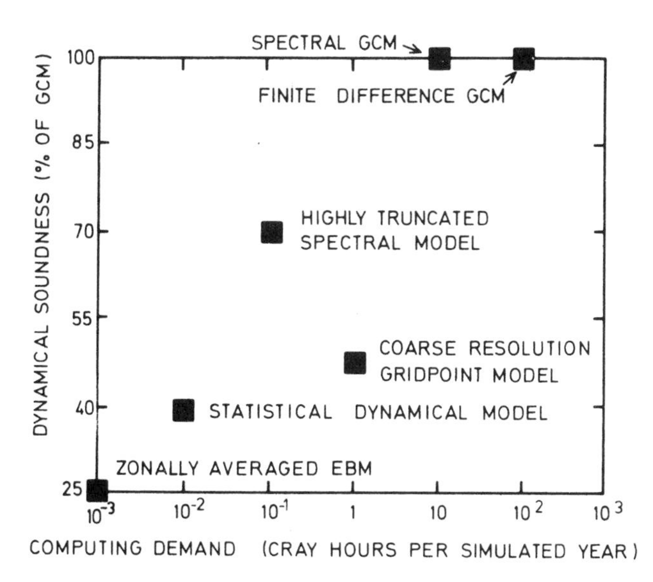

Figure 4.18 Variation of the 'dynamical soundness' of atmospheric models versus their computational requirements. The scale of dynamical soundness is subjective (reproduced by permission of the American Meteorological Society from Semtner (1984b), *Journal of Climate and Applied Meteorology*, **23**, 353–374)

of energy conversion and dissipation. While physical modellers have deemed that a very detailed effort here would be inappropriate since the results would be unlikely to compensate for lost spectral resolution, those developing integrated assessment models (Chapter 6) are exploring and developing this type of truncated model.

4.8.4 Repeating sectors in a global 'grid' model

Another technique for reducing the complexity of three-dimensional models is the 'Wonderland' approach, which has been used by modellers at the Goddard Institute for Space Studies in New York. The layout of the model is shown in Figure 4.19. Here the model domain is made up of continents which mimic the distribution of land on Earth but that occupy only 120°. The remainder of the globe is made up by repeating the same geography. The model therefore allows simulations which are hundreds or thousands of years long with only modest computer resources. This approach is similar to that used by some of the early global models at GFDL. It is possible to argue that severely truncated models such as these are similar to two-dimensional models in that they approximate the complete climate system by highly parameterized representation of one or more aspects. On the other hand, as they retain three dimensions they are, perhaps, more appropriately termed two-and-a-half-dimensional models.

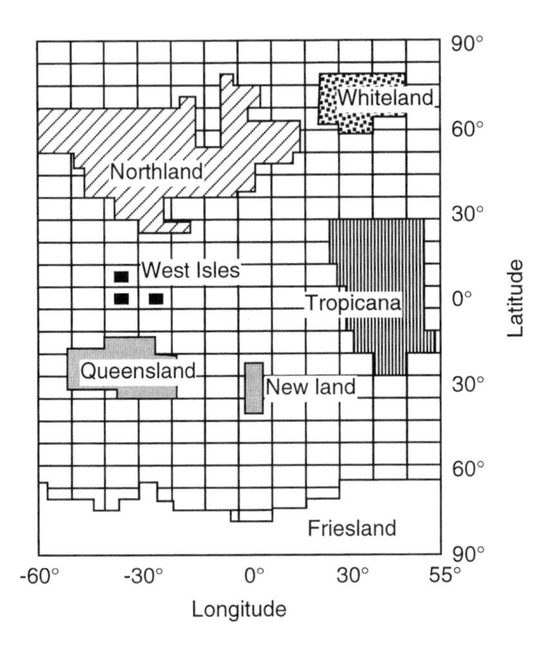

Figure 4.19 Distribution of 'Wonderland' continents in a global model with simplified geography (after Hansen *et al.*, 1992). The model's surface and atmosphere occupy only 120° of longitude, this sector being repeated around the globe

Some atmospheric dynamicists have chosen to solve a set of equations similar to equations (4.27)–(4.31) but in three directions (i.e. including longitudinal variations). However, they exclude much of the detail of the treatment of the radiative fluxes which is usual in one-dimensional RC and two-dimensional SD models. Thus, although they are strictly three-dimensional modellers, in choosing to neglect the complexities of radiative processes, these modellers have slipped a part-dimension and once again the result might be more correctly termed a two-and-a-half-dimensional model.

As the desire to increase the predictability obtainable with a climate model increases, so in general does the number of dimensions represented. In Chapter 5, the closest point yet attained by climate modellers to the apex of the climate-modelling pyramid, the coupled OAGCM, is described.

4.9 WHY ARE SOME CLIMATE MODELLERS FLATLANDERS?

In the introduction to Chapter 3, we noted that two-dimensional beings cannot understand three dimensions and must describe what they see in terms of the only dimensions they recognize. Since most climate modellers of our experience are three-dimensional, why is it that some of them choose to represent the climate system in terms of highly parameterized formulations which are essentially a reduction of the dimensionality of the real system?

One answer is, of course, that by simplifying, they aim to represent important mechanisms without overburdening their computational resources. In this way many of them can watch their simplified systems develop for longer time periods than would be possible if they were to use fully coupled OAGCMs operating on even the best available computers of today. This method can be carried further. For example, EBMs have been used to investigate 'Milankovitch' variations. However, as we tried to show in Section 4.8, the demarcation between two- and three-dimensional models is often rather arbitrary. The other reason is that by stripping away 'unnecessary complexities', the workings of any system are more readily viewed and, hence, hopefully more easily understood. Consequently, some questions, especially those pertaining to basic mechanisms and to long time periods, are better tackled with simpler (i.e. more highly parameterized) models.

The question of how far two-dimensional models should be used to represent the climate is a difficult one. Recent work has suggested that the latitudinal response to an increase in CO_2 and other trace gases is similar whether the perturbation is applied as a single 'jump' or as a transient change.

One message that we hope you have gained from this chapter and that will be underlined in Chapter 5 is that it is not scientifically short-sighted to choose to parameterize some aspects of the climate system and thereby reduce the number of dimensions; it is valuable and probably inevitable. The skill of a good modeller lies in the ability to identify results which are characteristic of features of the real system as opposed to those which are facets of the constrained framework of the model. Thus, climate modellers can choose to dwell happily in Flatland but they must not begin to think like natives.

RECOMMENDED READING

Chylek, P. and Kiehl, J.T. (1981) Sensitivities of radiative–convective models. *Journal of the Atmospheric Sciences* **38**, 1105–1110.

De Haan, B.J., Jonas, M., Klepper, O., Krabec, J., Krol, M.S. and Olendrzynski, K. (1994) An atmosphere–ocean model for integrated assessment of global change. *Water, Air and Soil Pollution* **76**, 283–318.

Garcia, R.R., Stordal, F., Solomon, S. and Kiehl, J.T. (1992) A new numerical model of the middle atmosphere. 1. Dynamics and transport of tropospheric source gases. *Journal of Geophysical Research* **97**, 12967–12991.

Hansen, J.E., Johnson, D., Lacis, A.A., Lebedeff, S., Lee, P., Rind, D. and Russell, G. (1981) Climate impact of increasing atmospheric CO_2. *Science* **213**, 957–966.

MacKay, R.M. and Khalil, M.A.K. (1991) Theory and development of a one dimensional time-dependent radiative–convective model. *Chemosphere* **22**, 383–417.

Manabe, S. and Wetherald, R.T. (1967) Thermal equilibrium of the atmosphere with a given distribution of relative humidity. *Journal of the Atmospheric Sciences* **24**, 241–259.

Potter, G.L. and Cess, R.D. (1984) Background tropospheric aerosols: Incorporation within a statistical–dynamical climate model. *Journal of Geophysical Research* **89**, 9521–9526.

Potter, G.L., Ellsaesser, H.W., MacCracken, M.C. and Mitchell, C.S. (1981) Climate change and cloud feedback: The possible radiative effects of latitudinal redistribution. *Journal of the Atmospheric Sciences* **38**, 489–493.

Ramanathan, V. and Coakley, J.A. (1979) Climate modelling through radiative–convective models. *Reviews of Geophysics and Space Physics* **16**, 465–489.

Rossow, W.B., Henderson-Sellers, A. and Weinreich, S.K. (1982) Cloud-feedback – a stabilizing effect for the early Earth. *Science* **217**, 1245–1247.

Saltzman, B. (1978) A survey of statistical–dynamical models of terrestrial climate. *Advances in Geophysics* **20**, 183–304.

Schneider, S.H. (1972) Cloudiness as a global climatic feedback mechanism: the effects on the radiation balance and surface temperature of variations in cloudiness. *Journal of the Atmospheric Sciences* **29**, 1413–1422.

Sellers, W.D. (1973) A new global climate model. *Journal of Applied Meteorology* **12**, 241–254.

Sellers, W.D. (1976) A two-dimensional global climate model. *Monthly Weather Review* **104**, 233–248.

Taylor, K.E. (1980) The roles of mean meridional motions and large scale eddies in zonally averaged circulations, *Journal of the Atmospheric Sciences* **37**, 1–19.

Thompson, S.L. and Schneider, S.H. (1979) A seasonal zonal energy balance climate model with an interactive lower layer. *Journal of Geophysical Research* **84**, 2401–2414.

Wang, W.-C. and Stone, P.H. (1980) Effects of ice-albedo feedback on global sensitivity in a one-dimensional radiative–convective model. *Journal of the Atmospheric Sciences* **37**, 545–552.

CHAPTER 5

General Circulation Climate Models

Predictability is to prediction as romance is to sex.

K. Miyakoda (public communication, 1985)

5.1 THREE-DIMENSIONAL MODELS OF THE CLIMATE SYSTEM

The most 'complete' models of the climate system are constructed by discretizing and then solving equations which represent the basic laws which govern the behaviour of the atmosphere, ocean and land surface. These three-dimensional models of the general circulation of the atmosphere and ocean have come to be known as GCMs. The term GCM is often used loosely and can be thought of as referring to a global climate model or to a general circulation model. In the latter case, the acronym is often qualified by the addition of an 'A' when speaking of strictly atmospheric models or an 'O' when talking of ocean only models. Sometimes the term 'coupled' and the acronym CGCM is also used. CGCMs and some AGCMs can include complex representations of the land surface. The history of GCMs, which was outlined in Chapter 2, is such that they are still often thought of as atmospheric models with additional 'add-on' components. However, as the models have developed, the amount of computational effort for the 'add-on' components has come to rival and even exceed the effort required to simulate the atmosphere alone.

The differences in geometry and composition of the atmosphere and ocean mean that there are a number of significant differences in the modelling approaches taken. The atmosphere is a spherical shell of compressible gas which covers the whole Earth. It is, oddly, heated mostly from the Earth's surface below. The ocean, on the other hand, is relatively incompressible, is heated only at its top surface and is confined to particular parts of the Earth's surface: the ocean basins. In this chapter, some of the ways in which the atmosphere and ocean are modelled in three dimensions are described. The ways in which these atmospheric and oceanic models are coupled to cryospheric and biospheric models and the ways in which the models are used are also considered in this chapter.

5.2 ATMOSPHERIC GENERAL CIRCULATION MODELS

In this section the basic formulation of atmospheric GCMs (AGCMs) is considered with particular references to the differences between so-called 'spectral' and 'finite grid' models. A schematic of an atmospheric model is shown in Figure 5.1. The dynamics comprise the numerical schemes by which large-scale atmospheric transports are accomplished. These dynamics are computed in either physical space or in spectral space, as described below. The equations which are solved in these models are similar to those first formulated for numerical weather forecast models, although in the early stages of climate modelling there was a requirement for increased emphasis on conservation (of energy and matter) which was less important for short-term

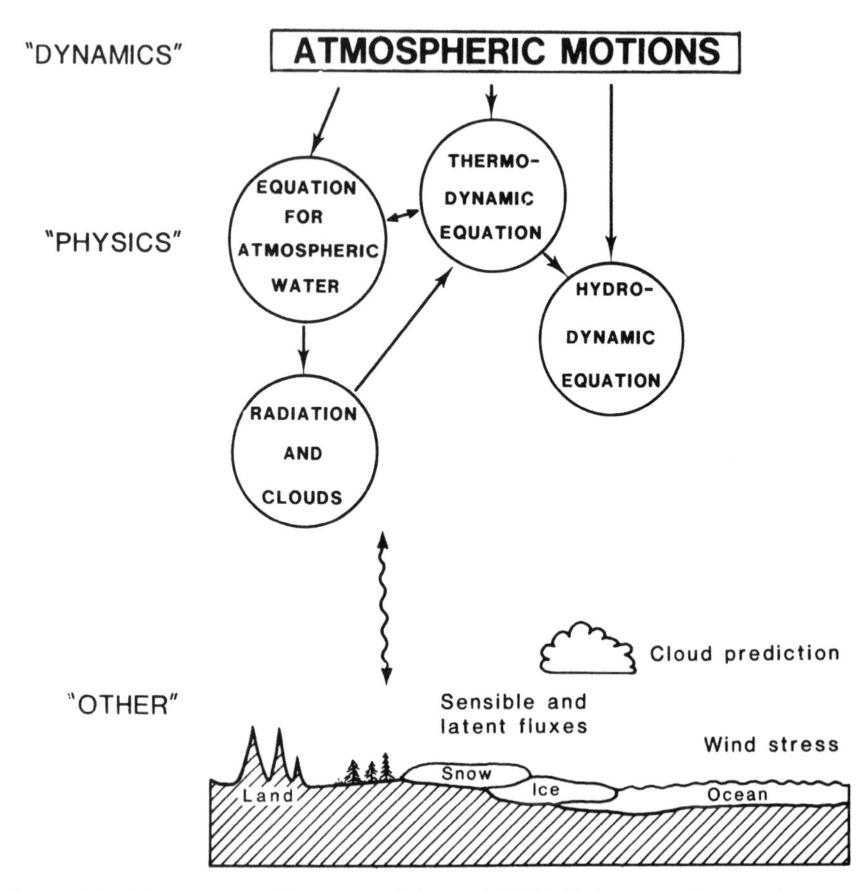

Figure 5.1 The processes incorporated in an AGCM. It is generally true that more computational effort is expended on the dynamics and the physics than on the other processes incorporated in AGCMs

weather forecasts. Any AGCM must be formulated with some fundamental considerations for:

1. conservation of momentum

$$\frac{\mathrm{D}\mathbf{v}}{\mathrm{D}t} = -2\boldsymbol{\Omega} \times \mathbf{v} - \rho^{-1}\nabla p + \mathbf{g} + \mathbf{F} \tag{5.1}$$

2. conservation of mass

$$\frac{\mathrm{D}\rho}{\mathrm{D}t} = -\rho\nabla \cdot \mathbf{v} + C - E \tag{5.2}$$

3. conservation of energy

$$\frac{\mathrm{D}I}{\mathrm{D}t} = -p\,\frac{\mathrm{d}\rho^{-1}}{\mathrm{d}t} + Q \tag{5.3}$$

4. ideal gas law

$$p = \rho RT \tag{5.4}$$

where $\mathbf{v}$ = velocity relative to the rotating Earth,
t = time,

$\frac{\mathrm{D}}{\mathrm{D}t}$ = total time derivative $\left[= \frac{\partial}{\partial t} + \mathbf{v} \cdot \nabla\right]$

$\boldsymbol{\Omega}$ = angular velocity vector of the Earth,
ρ = atmospheric density,
$\mathbf{g}$ = apparent gravitational acceleration,
p = atmospheric pressure,
$\mathbf{F}$ = force per unit mass,
C = rate of creation of atmospheric constituents,
E = rate of destruction of atmospheric constituents,
I = internal energy per unit mass [$=c_{\mathrm{v}}T$],
Q = heating rate per unit mass,
R = gas constant,
T = temperature,
c_{v} = specific heat of air at constant volume.

An atmospheric model needs also to conserve enstrophy (the root mean square of the vorticity). Failure to conserve enstrophy means that the energy of motion is transferred unrealistically to smaller and smaller space-scales. Very early model structures did not conserve enstrophy and, consequently, became unstable after short integration times.

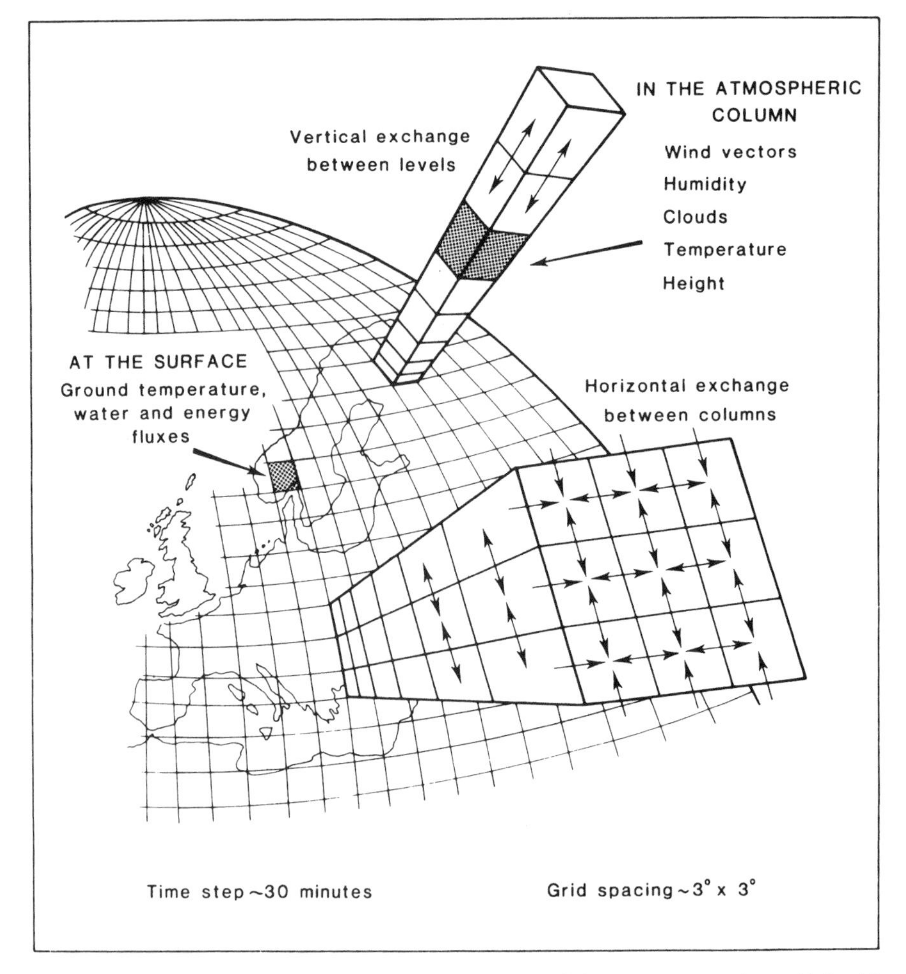

Figure 5.2 The construction of: (a) a finite grid AGCM (a rectangular grid in this case); (b) a spectral AGCM. In a finite grid AGCM, the horizontal and vertical exchanges are handled in a straightforward manner between adjacent columns and layers. Recently, modellers have begun experimenting with icosahedral grids. In a spectral AGCM, vertical exchanges are computed in grid point space (i.e. in the same way as for finite grid models), while horizontal flow is computed in spectral space. The method of transfer between grid space and spectral space can be followed with reference to the text and by reading around (b) from Point 1 to Point 4

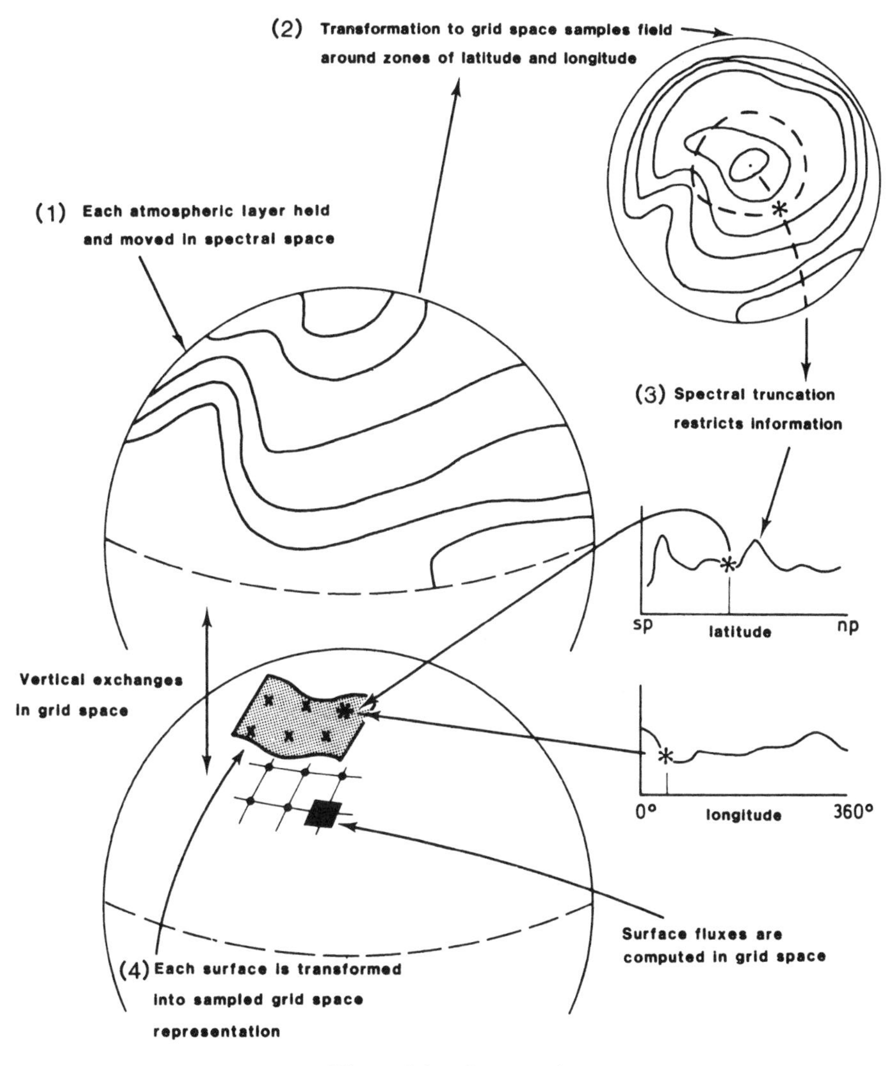

Figure 5.2 *Continued*

5.2.1 Finite grid formulation of atmospheric models

In order to simulate the atmospheric processes, the equations describing the atmosphere's behaviour have to be discretized. Modelling the atmosphere by dividing it into a series of 'boxes' is probably the easiest method to visualize. The atmosphere is reduced to a matrix of numbers, usually evenly spaced in latitude and longitude. Depending on the complexity of the model, values of a number of variables may be stored at each point. Care must be taken in the discretization process to preserve the properties of the original equations. In Figure 5.2(a) the structure of a cartesian grid AGCM is illustrated. The grid boxes are usually regularly spaced horizontally. The atmospheric column above each surface point is divided into a prespecified number of layers. Finer vertical resolution is often employed near the tropopause and the surface. The near-surface layers are influenced by boundary layer processes. High resolution near the surface is required to determine temperature and humidity gradients and atmospheric stability, needed for modelling surface fluxes.

In a finite grid AGCM, the structure is as might be expected: variables for a particular location are moved into the computer memory and all computations undertaken. Different grid structures involve the retaining of different variables at different points of the grid. Besides physical considerations, there are numerical influences on the grid spacing and timestep of an AGCM. The timestep must be short enough that the maximum speed of propagation of information does not permit any transfer from one grid point to another within one timestep. The timestep Δt is therefore governed by the restriction

$$\Delta t \leqslant \Delta x / c \tag{5.5}$$

where Δx is the grid spacing and c is the fastest propagation velocity, which in atmospheric GCMs is the speed of gravity waves. In the case where the model grid is rectangular, the grid spacing becomes small at the poles leading to the need for very short timesteps of the order of a few seconds. In this case, incorporating filtering procedures can overcome the numerical instability caused by not reducing the timestep to match this requirement. The semi-implicit timestepping scheme involves a special treatment of the motion of gravity waves (the main constraint on the timestep), with the result that longer timesteps are possible.

Finite grid models are considered more satisfactory than spectral AGCMs by many people. There are a number of reasons, including the advantage that transfer between two computational systems is not required and the fact that finite grid models do not produce the rather disconcerting predictions of, for example, negative masses of chemical species generated by spectral models. These non-physical phenomena occur in spectral AGCMs particularly where the considered variable is not smoothly varying. Steep gradients produce negative masses, humidities etc., locally. In an effort to improve the transport of 'tracers' such as water vapour in spectral models, modellers have developed 'semi-Lagrangian' schemes for water vapour which offer some improvement over the spectral method for transport. On the other hand, it has been suggested that spectral models require less computational effort than cartesian models of a similar resolution, as is illustrated in Figure 4.18, and have the advantage that the computation of gradients is more straightforward.

It has also been suggested that as available computational power increases through the use of massively parallel architectures and costs decrease, grid models will be favoured over spectral models. However, continued improvements to spectral models are retaining their competitive advantage. The next generation of grid models will likely be based on locally correct, icosahedral grids, thus overcoming the gradient-produced anomalies of spectral models and the coordinate selection difficulties of current cartesian models. At present, however, the computational advantage gained by using a spectral model has resulted in the rapid development of this model type.

5.2.2 Spectral formulation of atmospheric models

Spectral AGCMs are formulated in a fundamentally different way from finite grid AGCMs. Although the surface is retained as a grid, the atmospheric fields are held and manipulated in the form of waves (Figure 5.2(b)). The advantage of this is that computation of gradients is easier and hence computation times are reduced. Spectral models are, however, not usually formulated in all directions using waves: a rectangular grid is used for vertical transfers, and radiative transfer and surface processes are modelled in this grid space. The flow of a spectral AGCM is illustrated in Figure 5.3. The data fields are transformed to grid space at every timestep via fast Fourier transforms and Gaussian quadrature (a form of numerical integration) and back to spectral space via Legendre transform and Fourier transform. The timestepping is performed with the waveform representation and grid-point physics is incorporated after the transformation into grid space.

Representing the atmosphere with waves

Fourier's theorem states that any periodic function can be represented as a summation of sine and cosine waves. The variation of any quantity around a latitude zone is necessarily periodic and can therefore be represented as a summation of a number of waves. These waves are the Fourier transform of the original function. In the same way that the logarithm of a number contains the same information as the number, so the Fourier transform of a function contains the same information as the original grid representation. Analogously to logarithms, Fourier transforms allow mathematical operations to be performed more easily than if the original form were worked upon.

The same principle of increased computational time being required for increased resolution applies to spectral representation as well as to cartesian grid

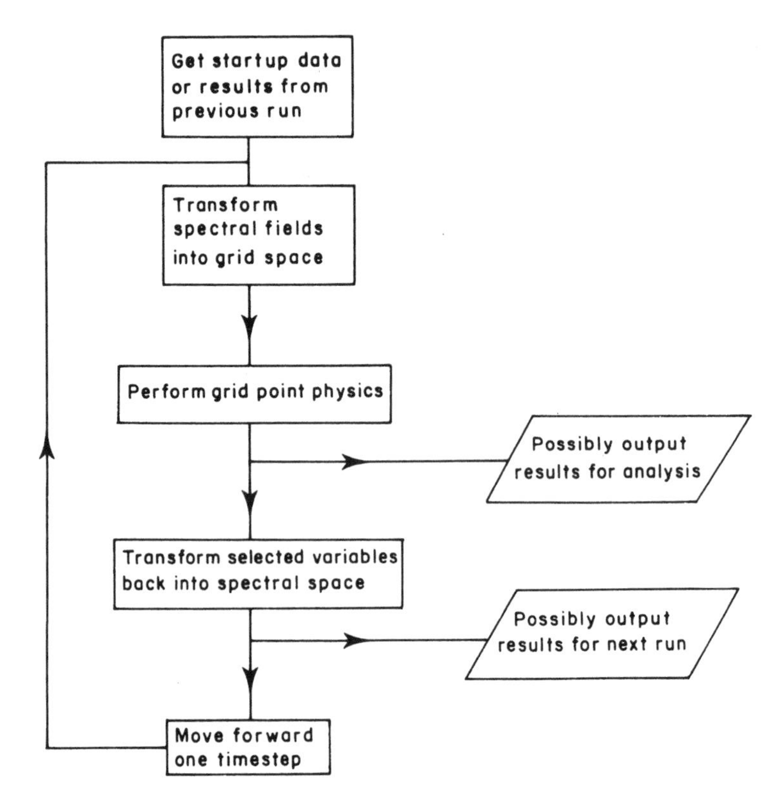

Figure 5.3 The flow through the computational scheme of a spectral AGCM. In this case there is only one transformation loop, although models with two transformation loops have also been developed

models. Resolution in a spectral AGCM is governed by the wavenumber of truncation. Early climate modelling applications used 15 waves to represent each variable in each latitude zone at each vertical level. If a model has 15 zonal waves then it is said to be truncated at wavenumber 15 (often referred to as R15 – the R standing for rhomboidal truncation, see below). More recently, it has become possible to compute and retain a larger numbers of waves and many current models use a triangular truncation (see below) with 42 zonal waves (termed T42). An example of the effect of truncation is the smoothing of the real orography of the Alps for two versions of the European Centre for Medium Range Weather Forecasting (ECMWF) model. This model is also spectral in nature, but as it is used for 10–20 day weather forecasts rather than climatic simulations it is truncated at much higher wavenumbers. Figure 5.4 compares the orography for model versions truncated at wavenumbers 63 and 106.

When considering the processes in grid space, an array of points, termed the Gaussian grid points, is defined, the number of these points being governed by

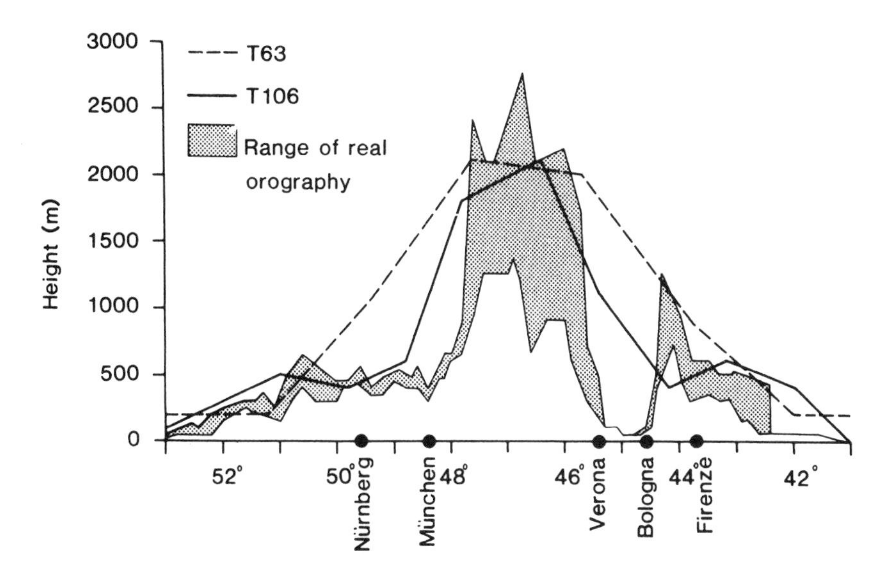

Figure 5.4 Comparison of the range in the real orography of the Alps (shown in cross-section) with the representation used in two versions of a spectral AGCM truncated at wavenumbers 63 and 106 (reproduced by permission from Simmons and Bengtsson, 1988)

the particular truncation level of the model. Overspecification of these grid points results in excessive computation times, whereas underspecification results in aliasing (the transferring of energy from high frequencies to low frequencies through poor sampling) of the high frequencies. Recent developments have allowed the use of different grids for the dynamics and the physics of a spectral model and the use of a reduced grid (fewer points in longitude) near the poles.

Structure of a spectral model

We have discussed how a spectral AGCM handles information both in wave form and in grid point form. In this section, the manner in which information is transferred between these two spaces, being the principal feature of such models, will be described. The full procedure for a single timestep of a spectral model is outlined below.

An arbitrary variable, X, which has values over the surface of a sphere (e.g. Figure 5.2(b)) can be expanded as

$$X = \sum_{m=-M}^{M} \sum_{n=|m|}^{|m|+j} X_n^m Y_n^m \tag{5.6}$$

with X_n^m, the spectral coefficients, being complex (i.e. containing an imaginary part premultiplied by $i=\sqrt{-1}$ as well as a real part). The spherical harmonic, Y_n^m, is a function of longitude, λ, and latitude, ϕ, such that

$$Y_n^m = P_n^m(\sin\phi)\exp(im\lambda) \tag{5.7}$$

where P_n^m is a Legendre polynomial of degree n and zonal wavenumber m. This is the case at point 1 on Figure 5.2(b). The nature of this functional representation of the atmospheric variables is governed by the spherical coefficients. In the next stage of the model cycle (point 2), the spectral coefficients of the dynamics variables (vorticity, divergence, wind components, water vapour mixing ratio and pressure) are transformed to a latitude–longitude grid (called the Gaussian grid). First a Legendre transform is evaluated for each spectral variable at each of the Gaussian latitudes, ϕ_j. These latitudes are related to the resolution of the model such that they are the roots of the associated Legendre polynomial of order zero. For a variable X,

$$X_n^m \rightarrow X_{(m)}(\phi_j; t) = \sum_{n=|m|}^{|m|+j} X_n^m P_n^m \tag{5.8}$$

The resulting Fourier harmonics, $X_{(m)}$, at each of the Gaussian latitudes ϕ_j and at time t are transformed via a fast Fourier transform (FFT) algorithm to longitudes $\lambda_l = 2\pi l/L$, $1<l<L$

$$X_n^m \rightarrow Z(\lambda_l, \phi_j; t) = \sum_{m=-M}^{M} X_{(m)} \exp(im\lambda_l) \tag{5.9}$$

By these two steps we have achieved a spatial distribution of values at a series of grid-points. Physical processes, such as radiative transfer, convection and the ground energy budget, can now be calculated in this rectangular grid (point 3 in Figure 5.2(b)). Subsequent to such calculations, variables are transformed back into wave space via fast Fourier transform, and then the inverse Legendre transformations are built up, one latitude at a time, using Gaussian quadrature (Figure 5.3) such that

$$Z_n^m = \sum_{j=1}^{k} w_{j,k}(\phi_j) Z_{(m)}(\phi_j) P_n^m(\sin\phi_j) \tag{5.10}$$

With the model reassembled in spectral form and appropriate expressions for the time rate of change of the variables incorporated, some form of timestep procedure is applied to find the values of the spectral fields at the advanced time point. This returns us to point 1 in Figure 5.2(b). Timestepping can be accomplished in several different ways, but all rely on approximating the time differential in a finite difference form. This gradient is applied for the appropriate time period and a value for each variable is derived at the next time point.

Truncation

The type of truncation of a spectral AGCM is often quoted in model descriptions. The simplest explanation is that the truncation number represents the number of waves which are resolved around a latitude zone. This is the governing factor in determining the resolution of a spectral AGCM. More completely, it is a description of the relationship between the largest Fourier wavenumber, the highest degree of the associated Legendre polynomial and the highest degree of the Legendre polynomial of order zero. These are termed M, K and N, respectively. With reference to Figure 5.5, the truncation types are defined as:

Triangular: $M = N = K$ (Figure 5.5(b))
Rhomboidal: $K = N + M$ (Figure 5.5(c))
Trapezoidal: $N = K > M$ (Figure 5.5(d))

These are all subsets of the pentagonal case illustrated in Figure 5.5(a). These truncation types carry with them requirements for the resolution of the Gaussian grid of the model. For example, for triangular truncation

$$N_{\text{long}} \geqslant 3M + 1; \qquad N_{\text{lat}} \geqslant (3M + 1)/2 \tag{5.11}$$

Such a specification is sufficient to remove most of the aliasing in the model.

The two most common truncations used in atmospheric GCMs are the triangular (T) and rhomboidal (R) methods. The choice of truncation is somewhat arbitrary. Although rhomboidal truncation gives higher resolution in high latitudes, has some advantages for vectorization on some supercomputers and can be considered a more 'exact' representation, triangular truncation is generally considered satisfactory and has special properties which result in a more uniform representation. Owing to the divergence of solutions at alternate timesteps, a time filter may need to be employed with certain timestepping schemes to prevent computational instability. Additionally, in rectangular grid schemes a multipoint filter is applied in the east–west direction in the polarmost rows. The use of triangular or icosahedral grids obviates the need for these filters.

5.2.3 Atmospheric GCM components

In AGCMs, three-dimensional, time-dependent equations govern the rate of change of the six basic model variables: surface pressure, two horizontal wind components, temperature, moisture and geopotential height. The six basic equations solved to derive these variables are, as described in Sections 2.2.4 and 5.2.1, the hydrostatic equation, two equations for horizontal motion (equation (5.1)), and the thermodynamic (equation (5.3)), water vapour and mass continuity (equation (5.2)) equations. The use of σ coordinates combined with the assumption that the atmosphere can be considered to be in

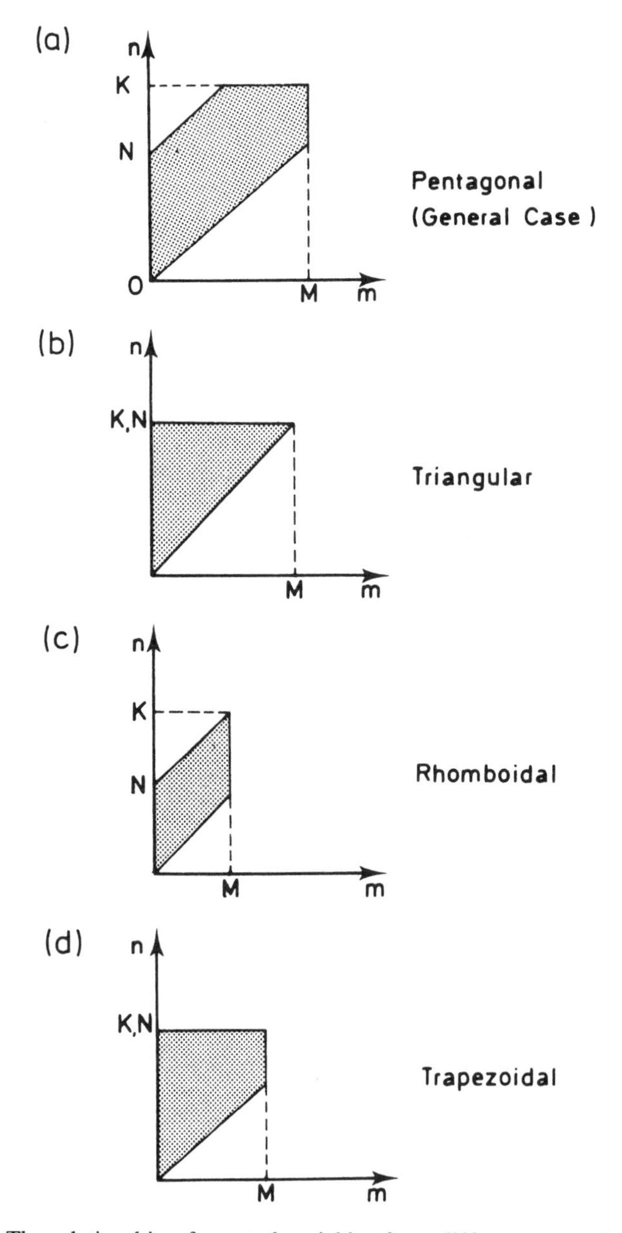

Figure 5.5 The relationship of spectral variables from different truncation types used in spectral AGCMs. The most common truncation types are triangular and rhomboidal

hydrostatic equilibrium permits the vertical wind component, w, to be obtained diagnostically from the convergence of horizontal wind components, u and v. The equations for the dynamics are solved in either wave (spectral) space or grid space, as explained above. The horizontal grid used in model integrations varies, but a typical resolution is 3–5° latitude by 4–7° longitude. Modelling groups often use a range of model resolutions so that, for example, a coarser version may be developed for the coupled ocean–atmosphere experiments. A typical timestep length is 10–30 minutes.

For the purpose of computation of the physics (as opposed to the dynamics), all GCMs can be visualized as consisting of an array of columns, which extend into both the atmosphere and the surface, distributed over the globe on a grid (Figure 5.2(a)).

The physical processes typically include the radiation scheme, the boundary layer scheme, the surface parameterization scheme, the convection scheme (including convective precipitation) and the large-scale precipitation scheme. All of these schemes, with the possible exception of the radiation scheme, are used at each location at each timestep.

Radiative transfer

The radiation scheme of an AGCM is likely to incorporate both daily and annual solar cycles. The radiation schemes are, in general, less complex than those incorporated in one-dimensional RC models (see Section 4.3), but are broadly similar. The shortwave and longwave fluxes are, of course, treated separately. For computational economy the radiation scheme is not called at every timestep, but rather only once every 1–3 hours. The radiation fluxes are usually held fixed over the intervening timesteps. Solar radiation is absorbed and scattered by atmospheric gases, clouds and the Earth's surface. Most AGCMs have shortwave radiation schemes which explicitly consider scattering and absorption by clouds as a function of zenith angle (e.g. the Delta–Eddington scheme used in the NCAR CCM2). Scattering properties are usually based on physical characteristics of the cloud, such as liquid water content and droplet size distributions, although early models used specified albedos for clouds (e.g. Table 4.1). The scattering and absorption of solar radiation is typically considered in four or five spectral regions (depending on the GCM). Usually the only gaseous absorbers considered at solar wavelengths are ozone and water vapour.

Layer cloud is usually assumed to occupy the whole depth of the model layer in which it occurs, while convective cloud is modelled as a vertical cloud column, sometimes extending through multiple layers and taking up less space horizontally than layer clouds. The sides of convective clouds are not usually allowed to interact thermodynamically with the surrounding clouds or air, or to reflect radiation. The radiative fluxes are calculated explicitly from the temperature, humidity, cloud and ozone distributions. Mixing ratios for ozone are either derived from a diagnostic scheme or from climatological values and prescribed and updated monthly.

The solar constant used in models varies a little among models, with 1370 W m^{-2} being the typical value. A simple treatment of enhanced surface absorption due to multiple reflection between clouds and the ground is likely to be included. The effective surface albedo can thereby be expressed as

$$\alpha_{\mathrm{eff}} = \alpha(1 - CR)/(1 - \alpha CR) \tag{5.12}$$

where α is the true surface albedo, C the mean fractional cloud cover and R the mean reflectivity of clouds as seen from the surface. The value of α_{eff} is calculated by weighting each cloud reflectivity by the amount of solar flux reaching the ground from that cloud.

The treatment of longwave radiation absorption is likely to be more complex than that for shortwave. For example, in one such AGCM scheme, seven spectral divisions are used. The first interval includes the 6.3 μm vibration–rotation water vapour band and the far infrared rotation band of water vapour; a further two intervals include the overlap between water vapour and the 15.0 μm carbon dioxide absorption bands. The fourth and fifth bands include the effect of the 9.6 μm ozone band and the water vapour continuum.

The emissivities and transmissivities for each gas in each interval at some temperature, e.g. 263 K, are sometimes stored in the computer code as 'look-up tables'. Carbon dioxide is usually assumed to be uniformly mixed and to have a mean concentration of ~330 ppmv (the Atmospheric Model Intercomparison Project (AMIP) used 345 ppmv), and in 'greenhouse' experiments this amount is increased. The surface and clouds are usually assumed to act as black bodies for longwave calculations, although in some schemes cloud properties are derived from the liquid water or ice content of the cloud.

Evaluation of the calculated radiative fluxes can be conducted by comparing the computed top-of-the-atmosphere fluxes with those observed by satellites. For example, data from the Earth Radiation Budget Experiment (ERBE) are widely used. Some aspects of model evaluation and intercomparison are considered in Chapter 6.

Boundary layer

The atmospheric boundary layer is the region in which surface friction has a large effect on the flow, typically the lowest kilometre or so. In particular, the near-surface wind is backed from the wind direction in the free atmosphere creating the Ekman spiral (Figure 5.6(a)). This layer can also suffer large fluctuations in temperature and humidity (Figure 5.6(b)) and its depth changes over the diurnal cycle. These features cannot be fully represented in most GCMs, primarily because the vertical resolution is inadequate and the parameterization schemes are unable to produce adequate approximations to the processes involved.

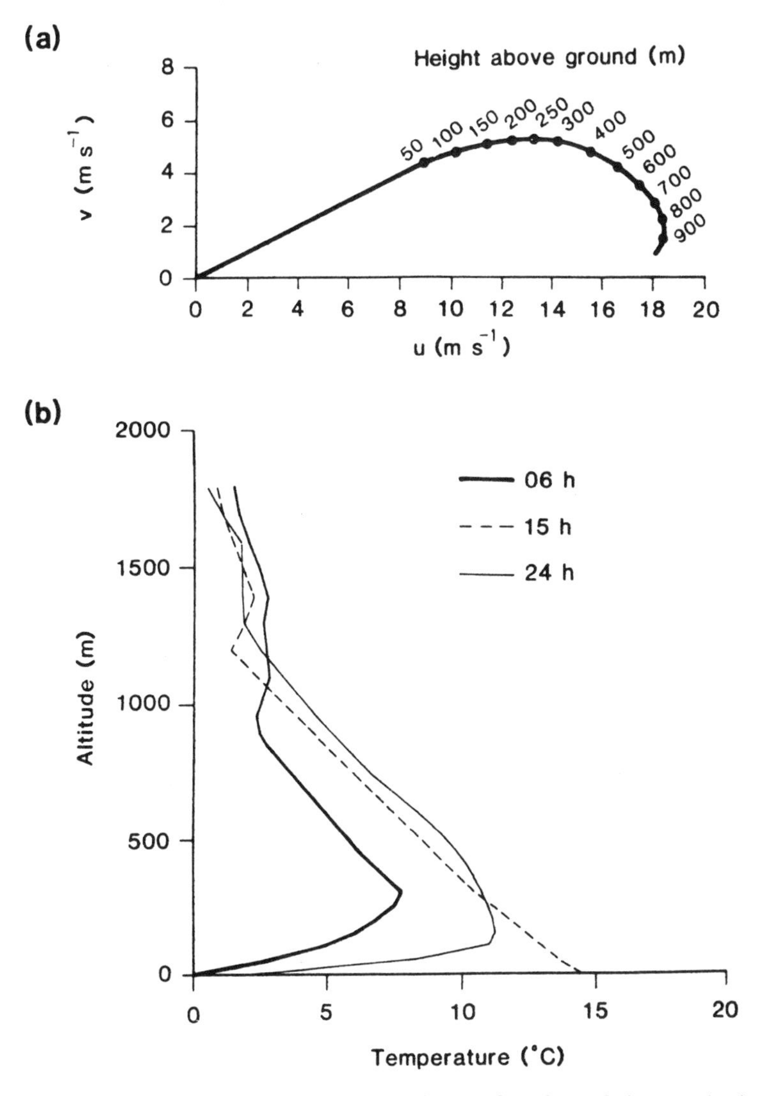

Figure 5.6 (a) The Ekman wind spiral plotted as a function of the two horizontal wind components; (b) a winter diurnal temperature variation in the planetary boundary layer. Typically neither of these phenomena can be fully represented in GCMs because the vertical resolution in the models near the surface is too coarse

An important characteristic of the boundary layer is its stability. This is calculated in terms of the potential temperature difference $\Delta\theta(z)$ between the surface and the lowest model level, at height z, and the difference $\Delta q(z)$ between the saturated specific humidity at the surface temperature and pressure and the specific humidity of the lowest layer. The bulk Richardson number, R_i, calculated as a function of $\Delta\theta(z)$, $\Delta q(z)$, the temperature of the lowest layer, T, and the wind speed, $V(z)$, of the lowest layer, such that

$$R_{\mathrm{i}} = \frac{gz[\Delta\theta(z) + 0.61T\Delta q(z)]}{TV^2(z)} \tag{5.13}$$

is used in conjunction with a specified surface roughness length to determine the bulk transfer coefficients for momentum, sensible heat and moisture. In AGCMs with a simple land-surface scheme, values of surface roughness are typically about 0.1 m over land and 10^{-4} m over ocean. The surface flux, F_x, of any variable x is given by

$$F_x = -C_x V(z)\Delta x(z) \tag{5.14}$$

where C_x is the bulk transfer coefficient and $\Delta x(z)$, the difference between the value of x at height z and at the surface. The surface fluxes are calculated. The temperature and moisture content of the surface and lowest atmospheric layer are updated accordingly.

Although there is a wide range of boundary layer schemes used in AGCMs, the interaction between the surface and the lowest model layer is similar in all these schemes. They differ in their treatment of the turbulent exchanges between model layers in the boundary layer. The turbulent exchanges between atmospheric layers in the boundary layer are usually modelled using the concept of eddy diffusivity, but higher-order closure schemes are becoming more common. In the eddy diffusivity approach, it is assumed that the turbulent flux, F'_x between adjacent model levels is proportional to the vertical gradient of that quantity and is given by

$$F'_x = -K_x(\mathrm{d}x/\mathrm{d}z) \tag{5.15}$$

where K_x is the diffusion coefficient for property x. The diffusion coefficient above can be expressed as

$$K_x = l^2(\mathrm{d}V/\mathrm{d}z) \tag{5.16}$$

where V is the horizontal wind speed and l is the mixing length.

Cloud prediction

The cloud amount is important to the radiation scheme, but there is no single, simple law which governs the formation of clouds. The considerable sensitivity of the radiation calculations to the cloud distribution means that cloud–radiation interactions have been recognized as critical to further model development. The early GCMs specified cloud amounts either zonally or globally according to climatological values. Modern climate models diagnose the cloud amount and type from other model variables. Most models have schemes which differentiate between convective clouds and stratiform clouds, relating predictions of the former to the result of the convection scheme and the latter to the large-scale condensation. Increasingly, models include prognostic schemes for cloud liquid water and ice, and hence are able to calculate cloud optical properties.

Early cloud prediction schemes simply related the cloud amount to the large-scale relative humidity. Such a technique can be as simple as assuming that the cloud cover is 100% for relative humidities above a certain threshold (usually 80–100%) and zero for humidities below this. There have also been schemes where the cloud amount, A_c, was some simple function of relative humidity, Q, such as the quadratic form

$$A_c = \left(\frac{Q - a}{b}\right)^2 \tag{5.17}$$

where a and b are empirical constants. Schemes have also been designed which are more general than this; for example, evaluating cloud on the basis of vertical velocity and atmospheric stability in addition to relative humidity.

The scheme used by the NCAR CCM2 diagnoses three types of cloud: convective cloud, layer cloud and marine stratus. Clouds are allowed to form in any layer except the lowest layer, and the minimum fraction for convective cloud is 20%. Low level frontal clouds form wherever there is upward motion. The relative humidity thresholds for upper and middle level clouds are functions of stability. Cloud liquid water paths (from which the cloud radiative properties are calculated) are derived from a time-dependent, meridionally and height-varying description of cloud liquid water density. The different cloud schemes used by different models and the differences in the underlying variables used for prediction results in a situation which is well illustrated by Figure 5.7. The very large differences in cloud distribution among these models (Figure 5.7(a)) are generally compensated for by assigning different cloud properties. In this way, the top-of-the-atmosphere (Figure 5.7(b)) and surface radiation fluxes are often much closer in model intercomparisons than the cloud fields themselves.

Many of the difficulties in cloud prediction arise from the requirement for sub-gridscale parameterization: cumuliform clouds are significantly smaller than the grid size of AGCMs, and stratiform condensation is likely to occur over smaller vertical distances than the vertical resolution of the AGCM. The difficulties of sub-gridscale parameterization are very hard to overcome, as improvements depend on gaining more detailed observational data and then developing and generalizing relationships. Increasingly, both shortwave and longwave radiative properties of clouds are being calculated as a function of a single variable such as cloud liquid (or frozen) water content. Although the quality of observational data which can be used for validation of models has improved immensely since the start of the International Satellite Cloud Climatology Project in 1985, we still do not have good models for cloud

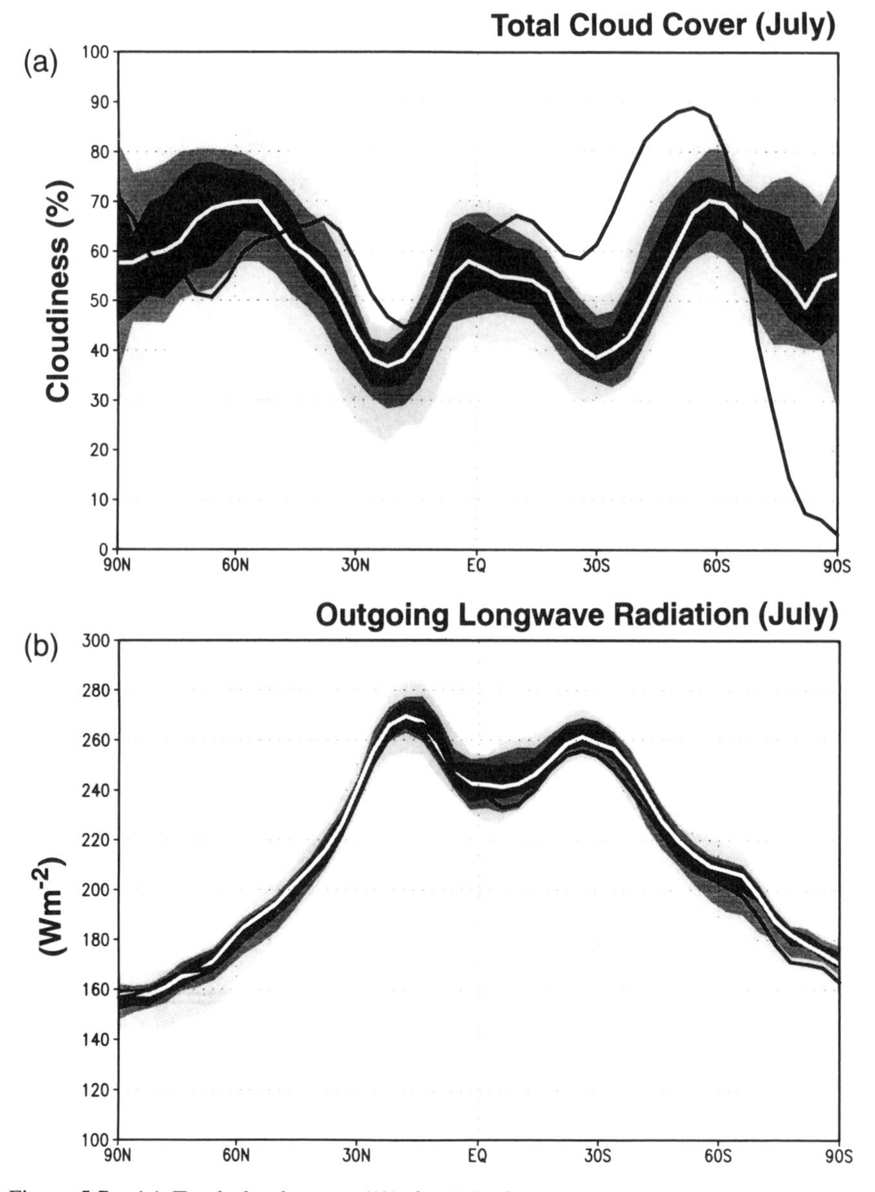

Figure 5.7 (a) Total cloud cover (%) for July from the models participating in the Atmospheric Model Intercomparison Project (AMIP), showing model mean and percentiles at 10, 20, 30, 70, 80 and 90 and observational data from the International Satellite Cloud Climatology Project (ISCCP). (b) As for (a), but for outgoing longwave radiation, with observations from NOAA polar orbiter satellite data. There is much greater agreement between model simulations of outgoing longwave radiation than there is for cloudiness. This is because models have been 'tuned' to match the observed radiation fluxes (reproduced by permission from Houghton *et al.*, 1996)

formation which can be incorporated in GCMs, and which can couple the radiative and hydrological role of clouds. The role of clouds in climate prediction remains one of the dominant sources of uncertainty in model predictions.

Convection processes

As discussed in Chapter 4 in relation to radiative–convective models, the thermal structure of the atmosphere which results from purely radiative processes is unstable and would result in convective motion. The major difficulty is parameterizing the sub-gridscale nature of convection. Often convection would be occurring in the real atmosphere over part of a $5° \times 5°$ area, but the average conditions would not satisfy the convection criterion. Similarly, the height of penetration of convection may often be less than the vertical distance between AGCM layers. The effects of these processes must be parameterized. Over the years, several schemes have been developed which accomplish this process. The earliest scheme, discussed in Chapter 4, is the simplest. This convective adjustment redistributes the energy within the column so that the lapse rate becomes some value assumed to be typical or average. Although the simplicity of this scheme is attractive, the instantaneous adjustment does not allow for the real life-cycle of cumulus clouds. Because of this, more complex schemes are generally employed in modern atmospheric models.

The Kuo scheme was once popular in climate models because of its relative simplicity. It relates the effects of cumulus convection to the rate of moisture convergence in the whole column. The Kuo scheme assumes that a fraction of the moisture which converges into the column is available to moisten the air and the remainder is condensed as rain. It has been largely superseded by various 'mass flux' schemes.

The mass flux approach is based on the notion that the grid box is populated by an array of cumulus clouds which have a spectrum of sizes. The mass flux scheme assumes that the behaviour of these many different clouds can be represented by a 'bulk' cloud. It tries to represent explicitly the fluxes of mass, energy and moisture within the cloud and the downward fluxes outside the cloud. The scheme is attractive because it can be argued to be more physically based and more amenable to increasingly complex treatments. The addition of new complex treatments like this into climate models must be carefully considered. One modelling group found that including a penetrative convection scheme in their model increased the sensitivity of the model climate (right-hand loop in Figure 5.8). It was discovered that the inclusion of the more complex convection treatment required a comparable upgrade in the cloud albedo scheme (left-hand loop in Figure 5.8) to avoid a mis-representation of the climate sensitivity.

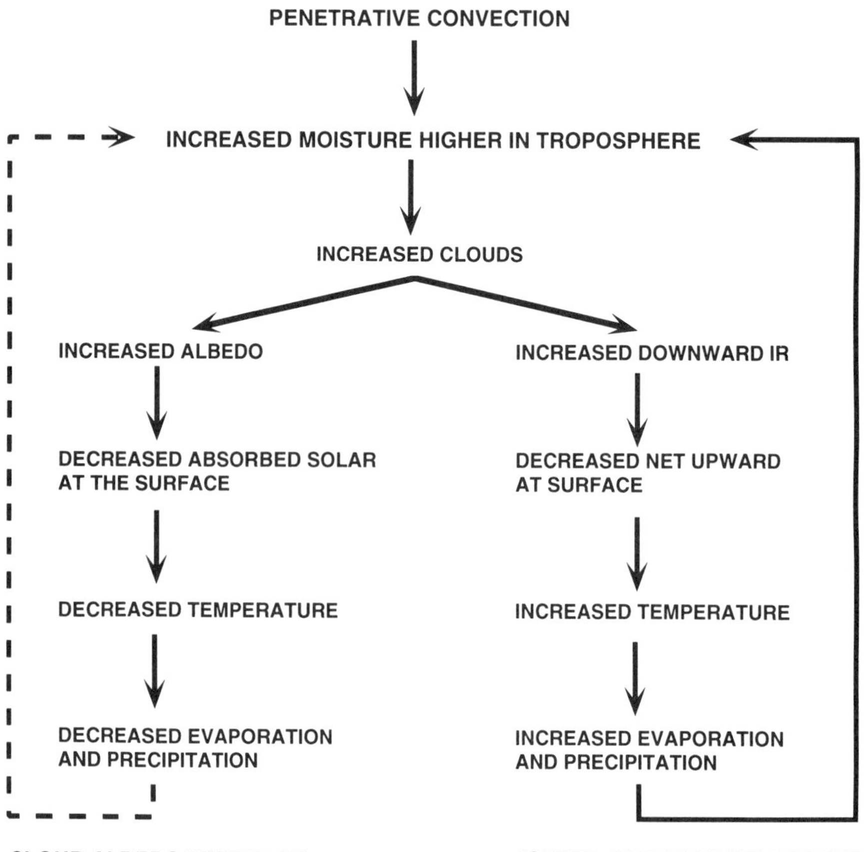

Figure 5.8 Feedback loops associated with enhanced penetrative convection in a global climate model. The right-hand loop results in increased sensitivity of the system through enhanced circulation of moisture, and the left-hand loop results in a decrease in sensitivity due to the enhanced cloud albedo (after Meehl and Washington, 1995)

Large-scale rainfall

Condensation can also occur in stable conditions when a layer in the model becomes supersaturated. The large-scale rainfall is usually based upon the saturation vapour pressure of each layer. Moisture is permitted to condense out of supersaturated air. Historically, if the temperature of the lowest atmospheric layer is less than 273 K the precipitation is classed as snowfall and the snow depth incremented accordingly, and if the lowest layer temperature exceeds 273 K, precipitation occurs as rainfall and soil moisture content is incremented. However, with the recent rapid development of complex land-surface schemes there has arisen a 'dispute' about whether the

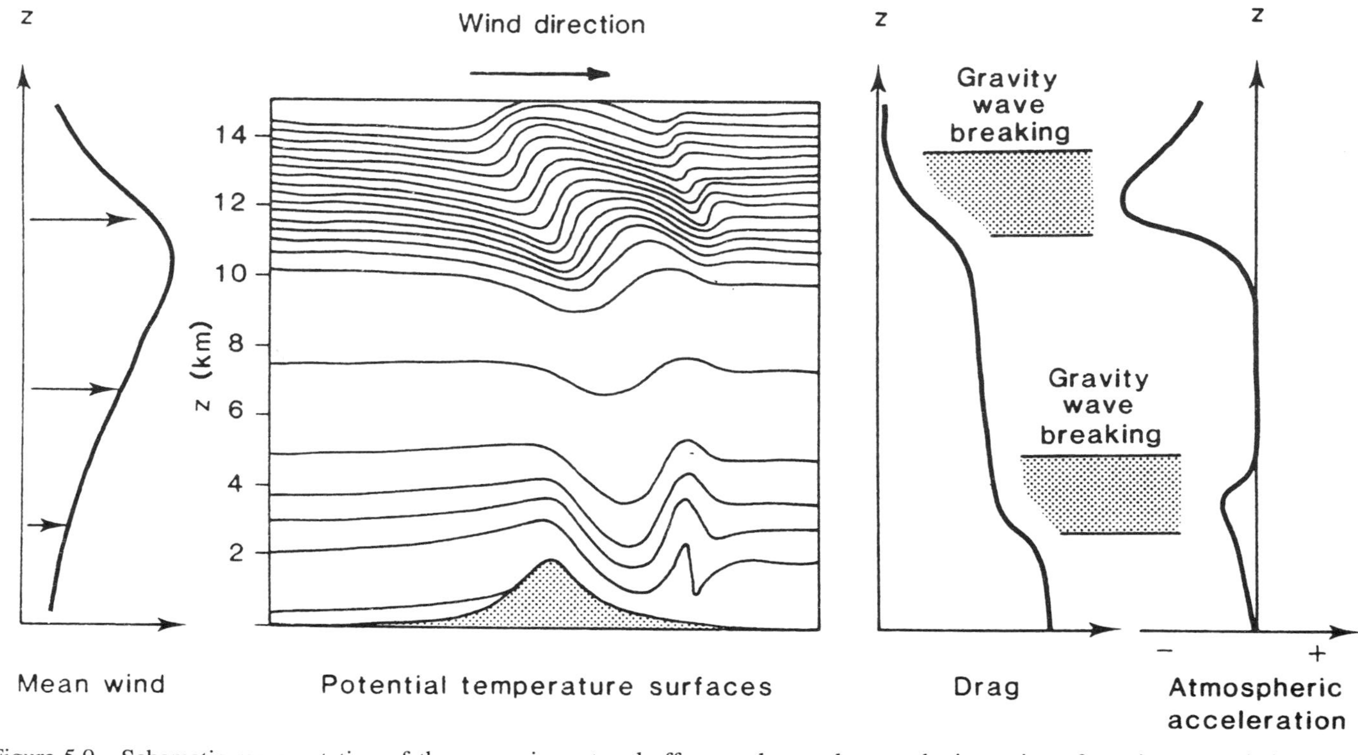

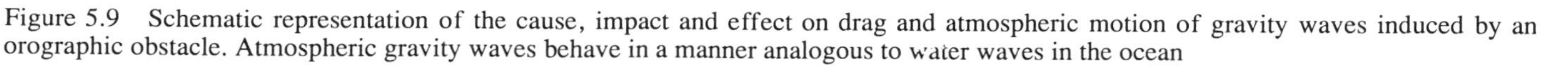
Figure 5.9 Schematic representation of the cause, impact and effect on drag and atmospheric motion of gravity waves induced by an orographic obstacle. Atmospheric gravity waves behave in a manner analogous to water waves in the ocean

AGCM or the land surface should determine the nature (solid or liquid) of precipitation.

Land-surface parameterization

The treatment of the land surface varies widely among different GCMs. The surface parameterization involves a surface type being prescribed for each grid area and the surface represented by one or more energy and moisture reservoirs. At each model timestep, the continental temperature and the moisture content are updated according to precipitation, evaporation and net radiation. Depending on the amount of precipitation and the nature of the scheme, some moisture is allowed to run off. The simplest form of this soil moisture 'book-keeping' is known as the 'bucket' or Budyko model. Many models include very detailed descriptions of moisture transport through the soil and through plant canopies. These models are discussed in Section 5.5.

Gravity wave drag

Gravity wave drag is the drag of the mountains on the atmosphere which is manifested in the production of gravity waves. This is illustrated schematically in Figure 5.9. These gravity waves can break in a manner analogous to waves on a beach and can transfer momentum from the large-scale flow at low levels to the flow at upper levels. Although neglected until the mid-1980s, gravity waves have come to be recognized as an important feature which requires inclusion in order to simulate the global- and regional-scale circulations. The richness of the gravity wave spectrum considerably complicates attempts to incorporate the effects into GCMs.

5.3 MODELLING THE OCEAN CIRCULATION

The ocean is driven mainly by the mechanical forcing of the winds and the net effect on the density and salinity of the water of surface exchanges of heat and moisture. It is confined to the ocean basins and governed by physical laws for the conservation of mass, momentum, energy and other properties. The currents which form are, as a result of these forcings and the rotation of the Earth, particularly narrow and strong at the western sides of the ocean basins. Many of the properties of the ocean, such as temperature, salinity (salt content), dissolved oxygen and other tracers, have maximum values in the cores of these strong currents. As a consequence of this, the proper representation of transports in an ocean model requires some very detailed calculations and careful assessment of the sensitivities of these transports to modifications in forcing. The challenge for ocean modellers is that many problems, from palaeoclimatic reconstruction to future climate prediction, require the correct simulation of these ocean responses.

The modelling of ocean processes by climate modellers has been a hierarchical procedure, and the coupling of ocean–atmosphere models can be thought of in terms of a hierarchy of oceanic components. The early, simple oceanic representations included the 'swamp' model (Figure 5.10(a)) with no heat storage capacity, and the fixed depth ocean surface layer models (slab models, Figure 5.10(b)) where there is a heat capacity but no dynamics. The latter has proved useful in helping to understand the processes acting at the air–sea interface and in providing a means for sensitivity testing of the results of coupled three-dimensional models. For example, utilizing a prescribed depth mixed layer permits the inclusion of a full seasonal cycle in the atmospheric GCM which is not possible with the oceanic 'swamp'.

The 'mixed layer' or 'slab' ocean model (Figure 5.10(b)) represents the ocean at each grid point as a slab of water with a prescribed depth, usually

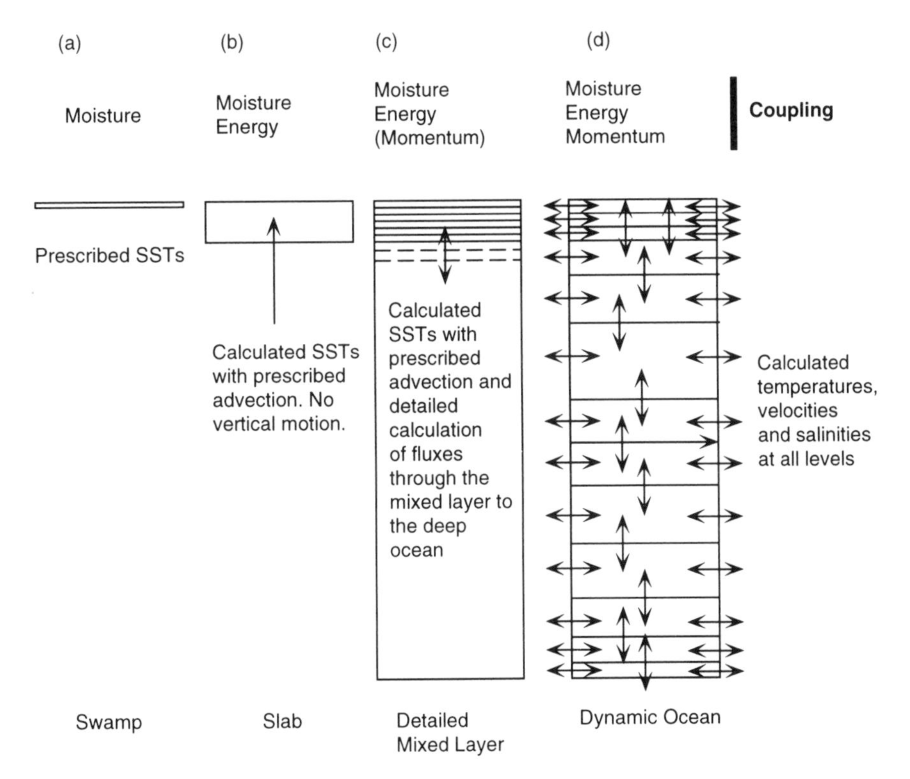

Figure 5.10 Schematic illustration of the level of complexity and coupling associated with various types of ocean model. (a) Swamp ocean with only moisture coupling; (b) a slab ocean with moisture and energy coupling; (c) a diffusive mixed-layer model with energy and moisture (and some momentum) coupling; (d) a full three-dimensional ocean model with momentum, moisture and energy coupling and full three-dimensional interactions in the water column

between 70 and 100 m. Occasionally the mixed layer depth may be geographically variable, but it is usually globally homogeneous. The use of prespecified mixed layer depth without any allowance for horizontal and vertical motion is inadequate for the simulation of the annual cycle of zonal heat storage, especially in the tropics, and most implementations of slab oceans do a very poor job at predicting the temperature and sea-ice distributions. The problem is that oceans advect a great deal of energy from the tropics to the poles and sequester heat into the deep ocean. The parameterization of dynamic ocean processes is very limited in slab models, but an adjustment of surface fluxes at every ocean point can be used as a surrogate for horizontal energy transfer. Some modellers have been successful in simulating reasonable temperature distributions and seasonality of temperatures and sea-ice type using this means of parameterization. However, the nature of predictions which can be made and questions which can be answered with this type of model is limited. For example, their response to enhanced greenhouse gases is simplistic, and they can offer no insight into phenomena such as El Niño since this requires coupling of the ocean model to the atmospheric wind field.

The modelling of the mixed layer has been enhanced by some modellers by the use of more complex treatments of the vertical diffusion of heat away from the surface (Figure 5.10(c)). It has been suggested that these more complex models are appropriate for middle latitudes since the sea-surface temperatures are governed largely by the local exchanges of heat and mechanical energy in these latitudes, whereas in equatorial latitudes, lateral advection of energy needs to be included. These more complex treatments of heat storage in one dimension have also found application in simplified global models such as those discussed in Chapter 3. Parallel development of three-dimensional ocean circulation models and mixed layer models has led to the need to embed the latter within the former to produce an efficient three-dimensional ocean circulation model.

The best known type of three-dimensional ocean model was developed at GFDL in the late 1960s (see Chapter 2). All of the early GFDL models used an assumption known as the 'rigid lid' approximation to support a longer timestep. In this kind of model, the height of the ocean surface is kept fixed to prevent the formation of external gravity waves. Motions which can be supported by the model ocean are therefore restricted to slow moving ocean currents (recall that the limitation on the timestep is related to the maximum speed of propagation through the medium). The earliest GFDL model was very slow to run and so was reformulated at a lower resolution (5° latitude × 5° longitude). At this resolution, ocean currents were broad and sluggish, but the model became very popular and a standard tool for climate modellers.

The scale over which ocean currents exist can be described by the internal radius of deformation, which varies from about 100 km in the tropics to less than 10 km in polar regions. The currents meander and produce cut-off features on scales similar to the radius of deformation. These mesoscale eddies, and the way in which they disperse the kinetic energy of the oceans, are difficult to describe theoretically. If currents and eddies are to be described in a numerical model of the ocean, then the grid size of the model must be suitably fine. The improved models of the 1980s, no longer relying on the rigid lid approximation, were able to make significant headway in the understanding of the origin and behaviour of ocean currents and eddies. Figure 5.11 shows results from a 1/6° eddy-resolving ocean model. The instantaneous salinity in the ocean region around southern Africa shows mesoscale eddies with great clarity. The CD which accompanies this book contains similar ocean model imagery.

Detailed regional investigation can be undertaken in ocean modelling as well as in atmospheric modelling. For example, Figure 5.12 shows the rate of poleward transport of heat in a model of the Atlantic basin in response to three different prescriptions of the flow of saline water from the Mediterranean through the Straits of Gibraltar. In the extreme case, representing the complete cessation of flow (characteristic of periods where the sea-level was much lower, e.g. 18 000 BP), the rate of energy transport by the Atlantic Ocean is greatly reduced.

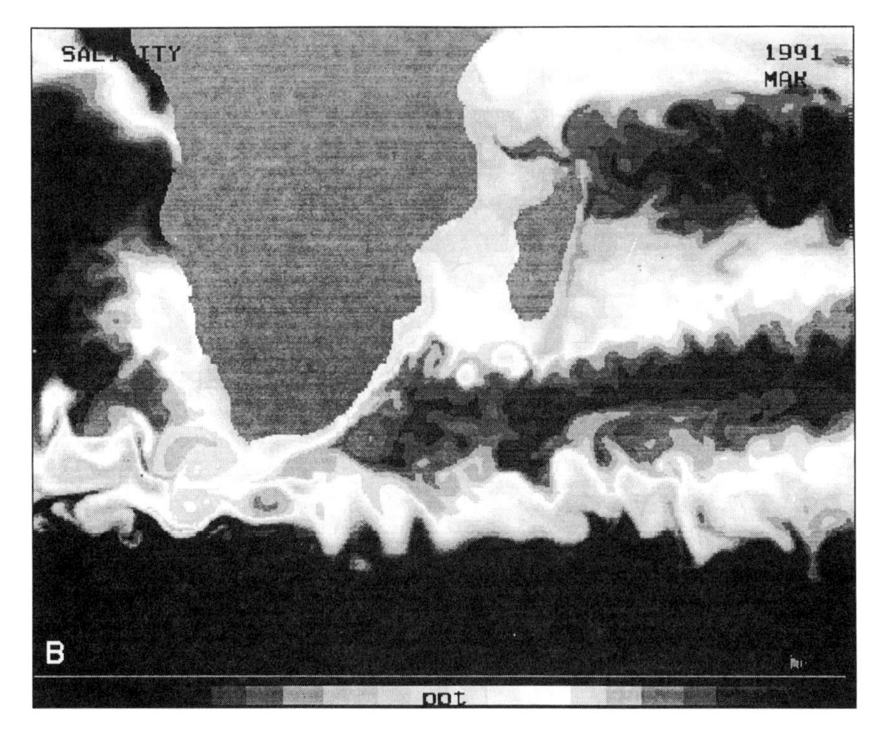

Figure 5.11 Instantaneous picture of salinity in a global 1/6° eddy-resolving ocean model (from Semtner, 1995)

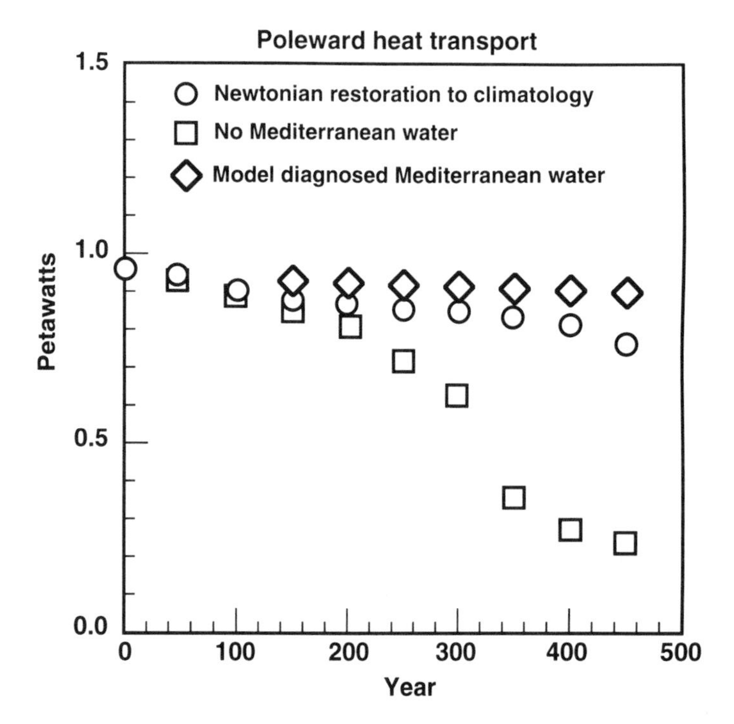

Figure 5.12 Maximum poleward heat transport simulated by a model of the North Atlantic Ocean under three different conditions of Mediterranean salt input. (a) Relaxation to climatology; (b) no input from the Mediterranean; (c) input from the results of a high-resolution model of the Mediterranean Sea (from Hecht *et al.*, 1996)

Formulation of three-dimensional ocean models

The model system adopted in 1969 for the GFDL model was a *z*-coordinate system, which can be reformulated slightly into an isobaric coordinate system. This model splits the ocean into a 3D array of points like those shown in Figure 5.13. The resolution of the model grid is higher near the boundaries of the ocean basin and the levels are unevenly spaced in the vertical, to allow for more detail near the upper and lower boundaries. There are problems with the isobaric or *z*-coordinate system caused by spurious transports across density surfaces. In response to this and other problems with the isobaric formulation, the isopycnal coordinate system was developed. Models with isopycnal coordinates have the equations formulated on constant-density surfaces analogous to the sigma coordinate system used in atmospheric models. In the real ocean, mixing processes are believed to be predominantly along constant-density surfaces. The isopycnal coordinate system therefore mimics, as much as possible, real structures within the ocean. The isopycnal coordinate system has the advantage of formulating the model in a manner which rigorously

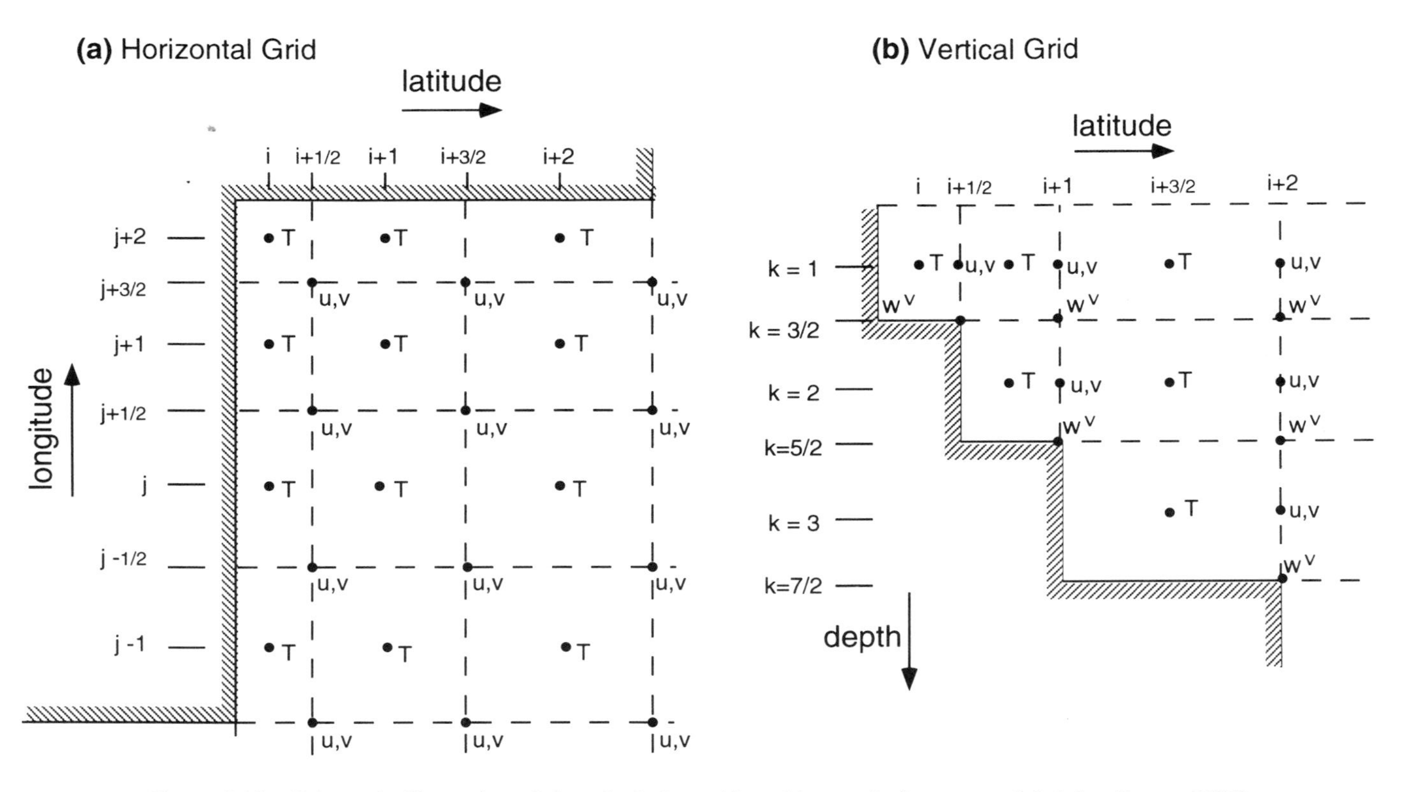

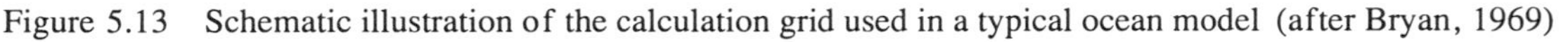

Figure 5.13 Schematic illustration of the calculation grid used in a typical ocean model (after Bryan, 1969)

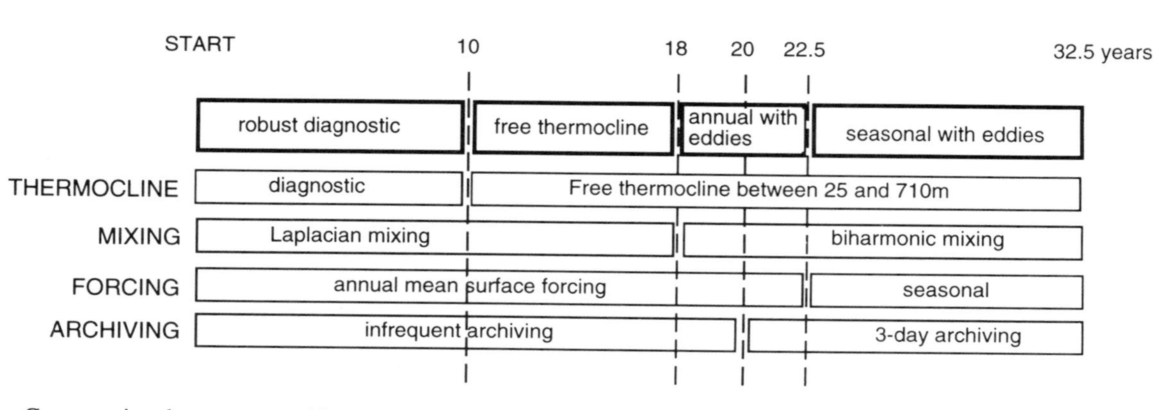

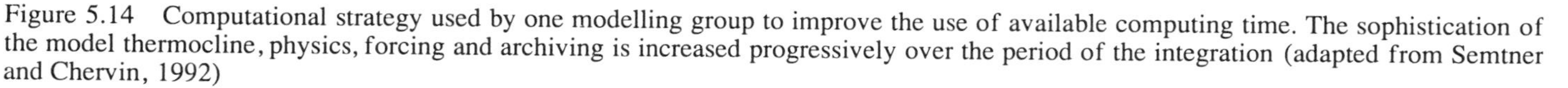

Figure 5.14 Computational strategy used by one modelling group to improve the use of available computing time. The sophistication of the model thermocline, physics, forcing and archiving is increased progressively over the period of the integration (adapted from Semtner and Chervin, 1992)

preserves potential vorticity. On the other hand, problems arise when the thickness of isopycnal layers drops to near zero or when they intersect with the surface. Hybrid coordinate systems and model schemes have been developed to overcome some of these problems. Topography is as important for ocean modelling as for atmospheric modelling, with basin geography being more important than bottom topography. Successful parameterization of mixing processes is also very important for the success of an OGCM.

Considerable effort in the development of ocean models has been in the adaptation of the model code to make effective use of available computer resources. Ocean modellers have been among the first to utilize the latest high performance architectures and have pushed supercomputer performance to very high levels. Ocean modellers have also adopted innovative strategies for the running of their models, as illustrated in Figure 5.14. Here a 32-year integration is split into several different stages, of increasing realism, in order to make the best use of computational and archiving resources.

5.4 MODELLING THE CRYOSPHERE

Any complete model of the climate must include prediction of snow cover, sea-ice, glaciers and possibly permafrost. General circulation models attempt to simulate all aspects of the climate system, although the accuracy with which they can do this is limited by several factors. The most important of these for the cryosphere is the very long time-scales associated with some elements (e.g. Table 1.1). Permafrost is not widely modelled in GCMs, with only a few models having the capability to represent this feature of polar hydrology. Sea-ice is modelled in GCMs but, as yet, with only limited success because of the incomplete representation of the three-dimensional ocean component. The prediction of snowfall in GCMs usually occurs if the lowest layer of the atmosphere has a temperature below 0 °C. If the surface is above 0 °C then the snow melts, cooling the surface and adding to ground water or runoff. Snow prediction is therefore dependent on the ability of the model to simulate the hydrological cycle and the distribution of surface and air temperatures. The effect of snowfall depends on incorporation of the effect of vegetation. For example, snow falling on tundra can raise the albedo from 0.2 to 0.8, whereas the same snow cover on a dense coniferous forest may not raise the albedo at all in the long term because the forest canopy shades the surface snow. The modelling of snow surface albedos in GCMs differs, and thus even similar predictions of snowfall and snow extent could lead to different radiative effects.

The major glaciers of Antarctica and Greenland are represented in GCMs as surface features which extend vertically. Snow falls on to them and is allowed to melt or sublime, but the major mode of ablation of these glaciers, the formation of icebergs, is not represented in GCMs (as discussed in Chapter 3). Temperate glaciers and all glacier dynamics are neglected in GCMs, although some modellers have forced glacier models with the output from GCMs.

As the dominant component of the summer-time surface energy balance in cryospheric regions is solar radiation, it is essential that the large-scale surface albedo be parameterized correctly. The albedo of sea-ice is predicted variously in GCMs. For example, one parameterization is $\alpha = 0.5$ if latitude <55°, $\alpha = 0.7$ if latitude >66.5°, with linear interpolation for intermediate locations. If the ice is melting, the albedo is reduced to 0.45, and when the thickness is less than 0.5 m the albedo decreases as a square-root function of thickness until it equals the albedo of the underlying surface. However, capturing all aspects is difficult to achieve: for example, the sudden decrease observed in Arctic sea-ice albedo when melt puddles form does not occur on the Antarctic pack ice. This means that the summer-time decrease of albedo in the Southern Ocean is much less than near the North Pole, and thus a globally applicable albedo parameterization is hard to develop.

Many of the difficulties associated with successful incorporation of cryospheric elements into GCMs result from the fact that parameterization depends upon successful prediction of other features such as oceanic heat transports and atmospheric wind fields. The strong influence of dynamics on sea-ice growth and decay in the Antarctic region inhibits successful modelling of these features of the cryosphere. The behaviour of Arctic sea-ice is very sensitive to the polar wind field. An example is in the representation of sea-ice processes. In the 1980s, GCMs used a thermodynamic model of the sea-ice which was developed in the mid-1970s for climate model applications. The Semtner model is almost a standard formulation for the thermodynamic behaviour of sea-ice and can be operated in a three-layer or one-layer mode. It predicts the accumulation and ablation of the sea-ice and snow-pack combination and predicts temperature in the snow-pack and at various levels within the ice. Ice fraction can be made to be a function of ice thickness predicted by these models. This allows for the inclusion of the effect of leads and other open areas in the modelled ice-pack. The motion of sea-ice under the influence of winds and ocean currents is rarely incorporated in GCMs, although a number of modelling groups are beginning to include at least the dynamic effect of winds on the ice in their models.

Where dynamic sea-ice is included, it is usually using the 'cavitating fluid' model developed by Flato and Hibler. This model allows ice to be advected across the model grid in response to the model wind fields. The sea-ice fraction and thickness around Antarctica, as predicted by the GENESIS model implementation of the Flato and Hibler sea-ice model, is shown in Figure 5.15. The vertical growth and decay of the ice in such models is still accomplished using the thermodynamic approach developed in the 1970s.

Two further aspects of the climate system are of considerable interest to most GCM modelling groups. The continental vegetation and the oceans had been neglected until very recently, but both areas have undergone significant developments in recent years (Section 5.3). The ocean, despite covering over two-thirds of the Earth's surface, has until recently been parameterized as a swamp or at best a relatively shallow mixed layer without horizontal advection. Land vegetation had received still less attention until model simulations of, for example, the effects of increasing atmospheric CO_2 focused attention upon the possibilities of water stress, altered growing season length and even altered vegetation distributions. In Sections 5.5 and 5.6, some of the recent developments in vegetation parameterization and in coupling ocean and atmospheric models are described.

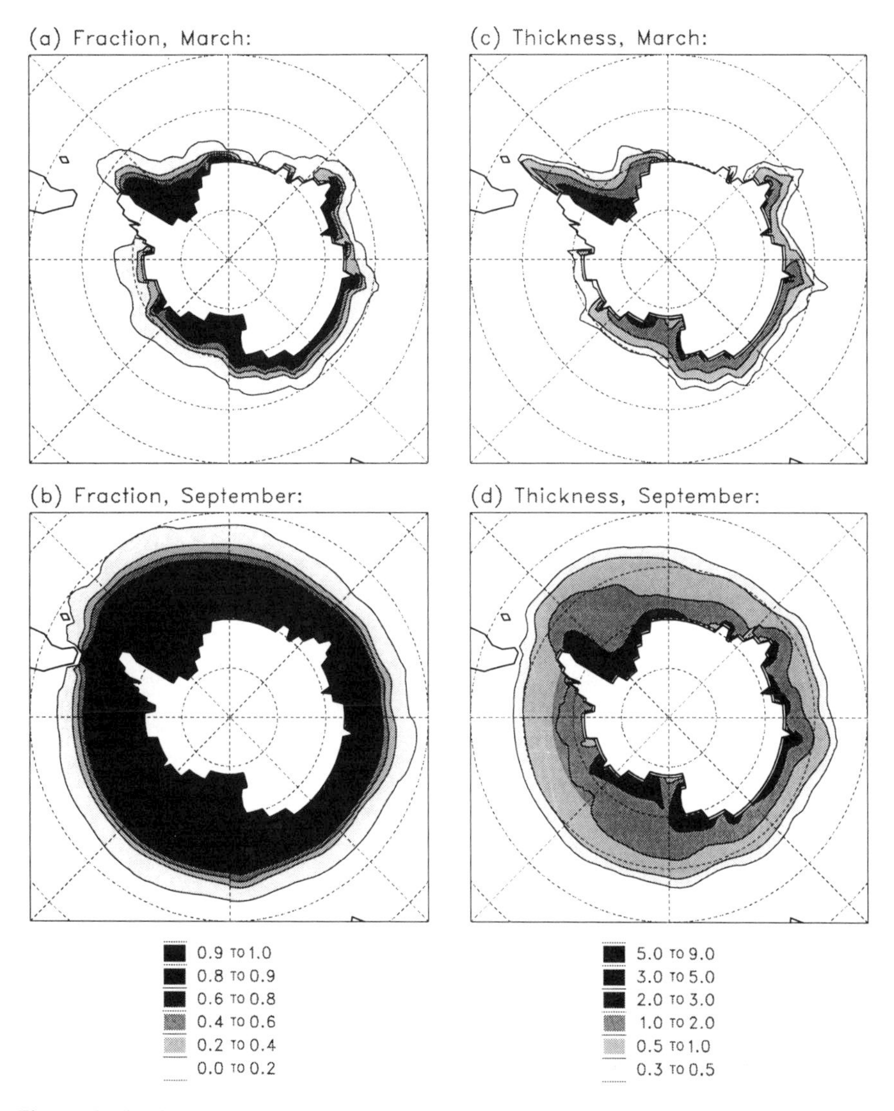

Figure 5.15 Seasonal range of sea-ice fraction and thickness for the Antarctic predicted using the GENESIS global climate model with a dynamic sea-ice model

5.5 INCORPORATING VEGETATION

Over 70% of the energy absorbed into the climate system is absorbed by the surface, and experiments with many climate models have underlined the sensitivity of the climate to the continental surface hydrology and to the vegetation cover. The treatment of the land surface has changed markedly over the history of climate modelling (Chapter 2). The Budyko or 'bucket' model dates back to 1969. This 'bucket' (Figure 5.16(a)) has a maximum depth, usually termed by the modellers 'field capacity'. The bucket fills when precipitation exceeds evaporation and, when it is full, excess water runs off. The bucket model has been demonstrated to be inadequate, particularly when the host model includes a diurnal cycle. It is also important to note that runoff in early GCMs did not play any further role in the hydrological cycle of the model, although in the late 1980s parameterizations of river routing have permitted the computation of freshwater inflow into the ocean basins.

Various GCM studies have identified the importance of the surface hydrology for climatic simulations. For example, Figure 5.17, from a classic paper, shows the considerable impact on rainfall of modifying surface evaporation. In this experiment, evaporation from the land surface was forced

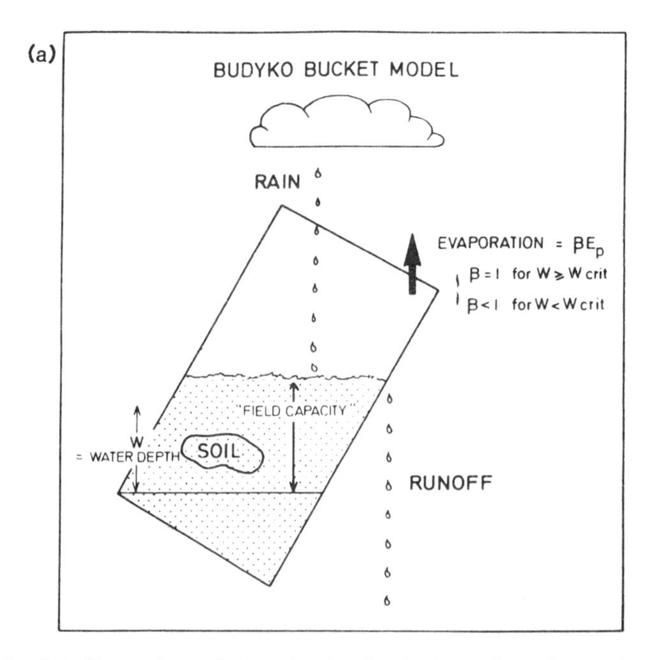

Figure 5.16 (a) Illustration of the simple 'bucket' land surface scheme; (b) an illustration of the processes included in more complex SVATS. This example scheme controls the radiative, latent and sensible heat fluxes occurring at the surface, and models the movement of soil moisture below the ground and through the plants

to be equal to the potential evaporation (free evaporation) or set equal to zero at all land-surface points. As can be seen, for the month of July the resulting precipitation is vastly reduced (Figure 5.17(b) cf. Figure 5.17(a)). As well as the considerable reduction, there is also a shift in the position of the remaining maxima of rainfall over the continental areas. This experiment, although extreme, suggests that the modelling of evapotranspiration from the land surface may be crucial to the accurate modelling of the global hydrological regime. As schemes for energy and moisture exchanges have become more

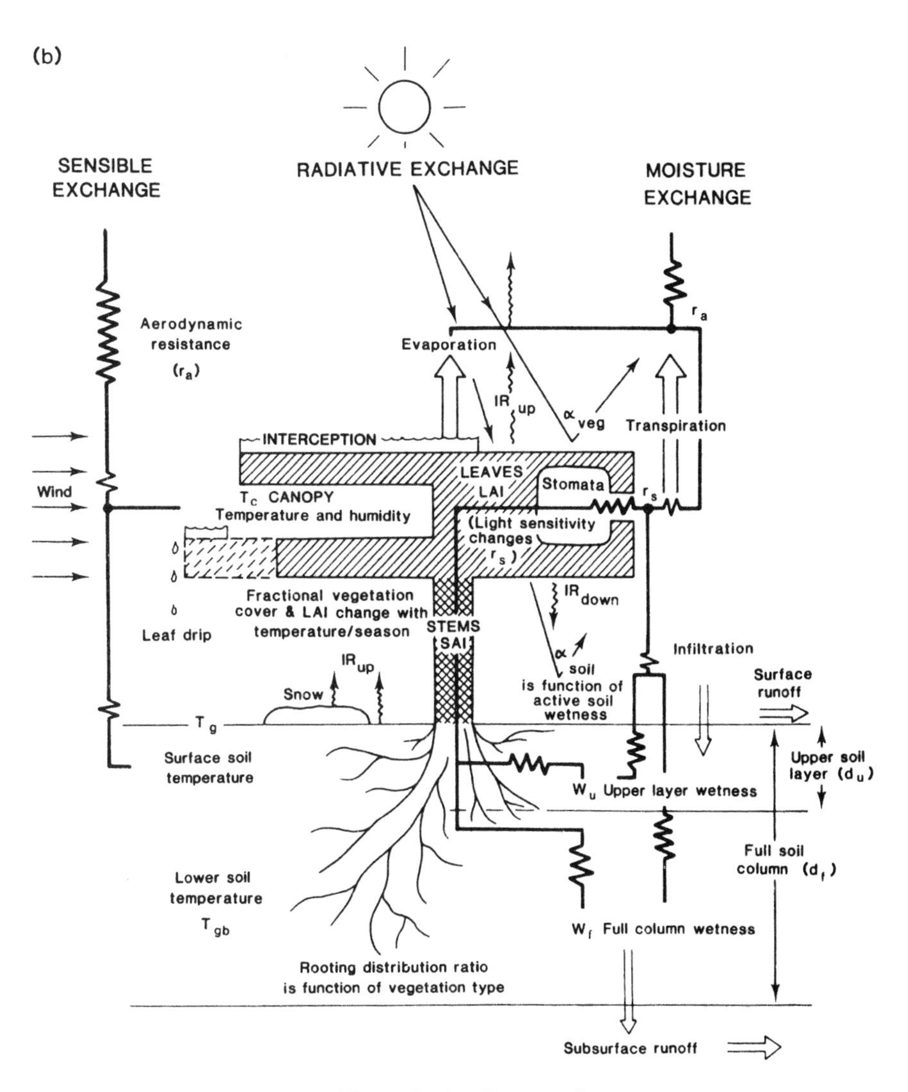

Figure 5.16 *Continued*

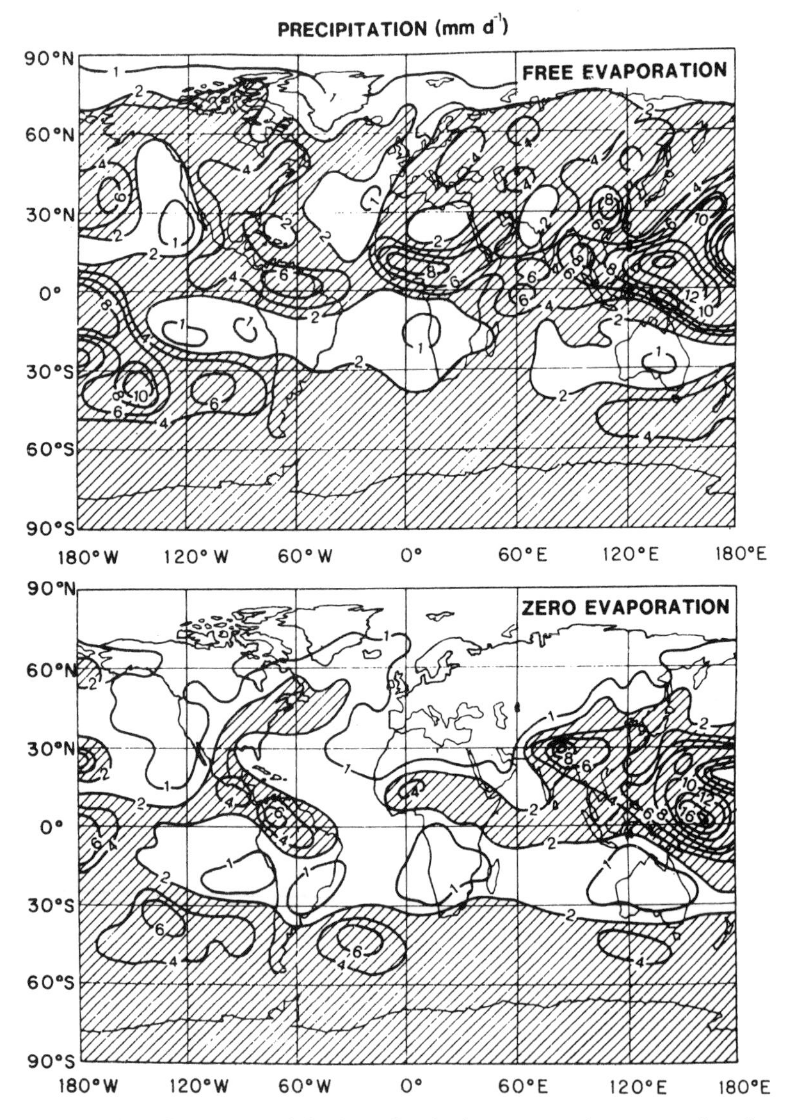

Figure 5.17 The effect on precipitation of reducing to zero the evaporation from the continental land surface as compared with permitting evaporation at the potential rate, as simulated by a general circulation climate model. Results are averaged from the last 30 days of two 60-day 'perpetual July' simulations (after Shukla and Mintz, 1982)

realistic (Figure 5.16(b)) a major problem in land-surface modelling is the reconciliation of basin hydrological studies with the resolution typical of general circulation climate models: 3° × 5° grid elements. These difficulties are linked to the problems of 'downscaling' GCM results to smaller areas, a topic discussed in Chapter 6. The second, and equally acute, problem is the dearth of hydrological data with which to initialize and validate global models. Current observational and modelling programmes are making some headway on these problems. The Project for Intercomparison of Land-surface Parameterization Schemes (PILPS), which is discussed in Chapter 6, has been successful in bringing together land-surface modellers and in fostering the use of datasets for intercomparison and validation.

From being a surface of uniform roughness with very limited hydrological capabilities (e.g. Figure 5.16(a)), the land surface as represented in GCMs has evolved into a complex sub-component of the fully coupled atmosphere-ocean-biosphere GCM (AOBGCM). Plants, the dominant component of the land-surface climate, have been modelled using three main strategies: (i) the physically based approach, which has resulted in the development of complex soil–vegetation–atmosphere transfer schemes (known generically as SVATS (e.g. Figure 5.16(b))); (ii) biogeochemical models of vegetation and soil processes which emphasize exchanges of carbon, nitrogen, phosphorus and sulphur; (iii) equilibrium biospheric prediction models, which either define the nature of an 'equilibrium' vegetation based on a simple classification scheme or use succession models to simulate the nature of a biome as a combination of a range of species, all of which have different growth functions. These different scheme types therefore emphasize (i) energy and moisture exchanges on time periods of minutes to months, (ii) chemical storage and exchange on time periods of months to decades, and (iii) ecosystem dynamics over very long time periods of decades to millennia. These strategies, developed from very different positions, are only recently beginning to converge to meaningful models of the biosphere and its response to, and role in, the overall climate system.

5.6 COUPLING MODELS: TOWARDS THE AOBGCM

The notion of coupling models is not new. The nature of climate model construction has been that of a modular approach. Modellers constructed routines to deal with clouds, land-surface processes, ocean thermal response and sea-ice. The complexity of the schemes is rivalled by the complexity of the coupling.

Land-surface models, for example, have grown from very simple schemes to the exceedingly complex SVATS discussed in Section 5.5. These are coupled closely with the atmospheric model, and exchange fluxes every timestep. On longer time-scales, modellers are beginning to use schemes to predict the characteristics of the vegetation and soils and the exchanges of chemical elements. The time-scales of coupling range from minutes to millennia.

A most important coupling is that between atmosphere and ocean models. In early climate models, the atmosphere was driven by prescribed climatological sea-surface temperatures and sea-ice distributions. Most climate predictions of the early 1980s were based on atmospheric models with prescribed (seasonally varying but present-day) sea-surface temperatures and sea-ice. By the late 1980s, mixed layer ocean models were used in which the meridional energy transport of the oceans was prescribed at present-day values. This latter

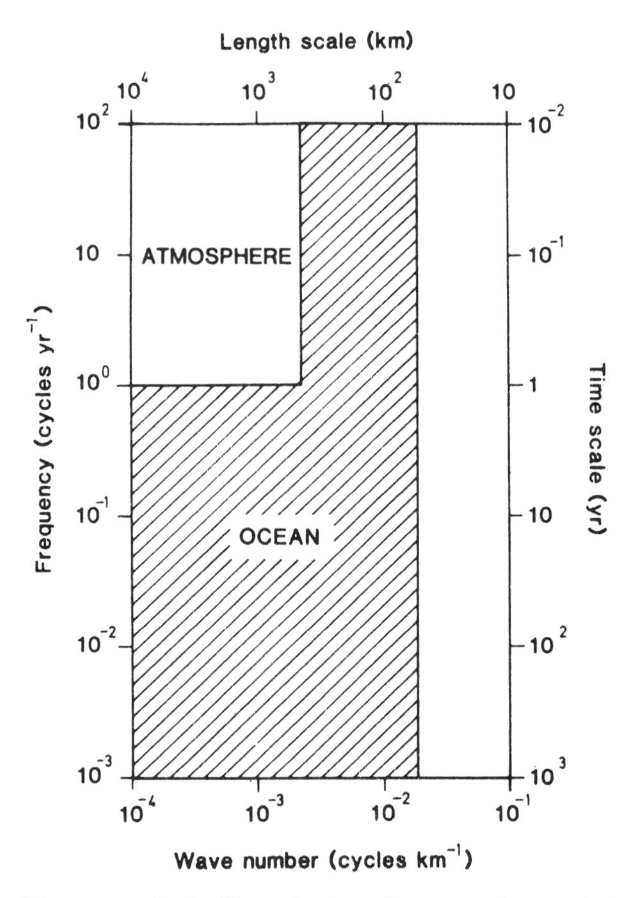

Figure 5.18 The range of significant horizontal wavenumbers and frequencies in climate models of the atmosphere and ocean. The lowest frequency is determined by the thermal relaxation times of the subsystem and the highest by gravity waves. The lowest wavenumber is determined by the planetary scale and the highest by the radius of deformation of the subsystem medium. The inverse of the scales gives the more familiar time and length scales. Note that the ocean spans a greater range of *both* fundamental scales than the atmosphere

approach allows the temperature of the ocean to change in response to changed forcings (such as enhanced CO_2) but clearly constrains the simulation by prescribing present-day transports. The full ocean system with deep ocean processes as well as those in the upper mixed layer is now included in three-dimensional climate models.

The difficulties inherent in ocean–atmosphere coupling are identified in Figure 5.18. This diagram underlines the different response times associated with the atmosphere and the ocean subsystems, and emphasizes that the ocean subsystem spans a greater range in both time and space than the atmosphere subsystem.

The considerable discrepancy in response (or equilibration times) of the atmosphere and the ocean, including the deep ocean, have already been described (see Section 1.3 and especially Table 1.1). Since the time to reach equilibrium is much longer for the ocean than for the atmosphere, linking of an ocean model with an atmospheric model is very difficult. Ideally the linkages should be between the thermodynamic systems, among the variables represented in the equations of motion, and in terms of the parameters and variables of the water cycle (Figure 5.19). Although there are eddy-resolving ocean models with very high resolution (e.g. Figure 5.11), they have not yet been coupled to atmospheric models. The models which are generally coupled to atmospheric models are of the GFDL type described in Section 5.3 and are usually at a resolution similar to that of the atmospheric model, with model components communicating at least once each model day.

The difference in response time between atmosphere and ocean means that for effective use of computer resources, the models are not always run in a continuously coupled mode; rather, they are sometimes coupled asynchronously.

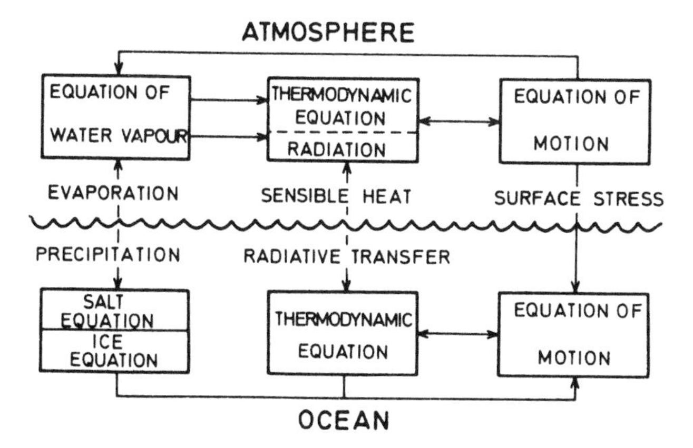

Figure 5.19 Box diagram illustrating the major components of a joint ocean–atmosphere model and the interaction among the components (reproduced by permission from Manabe *et al.*, 1979)

The coupling scheme used in the GENESIS model links a T31 atmospheric model to a 2° × 2° land-surface grid and a 2° × 2° ocean model (Figure 5.20). Initially, they are run together, fully coupled, for a period given by τ_{coupled}. The latter part of this period, τ_{ave}, is used to derive an average atmospheric climate with which to force the ocean model during the $\tau_{\text{asynchronous}}$ period. During this second period, the atmospheric model is not operated. The cycle is then repeated. This saves considerable amounts of computer time since the cost of computing the atmospheric model is typically many times that of the ocean model per year. In the GENESIS asynchronous coupling, τ_{coupled} is around 15 years, τ_{ave} is around 10 years and $\tau_{\text{asynchronous}}$ is around 85 years.

Another problem associated with modelling the ocean is that the response time of the deep ocean to climate changes is several thousand years. To avoid running the entire model for thousands of years to 'spin up' the deep ocean, a technique called 'distorted physics' is used. The specific heat capacity of the deep ocean water is reduced by a factor of up to 10 so that the deep ocean temperatures respond more rapidly than the surface layers (an analogous distortion is applied to salinity). These alterations distort the dynamical behaviour of the ocean to some extent, so that at the end of a long 'distorted physics' run, a period of several decades without distorted physics is needed. Typical results from such a coupled model are shown in Figure 5.21 – in this case the model is Version 0 of the NCAR Community Climate Model. In this June–July–August cross-section of the atmosphere and ocean, the cold deep waters in the Arctic are clearly shown, as is the variation in mixing depths with latitude.

Climate drift

On a local scale, the fluxes between the atmosphere and ocean are such that the ocean is warmed in the summer and cooled in the winter. The small imbalance

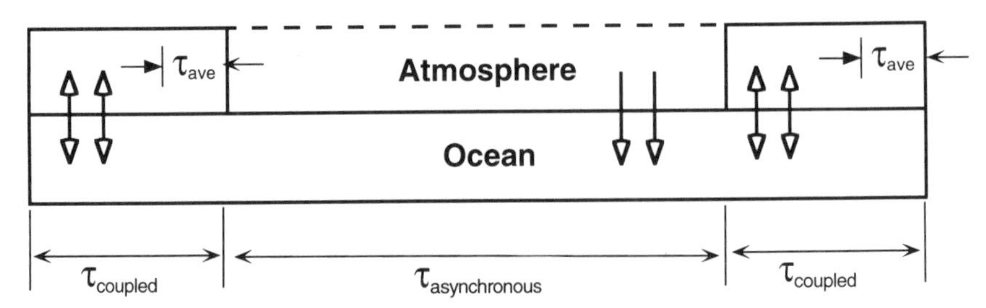

Figure 5.20 An illustration of an asynchronous coupling between atmospheric GCM and an oceanic GCM. The models are run together for only a short fraction of the total integration time. Most of the integration involves running the ocean model with only mean forcing information from the atmosphere

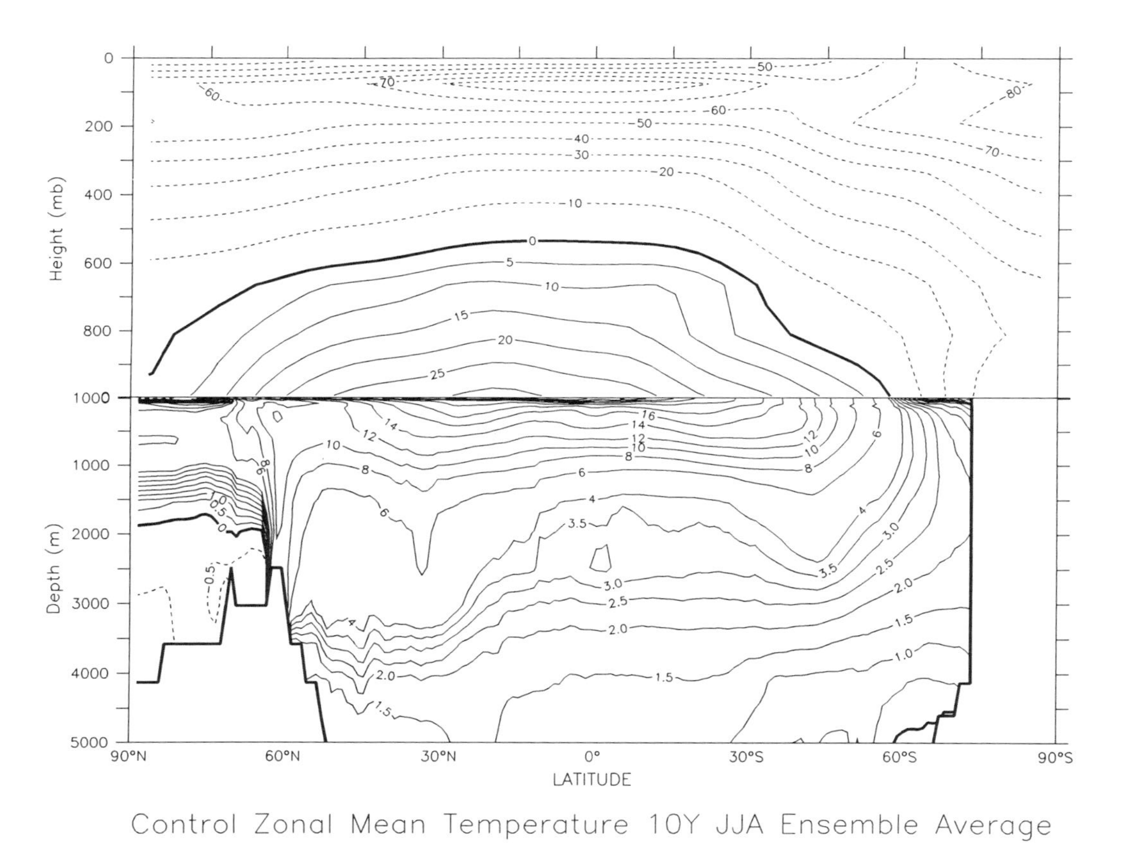

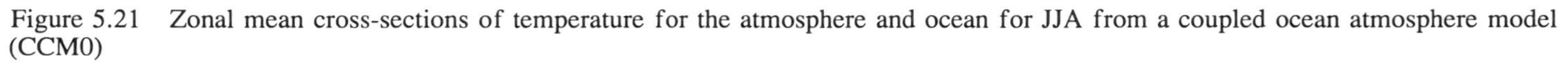
Figure 5.21 Zonal mean cross-sections of temperature for the atmosphere and ocean for JJA from a coupled ocean atmosphere model (CCM0)

drives the ocean circulation. The coupling of ocean and atmospheric models can highlight discrepancies in the fluxes calculated by the two models. A good estimate of the difference between two large and similar numbers is not easily obtained. Because the relative errors in this resultant forcing are large, they can cause problems in the ocean circulation. Some modellers have chosen to apply flux adjustments to their coupled model in order to prevent the climate drifting from present conditions. Other modellers prefer to run long climate simulations without the aid of such adjustments, but have had to accept a drift in climate. The model in Figure 5.21, which has no flux corrections, exhibits little, if any, drift, but is consequently left with some significant systematic errors in the simulation compared with observations. Although the process of flux adjustment has been viewed sceptically in some quarters, it, and the alternative of climate drift, should be thought of as a necessary stage in the process of

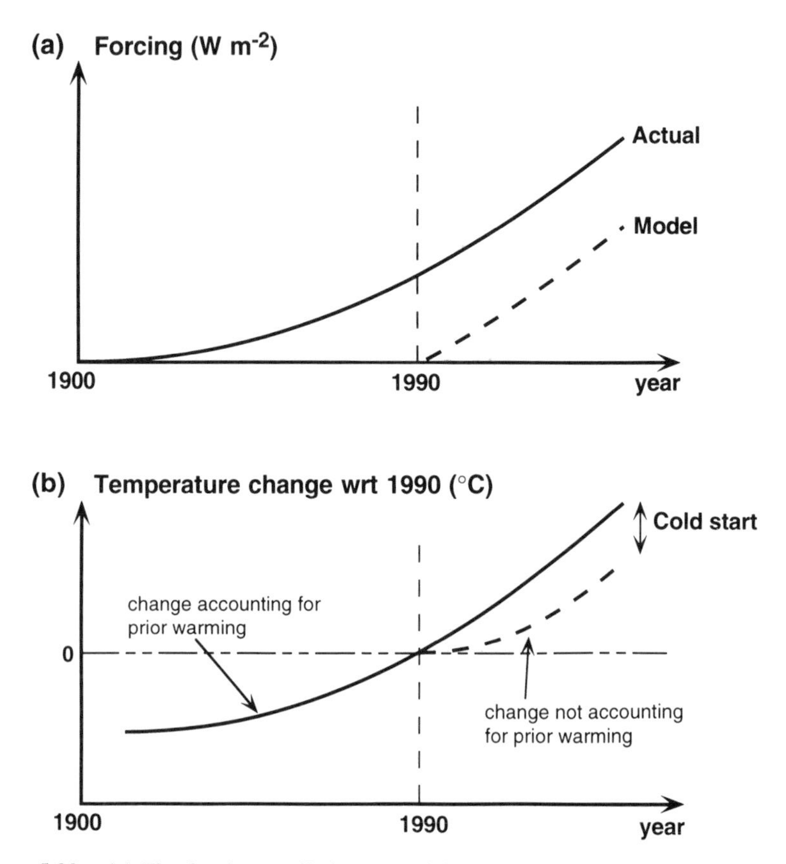

Figure 5.22 (a) The forcing applied to a model which starts running in 1990 matches the slope of the actual forcing which began before 1900. (b) The sequestration of heat by the oceans results in a slow start ('cold start') to the model warming

model development and improvement. The need for flux adjustment will lessen as model components are improved.

The 'cold start' phenomenon

Normally, model simulations of the effect of changing carbon dioxide concentrations are thought of as changes with respect to the 'present day'. If we start a coupled model from present-day (1990) conditions, and apply forcing to the model as shown in Figure 5.22(a), then because of the thermal inertia of the oceans, the warming rate is low at first (Figure 5.22(b)). Since the forcing in the real world has been applied since before 1900, that period of slow warming has passed. If model simulations do not account for this prior warming, then the projected warming path will be different. This slow initial warming is known as the 'cold start' phenomenon.

5.7 USING GCMs

Use of fully coupled climate models, because of the complications associated with 'cold start' and 'climate drift', necessitates careful attention to the impact of the coupling process itself. The concepts of different time-scales and subsystem equilibration times introduced in Chapter 1 are critical to the effective use and appropriate interpretation of results from GCMs. Three-dimensional climate models have developed over the years to form vital tools in studies of the climate system. In Chapter 6 we will examine how GCMs can be used to evaluate possible regional-scale impacts associated with tropical deforestation, predict the impact of changes in greenhouse gases and be used for the analysis of palaeoclimates. This range of applications of climate models and the diversity and complexity of the models has led to a set of projects aimed at the coordination, validation and interpretation of model results. These are also discussed in Chapter 6.

RECOMMENDED READING

Boer, G.J., McFarlane, N.A., Laprise, R., Henderson, J.D. and Blanchet, J.-P. (1984) The Canadian Climate Centre spectral atmospheric general circulation model. *Atmosphere–Ocean* **22**, 397–429.

Briegleb, B.P. (1992) Delta–Eddington approximation for solar radiation in the NCAR Community Climate Model. *Journal of Geophysical Research* **97**, 7603–7612.

Bryan, K. (1969) A numerical method for the study of the world ocean. *Journal of Computational Physics* **4**, 347–376.

Cess, R.D., Zhang, M.H., Minnis, P., Corsetti, L., Dutton, E.G., Forgan, B.W., Garber, D.P., Gates, W.L., Hack, J.J., Harrison, E.F., Jing, X., Kiehl, J.T., Long, C.N., Morcrette, J.-J., Potter, G.L., Ramanathan, V., Subasilar, B., Whitlock, C.H., Young, D.F. and Zhou, Y. (1995) Absorption of solar radiation by clouds: Observations versus models. *Science* **267**, 496–499.

Deardorff, J. (1978) Efficient prediction of ground temperature and moisture with inclusion of a layer of vegetation. *Journal of Geophysical Research* **83**, 1889–1903.

Dickinson, R.E. (1984) Modelling evapotranspiration for three-dimensional global climate models. In J.E. Hansen and T. Takahashi (eds), *Climate Processes and Climate Sensitivity*. Geophysical Monograph 29, Maurice Ewing Vol. 5. American Geophysical Union, Washington, DC, pp. 58–72.

Dickinson, R.E. (1995) Land processes in climate models. *Remote Sensing of Environment* **57**, 27–38.

Dickinson, R.E., Henderson-Sellers, A. and Kennedy, P.J. (1993) *Biosphere/* Atmosphere Transfer Scheme Version 1e (BATS1e) as Coupled to the NCAR Community Climate Model. NCAR Technical Note TN-387+STR, 80 pp.

Eagleson, P.S. (ed.) (1982) *Land Surface Processes in Atmospheric General Circulation Models*. Cambridge University Press, Cambridge, 560 pp.

Gordon, C.T. and Stern, W.F. (1982) A description of the GFDL global spectral model. *Monthly Weather Review* **110**, 625–644.

Flato, G.M. and Hibler, W.D. III (1992) Modelling sea-ice as a cavitating fluid. *Journal of Physical Oceanography* **22**, 626–651.

Hack, J.J. (1993) Parameterization of moist convection in the NCAR Community Climate Model (CCM2). *Journal of Geophysical Research* **99**, 5551–5568.

Hansen, J., Russell, G., Rind, D., Stone, P., Lacis, A., Lebedeff, S., Ruedy, R. and Travis, L. (1983) Efficient three-dimensional global models for climate studies: Models I and II. *Monthly Weather Review* **111**, 609–622.

Manabe, S. and Bryan, K. (1985) CO_2 induced changes in a coupled ocean–atmosphere model and its palaeoclimatic implications. *Journal of Geophysical Research* **90**, 1689–1707.

Manabe, S., Bryan, K. and Spelman, M.J. (1979) A global ocean–atmosphere climate model with seasonal variation for future studies of climate sensitivity. *Dynamics of Atmosphere and Oceans* **3**, 393–426.

Matthews, E. (1983) Global vegetation and land use: New high-resolution data bases for climate studies, *Journal of Climate and Applied Meteorology* **22**, 474–487.

Meehl, G.A. (1990) Development of global coupled ocean–atmosphere general circulation models. *Climate Dynamics* **5**, 19–33.

Mintz, Y. (1984) The sensitivity of numerically simulated climates to land-surface boundary conditions. *In* J.T. Houghton (ed.), *The Global Climate*. Cambridge University Press, Cambridge, pp. 79–106.

Mitchell, J.F.B., Davis, R.A., Ingram, W.J. and Senior, C.A. (1995) On surface temperature, greenhouse gases and aerosols: Models and observations. *Journal of Climate* **10**, 2364–2386.

Ramanathan, V. (1981) The role of ocean–atmosphere interactions in the CO_2 climate problem, *Journal of the Atmospheric Sciences* **38**, 918–930.

Rasch, P.J. and Williamson, D.L. (1990) Computational aspects of moisture transport in global models of the atmosphere. *Quarterly Journal of the Royal Meteorological Society* **116**, 1071–1090.

Sellers, P.J., Mintz, Y., Sud, Y.C. and Dalcher, A. (1986) A simple biosphere model (SiB) for use with general circulation models. *Journal of the Atmospheric Sciences* **43**, 505–531.

Semtner, A.J. (1995) Modelling ocean circulation. *Science* **269**, 1379–1385.

Shukla, J. and Mintz, Y. (1982) Influence of land-surface evapotranspiration on the Earth's climate, *Science* **215**, 1498–1501.

Simmons, A.J. and Bengtsson, L. (1988) Atmospheric general circulation models: Their design and use for climate studies. *In* M.E. Schlesinger (ed.), *Physically Based*

Climate Models and Climate Modelling. Proceedings of a NATO ASI, Reidel, Dordrecht.

Slingo, A. (1989) A GCM parameterization for the shortwave radiative properties of water clouds, *Journal of the Atmospheric Sciences* **46**, 1419–1427.

Washington, W.M. and Parkinson, C.L. (1986) *An Introduction to Three-Dimensional Climate Modelling*. University Science Books, Mill Valley, CA, 422 pp.

Washington, W.M., Semtner, A.J., Jr., Knight D.J. and Mayer, T.A. (1980) A general circulation model experiment with a coupled ocean, atmosphere and sea-ice model. *Journal of Physical Oceanography* **10**, 1887–1908.

Wells, N.C. (1979) A coupled ocean–atmosphere experiment: The ocean response. *Quarterly Journal of the Royal Meteorological Society* **105**, 355–370.

Wilson, M.F. and Henderson-Sellers, A. (1985) Land cover and soils data sets for use in general circulation climate models. *Journal of Climatology* **5**, 119–143.

CHAPTER 6

Evaluation and Exploitation of Climate Models

Scenario ... a sketch, outline, or description of an imagined situation or sequence of events; esp. (a) a synopsis of the development of a hypothetical future war, and hence an outline of any possible sequence of future events; (b) an outline of an intended course of action; (c) a scientific model or description intended to account for observable facts. ... The over-use of this word in various loose senses has attracted frequent hostile comment

Oxford English Dictionary, Compact Edition (OED, 1992, p. 1669).

6.1 EVALUATION OF CLIMATE MODELS

It is clear from the preceding chapters that there is a wide variety of climate models with different characteristics and different applications. Even within one particular climate model type (e.g. AGCMs) there are many different features and stages of development. Moreover, because climate models share a commonality of purpose, it is possible, and often useful, to apply different climate model types to the same prediction task. The result of this profusion of model types and model characteristics is a bewildering array of model predictions. This profusion of predictions and predictive capability has prompted the climate modelling community to initiate a series of model intercomparisons and evaluations of performance.

Evaluation of climate models can produce a range of outcomes which have been grouped as (a) predictions that are unreasonable, (b) predictions that are so reasonable as to be already known, (c) unexpected predictions, but which can be readily understood and accepted, or (d) model predictions that, whilst being reasonable, identify novel outcomes that challenge current theories. Normal practice in model development would screen out all developments producing unreasonable results, and there is little benefit in intercomparison of results which are totally reasonable and well known. Thus the intercomparisons and group evaluations tend to try to focus on results in categories (c) and (d): new predictions that are consistent with theory and those which challenge existing ideas.

The process of comparison of model predictions and group evaluation is complex as it has to encompass models and modelling groups from around the world and has to be organized so that comparable results are being compared. To facilitate the process of model evaluation and intercomparison, the WCRP's Working Group on Numerical Experimentation (WGNE) categorized intercomparisons into three levels (Figure 6.1(a)). Level 1, the simplest, uses any available model results and a common diagnostic set. The IPCC assessments are Level 1 intercomparisons. Level 2 requires that the simulations are made according to prespecified, identical conditions, that common diagnostics are employed and that there is a common diagnostic set against which all the predictions are evaluated. Level 3, the 'best' intercomparison process, requires, in addition to the requirements of the two lower levels, that all the models employ the same resolution and that the intercomparison includes the use of some common routines or code modules.

Until the 1990s, intercomparisons were conducted at Level 1. Recently, a series of initiatives has led to intercomparisons at Level 2 – some of these are described in the following sections. At the time of writing, there are no Level 3 intercomparisons, although some of the Level 2 intercomparisons are planned later to develop common code modules. For example, the Project for Intercomparison of Land-surface Parameterization Schemes' (PILPS) timelines (Figure 6.1(b)) show its Phase 1 as Level 1, and Phases 2 and 3 as Level 2, but has planned for part of Phase 4 (part (c)) to include a common coupling module.

6.1.1 Intercomparisons facilitated by technology

Most of the recent climate model intercomparisons have only been possible because of the advent of global telecommunications. The facilities of the 'information superhighway' are essential for a Level 2 or higher type of intercomparison. Typically, a coordinating group is identified and this group takes responsibility for the provision of the agreed model simulation instructions, including the experimental design and the forcing data. They also provide independent data against which to compare the model results and facilitate model intercomparisons by providing quality control procedures and a central electronic results and data 'library' which can be accessed by all the participating modelling groups.

The demands associated with providing these facilities are quite considerable. This is one of the reasons why Level 2 intercomparisons have so far been restricted to specific aspects of the whole climate system. The following sections review, as examples, intercomparisons of atmospheric models, land-surface schemes, ocean carbon models, vegetation models and radiation schemes.

6.1.2 Atmospheric Model Intercomparison Project (AMIP)

AMIP was established in 1989 and moved into its second stage (AMIP II) in 1996. It focuses on structured (Level 2) intercomparisons of the atmospheric

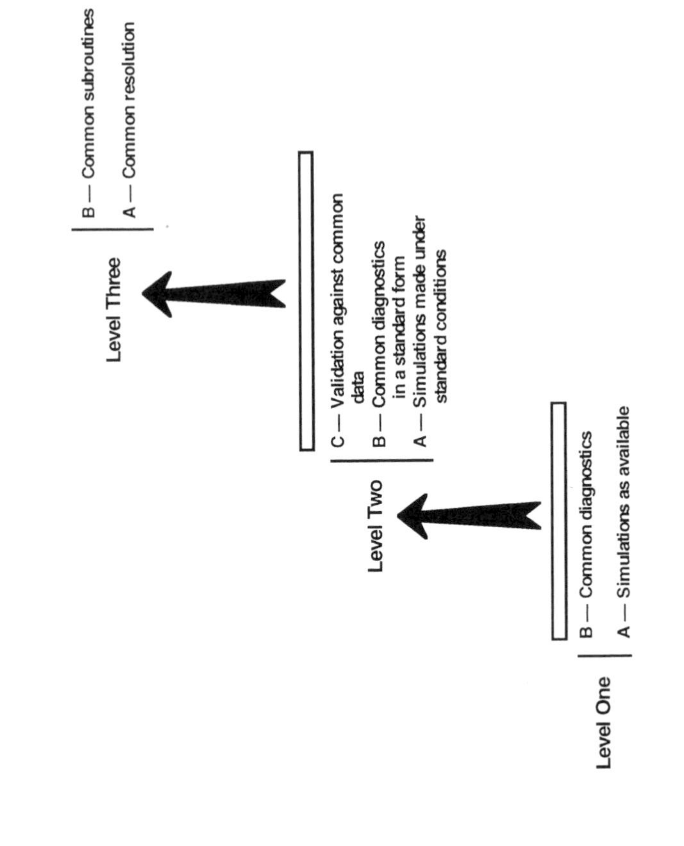
Level One
A — Simulations as available
B — Common diagnostics
Level Two
A — Simulations made under standard conditions
B — Common diagnostics in a standard form
C — Validation against common data
Level Three
A — Common resolution
B — Common subroutines

(a)

(b)

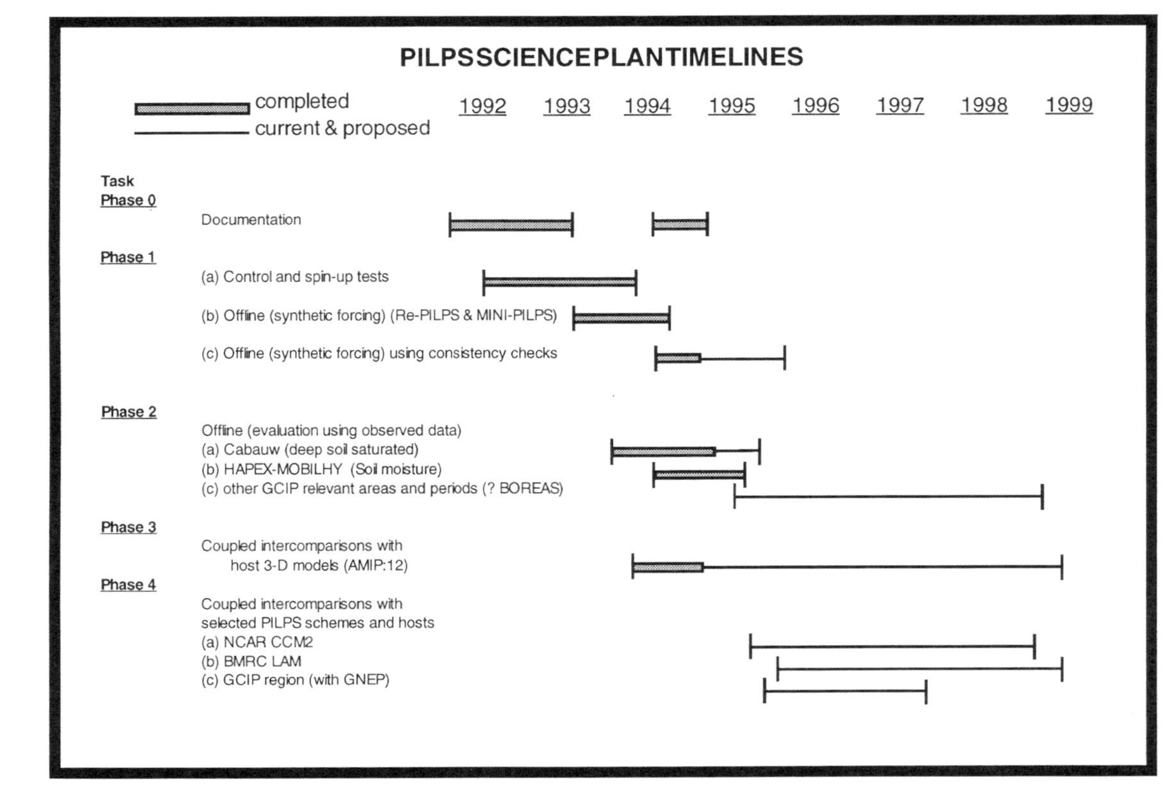

Figure 6.1 (a) Levels of model intercomparison as defined by the Working Group on Numerical Experimentation (WGNE) in support of the World Climate Research Programme (WCRP) (after Gates, 1992). (b) A Gantt chart for the Project for Intercomparison of Land-surface Parameterization schemes (PILPS). The PILPS project includes progressively higher levels of intercomparison

component of global climate models. Participating models use prescribed ocean surface temperatures and sea-ice extents, as well as agreed values of the solar constant (1365 W m^{-2}) and the atmospheric concentration of CO_2 (345 ppmv) as input to a fixed-length simulation. The simulation period for AMIP I was from 1 January 1979 to 31 December 1988, and that for AMIP II from 1 December 1978 to 1 December 1995. The prescribed forcings did not extend to the use of common surface elevation information nor, in AMIP I, to an agreed spin-up procedure, although for AMIP II there will be a recommended procedure.

All participating model groups (around 30) are required to submit output in an agreed format, but there is no requirement for a particular resolution. The results from these global atmospheric simulations have been reported in the IPCC Second Scientific Assessment as a partial demonstration of the validity of GCMs.

Table 6.1 illustrates the differences found between observed values and the mean of participating AMIP models. Some of these differences are fairly small (e.g. sea-level pressure and surface air temperature) but others, especially those associated with clouds and radiative forcing, can be seen to be rather large.

6.1.3 Radiation and cloud intercomparisons

Interactions between clouds and radiation are known to be the source of many of the differences among climate model predictions. This recognition has prompted two complementary intercomparisons: the Intercomparison of Radiation Codes in Climate Models (ICRCCM) and the Feedback Analysis of GCMs and Intercomparison with Observations (FANGIO). Both studies pre-date AMIP, ICRCCM being initiated in 1984 and FANGIO in 1988. They focus on different aspects of cloud–radiation interactions.

Table 6.1 Hemispheric mean seasonal root-mean-square differences between observations and the mean of the AMIP models (after Gates, 1996)

	DJF		JJA	
Variable	NH	SH	NH	SH
Mean sea-level pressure (hPa)	1.4	1.4	1.3	2.4
Surface air temperature (°C) (over land)	2.4	1.6	1.3	2.0
Precipitation (mm d^{-1})	0.80	0.71	0.62	0.77
Cloudiness (%)	10	21	14	16
Outgoing longwave radiation (OLR) (W m^{-2})	2.8	3.2	2.9	5.5
Cloud radiative forcing (W m^{-2})	9.1	20.5	16.2	6.5
Surface heat flux (W m^{-2}) (over ocean)	22.5	27.3	30.5	17.2
Zonal wind (m s^{-1}) (200 hPa)	2.4	1.8	1.8	2.4

ICRCCM has a straightforward mandate: to intercompare results from participating radiation codes in the long and short wavelength regions of the spectrum for the cases of clear and cloudy skies. Around 30–40 modelling groups participated in the first phase, representing both climate models and also radiative transfer algorithms employed in retrieval of fluxes from satellite measurements. The calculations of these schemes were compared with the most detailed radiative transfer calculations available, termed 'line-by-line' calculations. In the second phase, these line-by-line calculations are being augmented by observational data from satellite- and surface-based field programmes, including ERBE, ISCCP and the Surface Radiation Budget Climatology Programme.

The ICRCCM intercomparisons reveal very large differences among the predictions of radiation models. In the longwave region ranges of 30–70 W m^{-2} were discovered, while in the shortwave region ranges varied from 3% for the (simplest) pure water vapour cases, to 46% for the cases with thick cloud, and as high as 60% for situations with high aerosol loadings. An ICRCCM report summarizes these findings as showing that many algorithms have inherent, unknown but large errors which may significantly affect the conclusions of the studies in which they are used. However, it is noted, in the same summary, that it is difficult to draw conclusions regarding the accuracy, or otherwise, of climate model simulations overall because of their dependence on other compensating processes and adjustments of model parameters.

The FANGIO project seeks to improve understanding of the feedback processes in climate models involving cloud and radiation calculations. Defining a climate model sensitivity parameter, λ', as

$$\lambda' = \frac{1}{(\Delta F/\Delta T_s - \Delta Q/\Delta T_s)} \tag{6.1}$$

where ΔQ is the change in shortwave flux, ΔF is the change in infrared flux, and noting that for conditions typical of the present day Earth ($F = 240$ W m^{-2}, $T_s = 288$ K) the value of λ' in the absence of any feedbacks is 0.3 K m^2 W^{-1}, the FANGIO investigators have calculated the value of λ' for their models in the cases of clear skies, cloudy skies and for the global response overall (Figure 6.2). The range in the clear-sky values is very small, underlining the main conclusion of this intercomparison: the threefold variation among AGCMs' sensitivity to a prescribed climate change is due almost entirely to cloud feedback processes.

6.1.4 Project for Intercomparison of Land-surface Parameterization Schemes (PILPS)

This intercomparison was established by WCRP in 1992. As noted in connection with Figure 6.1(b), PILPS is predominantly a Level 2 exercise, but

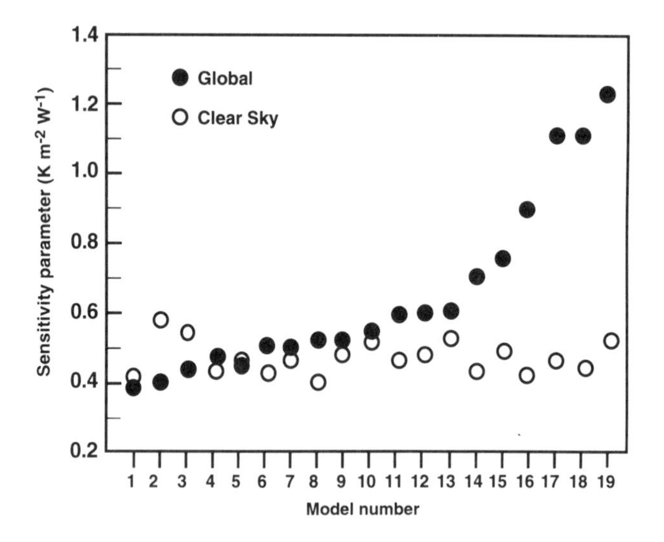

Figure 6.2 Clear-sky and global sensitivity parameters (K m^2 W^{-1}) for 19 GCMs participating in FANGIO (after Cess *et al.*, 1992)

its final stage, Phase 4(c), will incorporate the use of common code modules and is therefore a Level 3 intercomparison. The structure of PILPS is similar to that of AMIP: one component of the climate system is evaluated by prescribing a 'perfect' performance from the other components. For AMIP, the oceans and sea-ice are prescribed. In PILPS, the atmospheric forcing, soil and vegetation characteristics are identical across all simulations.

The largest differences discovered by PILPS occur in the treatment of energy and water partitioning by the land-surface models. The available net radiation at the land surface can be lost either as latent heat (evaporation) or sensible heat (by warming the air close to the surface and its subsequent removal by advection). Evaporation can only occur if there is moisture in the soil or on the vegetation. Similarly, precipitation falling on the land can be lost either as evaporation or runoff and drainage. These two sets of processes are coupled, since when the land is dry the radiative energy income must be balanced only by sensible heat loss. These coupled partitionings of net radiant energy (Figure 6.3(a)) and precipitation (Figure 6.3(b)) generated from PILPS Phase 2(a) are compared with observations from Cabauw in the Netherlands. The energy partitioning does not produce a perfect line or relationship because different schemes predict differing surface temperatures and hence differing emissions of infrared radiation from the surface and thus different values of net available radiant energy: the scatter (around the line). The range along the line is about 25 W m^{-2}, which is commensurate with the range in partitioning of precipitation (Figure 6.3(b)) of about 320 mm yr^{-1}.

It can be seen that the bucket model (BUCK) is an outlier in PILPS. On a monthly basis, Figure 6.3(c) shows that although the majority of schemes

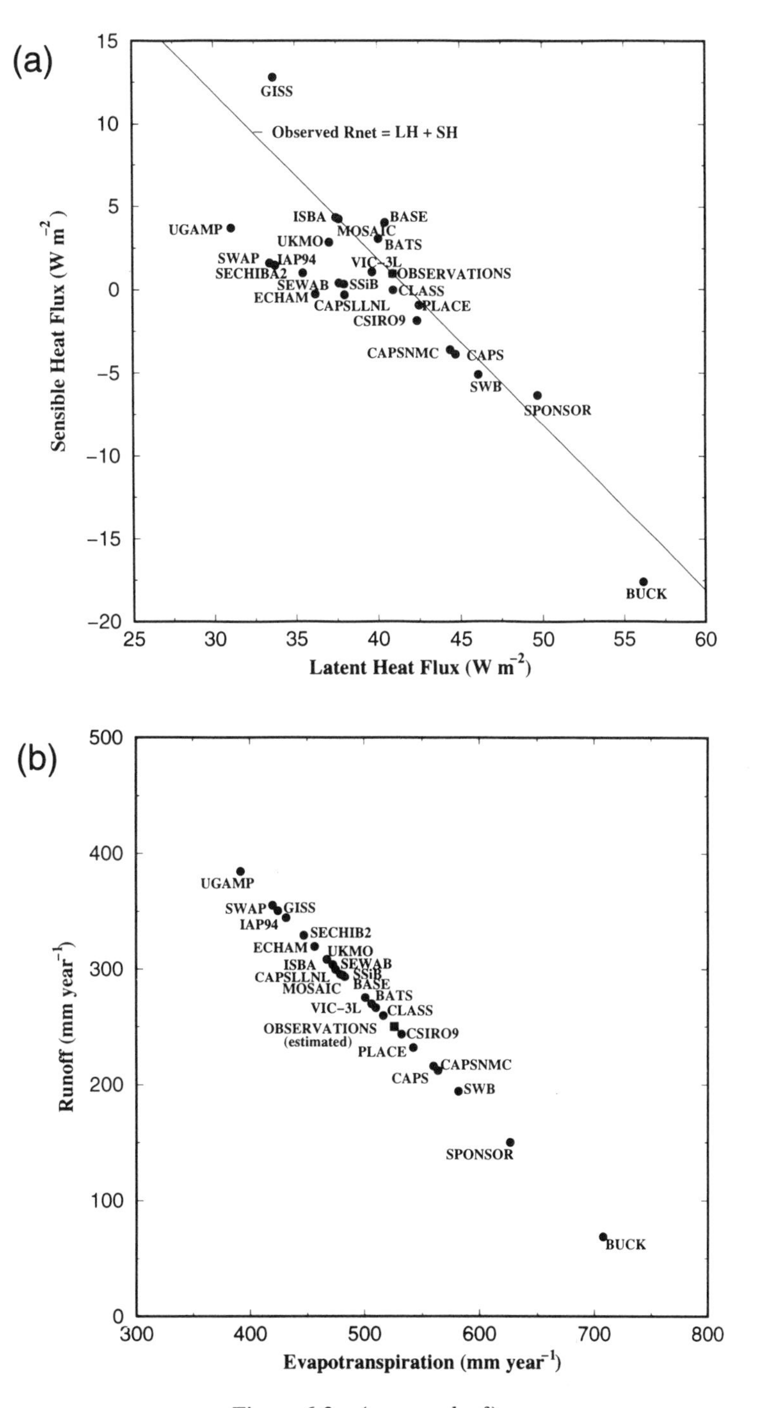

Figure 6.3 (*see overleaf*)

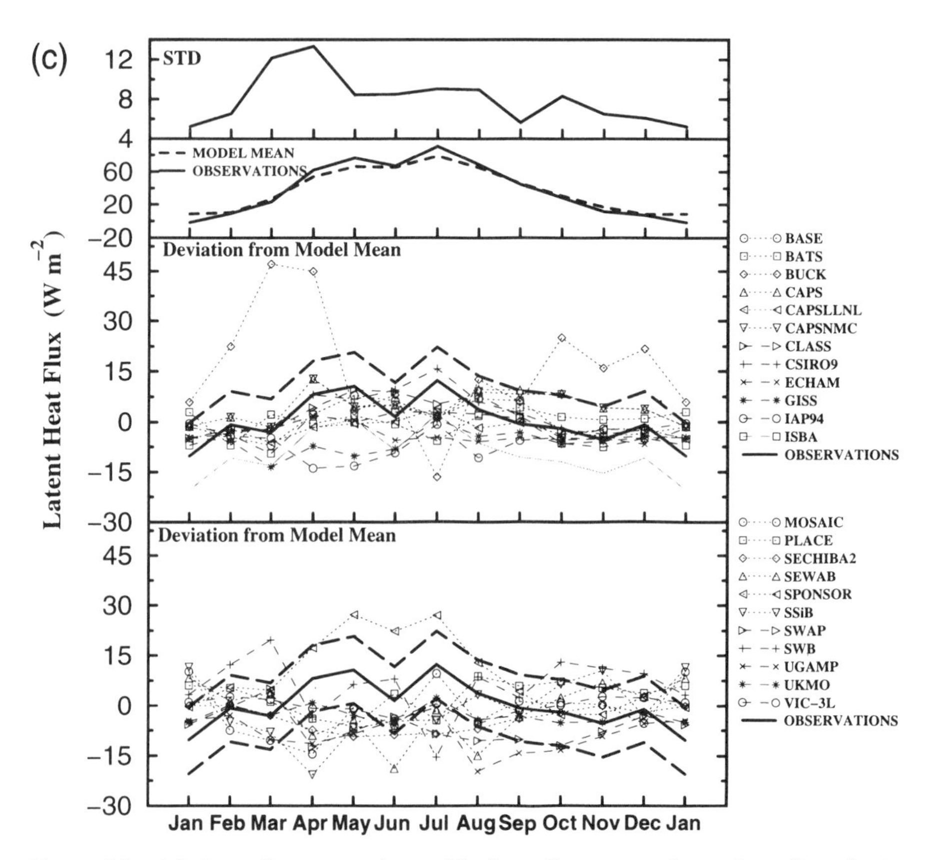

Figure 6.3 (a) Annually averaged sensible heat flux versus latent heat flux (both W m^{-2}) for the PILPS Phase 2(a) experiments. Observed annual mean net radiation must be the sum of the latent and sensible heat fluxes. (b) Annual runoff versus evapotranspiration (mm yr^{-1}) for the same experiment. (c) Seasonal variations of latent heat flux together with across-model standard deviation (STD) from PILPS Phase 2(a) (from Chen *et al.*, 1996)

produce predictions deviating from the group mean that lie within ±10 W m^{-2} of the observed latent heat flux, the group mean itself significantly outperforms any individual model.

6.1.5 Comparing carbon-cycle subcomponents of climate models

The uptake and release of carbon dioxide at the land and ocean surfaces controls its atmospheric concentration. The magnitude and extent of terrestrial and oceanic sources and sinks of CO_2 must be fully understood if predictions are to be made of the ultimate levels of atmospheric CO_2 and hence of the future climate. Two intercomparison projects are designed to evaluate the performance of ocean carbon simulation and of terrestrial vegetation exchanges of CO_2 with the atmosphere.

The Vegetation/Ecotype Modelling and Analysis Project (VEMAP) has the goal of intercomparing the performance of vegetation model simulations. These models simulate the influence of the physical environment on: (i) the availability of plant functional types (i.e. which plants can grow and reproduce); (ii) competition for resources; (iii) the emergent equilibrium vegetation cover. Associated models, termed terrestrial biogeochemistry models, have been used to simulate the flow of carbon and mineral nutrients within vegetation, surface litter and soil organic matter pools. These models have also been used to examine the global patterns of net primary production, carbon storage and mineral uptake and their sensitivity to climate change.

The emergence of divergent types of biosphere models makes it difficult to address complex issues of global change and terrestrial ecosystems. In particular, examining the response of ecosystems to multiple, and potentially interacting, factors and how the resulting changes in the terrestrial biosphere may influence the Earth system as a whole requires an integrated perspective. The VEMAP intercomparisons (Table 6.2) include the synergistic effects of different vegetation models, different biochemistry models and the different climates simulated by different GCMs. This intercomparison shows a very large range in the projected impact on the total terrestrially-stored carbon: estimates range from a predicted reduction of −39% when BBGC is run with the MAPSS vegetation and the UKMO climate, to an increase in stored carbon of +32% when TEM is run with MAPSS for both the OSU and GFDL climate projections (Table 6.2).

Table 6.2 Annual total carbon storage (10^{15} gC) and percentage change for the linkage of the three biogeochemistry models (BGC) (BIOME-BGC [BBGC], CENTURY [CEN] and the terrestrial ecosystem model [TEM]) with the vegetation distributions of the three biogeography models (VEG) (BIOME2, DOLY and MAPSS) for contemporary climate (CON) at 335 ppmv CO_2 and three GCM climates (OSU, GFDL and UKMO) at 710 ppmv CO_2 (after VEMAP, 1995)

			Models		
BGC	VEG	CON	OSU (%)	GFDL (%)	UKMO (%)
BBGC	BIOME2	122	−13.2	−9.5	−34.7
	DOLY	122	−18.1	−13.6	−36.4
	MAPSS	120	−8.3	−13.8	−39.4
CEN	BIOME2	125	−0.8	+12.6	−1.8
	DOLY	124	+9.8	+17.7	+7.8
	MAPSS	120	+17.0	+20.4	−1.5
TEM	BIOME2	114	+11.9	+25.7	0.0
	DOLY	114	+19.7	+25.3	+12.5
	MAPSS	109	+32.3	+32.2	+1.7

Oceanic uptake and release of CO_2 completes the global carbon system. Ocean–atmosphere carbon exchanges occur both as a result of the degree of solubility and as a function of the ocean biology. The Ocean Carbon Model Intercomparison Project (OCMIP) is evaluating the capability of models to predict both anthropogenic and natural CO_2 exchange by comparison with observations of radiocarbon data, and with determinations of the extent and type of ocean biology derived from satellite observations of ocean colour. It is anticipated that results from OCMIP will be a valuable resource for the further development of the 3D ocean components of OAGCMs and, ultimately, permit improved simulations of the global carbon system in AOBGCMs.

6.1.6 Benefits gained from climate model intercomparisons

The most important outcomes of the international intercomparisons of climate model performance described here are: (i) the identification of group outliers; (ii) the estimation of the range of confidence (or uncertainty) inherent in predictions of any one of the 'reasonable' models; (iii) the development and dissemination of datasets for continuing model evaluation. It has generally been found that:

- no one model performs well in all the evaluations employed;
- no one test evaluates all aspects of the participating models;

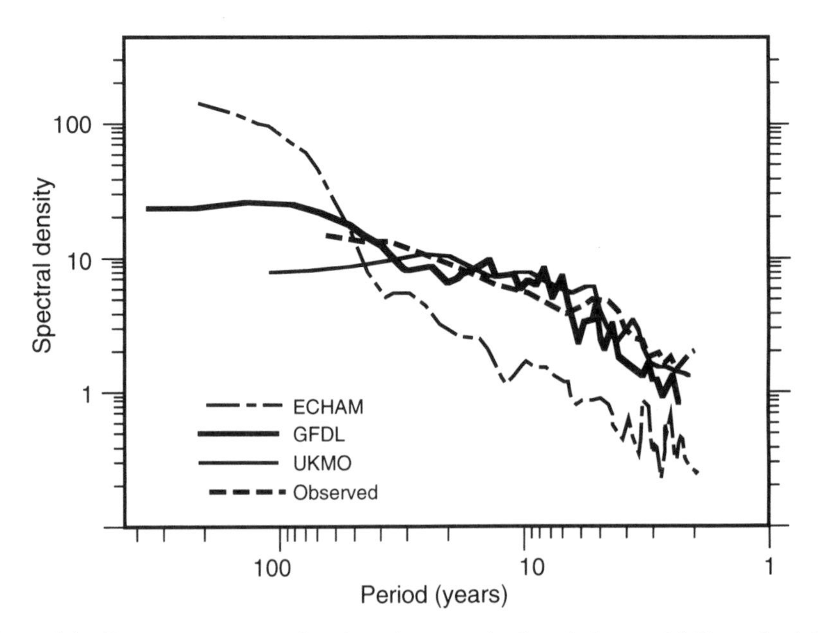

Figure 6.4 Power spectrum, showing the magnitude of the variability of global mean surface temperature from three global climate models and from observations (reproduced by permission from Houghton *et al.*, 1996)

- the model group mean (after excluding unreasonable results/outliers) outperforms any one individual model.

There are dangers as well as benefits associated with large intercomparisons. The most omnipresent is the trend towards group median values. However, if the intercomparisons are of Level 2 or 3 (Figure 6.1), the existence of observational data, especially if released only after the simulations, should counterbalance this central tendency. To date, considerable benefits have been derived from well-structured climate model intercomparisons.

The intercomparison of models also leads to the investigation of sets of statistics which can be used to characterize the behaviour of the climate. For example, the spectrum of variability of the global daily mean near-surface temperature (Figure 6.4) illustrates the differences (and agreements) between observational data and three different coupled ocean–atmosphere models. Since the observational data are of a relatively short period (140 years), estimation of long time-scale variability is not possible. Unfortunately it is this part of the spectrum where the models are in greatest disagreement.

6.2 EXPLOITATION OF CLIMATE MODEL PREDICTIONS

Climate models have the potential to develop information about future and past climates that have applicability to a wide range of human activities. For example, the search for 'safe' disposal sites for nuclear waste materials has involved not only geological evaluation of possible sites, but also climatological assessment using climate model predictions. Some mineral exploration companies have examined the results of past climate predictions in order to try to infer the likely locations of mineral deposits. The model predicted 'threat' of a 'nuclear winter' following a nuclear war is believed by some commentators to have contributed to the de-escalation in weapons development and holdings in the late 1980s. However, the most widespread application of climate model predictions currently is the evaluation of the impacts of greenhouse warming.

6.2.1 Expert assessment

During the 1980s, a growing body of evidence on the likely impacts of global climate change led to increased public concern. As discussed in Chapter 1, this growing body of evidence from both models and observations led to the establishment of the Intergovernmental Panel on Climate Change (IPCC) by the World Meteorological Organization and the United Nations Environment Programme (UNEP). Since it was set up in 1988, the IPCC has acted to focus efforts of climate scientists of all descriptions (e.g. modellers, observational meteorologists, data analysis experts, impacts assessors and economists) on the problem of climate change associated with the enhanced levels of greenhouse gases produced by human activities (particularly fossil fuel combustion). In the lead-up to the 1990 reports, the IPCC had three working groups, focusing on (i) the assessment of scientific information on climate change, (ii) the assessment of environmental and socio-economic impacts of climate change, and (iii) the formulation of response strategies. The IPCC also established a special panel on the participation of developing countries. An updated supplementary report was issued in 1992 and a second full assessment published in 1996.

Since 1992, the focus of the IPCC working groups has been changed slightly. Working Group I still focuses on the assessment of available scientific information on climate change, especially as related to human activities. Working Group II is charged with the assessment of environmental and socio-economic impacts and possible response options, while Working Group III is examining cross-cutting issues related to climate change, particularly socio-economic and technological issues. Although not a scientific research programme, the IPCC has acted as a focus for climate researchers. It has drawn heavily on the established research and intercomparison projects which were discussed in Section 6.1 and has also encouraged them.

The IPCC process has required modellers, and those who have examined records of past climates, to make an assessment of their confidence in the different aspects of their results, and this in turn has generated impetus in the research community towards model improvements. These assessments of confidence have been subject to change since 1990. In some cases, an enhanced understanding and implementation of processes in models has not led to an increase in confidence in the results. For example, the confidence in soil moisture predictions from GCMs has been reduced between 1990 and 1996 because we now know more about soil moisture processes: specifically, we know that they are more complex than early GCMs allowed. The information presented in the IPCC 1990 reports was used as a foundation for a global agreement to formulate policy on climate change: the Framework Convention on Climate Change. The ongoing process of scientific evaluation and assessment continues to feed into a political framework of commitments and targets.

6.2.2 GCM experiments for specific applications

Tropical deforestation

The possible impacts of tropical deforestation on the local, regional and global climate have received considerable attention from climate analysts and modellers in recent years since tropical forests provide the habitats of about half of the world's species and are an important natural sink of CO_2 and a source of tropical aerosols and trace gases. In the context of the global atmospheric circulation, since strong ascending branches of the Walker and Hadley circulation are located over tropical forest regions, it has been suggested that changing the land-surface characteristics in regions of tropical forest may affect the atmospheric circulation.

GCMs are one tool which can be used in an attempt to answer the many questions which have been raised with regard to the impacts of tropical deforestation. In experiments designed to examine the effects of processes such as tropical deforestation, modellers change the characteristics of the vegetation and soil surface at a number of points. These changes depend on the nature of the land-surface scheme employed in the model. The choice of parameters is difficult, since relatively little information is available on the characteristics of deforested regions. Modellers must choose values of roughness length, leaf area index, soil colour and vegetation type (trees, grass, shrub etc). Models do not yet attempt to simulate the gradual change which really occurs as tropical forest is removed and replaced. The best that can be done currently is to compare long-term means 'before' and 'after' an imposed deforestation.

Among the significant properties of tropical rainforests are that they have a very low surface albedo throughout the year, their leaf area and stem area are larger than those of any other vegetation and the trees are tall. Replacing the tropical rainforest with grassland leads to three main changes in the land-surface properties: (i) the surface albedo is increased, which directly causes a reduction in the surface net radiation; (ii) the reductions in the leaf area and stem area lead to a decrease in the water-holding capacity of the vegetation – thus the evaporation of the intercepted precipitation and, probably, the vegetation transpiration are decreased following deforestation; (iii) the grassland replacing the tropical forest is much shorter and smoother than the forest so that the surface roughness is dramatically reduced and the surface friction is reduced. The decreased surface roughness length has two competing effects on the evapotranspiration. Strengthened surface wind speed acts to enhance evaporation, whereas the effect of decreased surface roughness is to reduce evaporation. Most model simulations indicate that the surface evapotranspiration is decreased overall.

The changes in surface evapotranspiration, acting as the connection between the changes in the hydrological processes (determining the regional water recycling) and the changes in the land-surface and atmospheric energy budget (the sink of the surface energy budget and the source of the atmospheric energy budget), have a crucial role in determining the local impact of tropical deforestation. Moreover, the reduction in the vertical motion caused by less solar radiation being absorbed by the land surface and decreased latent heat from the land surface results in the reduction in the cloud cover amount over the deforested regions and may affect the regional atmospheric dynamics, especially the atmospheric moisture flow.

Results from one set of model experiments have shown that the impacts simulated in response to tropical deforestation in the Amazon Basin, SE Asia and tropical Africa are regionally dependent. Significant climatic changes are

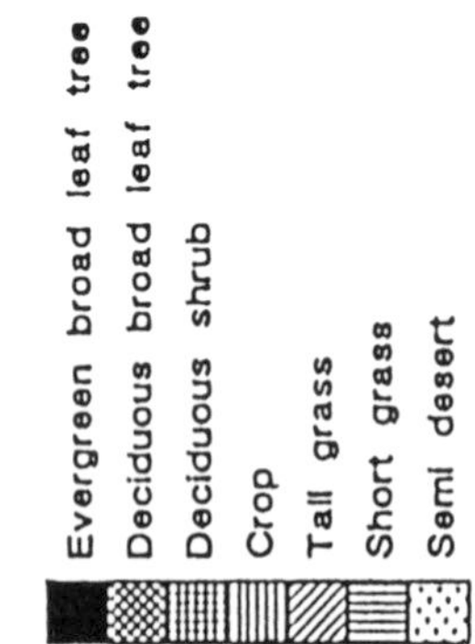

Evergreen broad leaf tree
Deciduous broad leaf tree
Deciduous shrub
Crop
Tall grass
Short grass
Semi desert

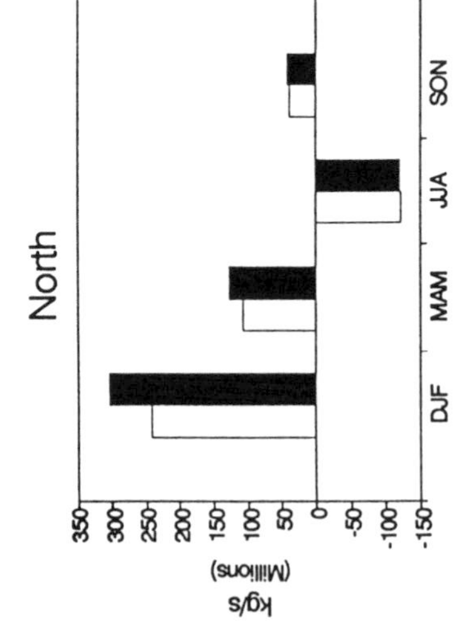

North
kg/s
(Millions)
350
300
250
200
150
100
50
0
-50
-100
-150
DJF
MAM
JJA
SON

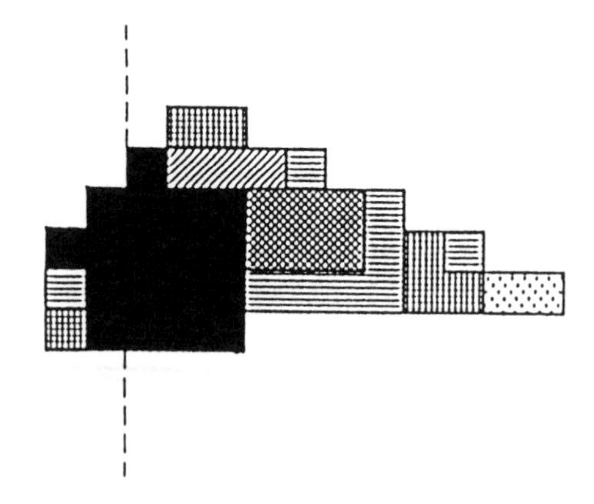

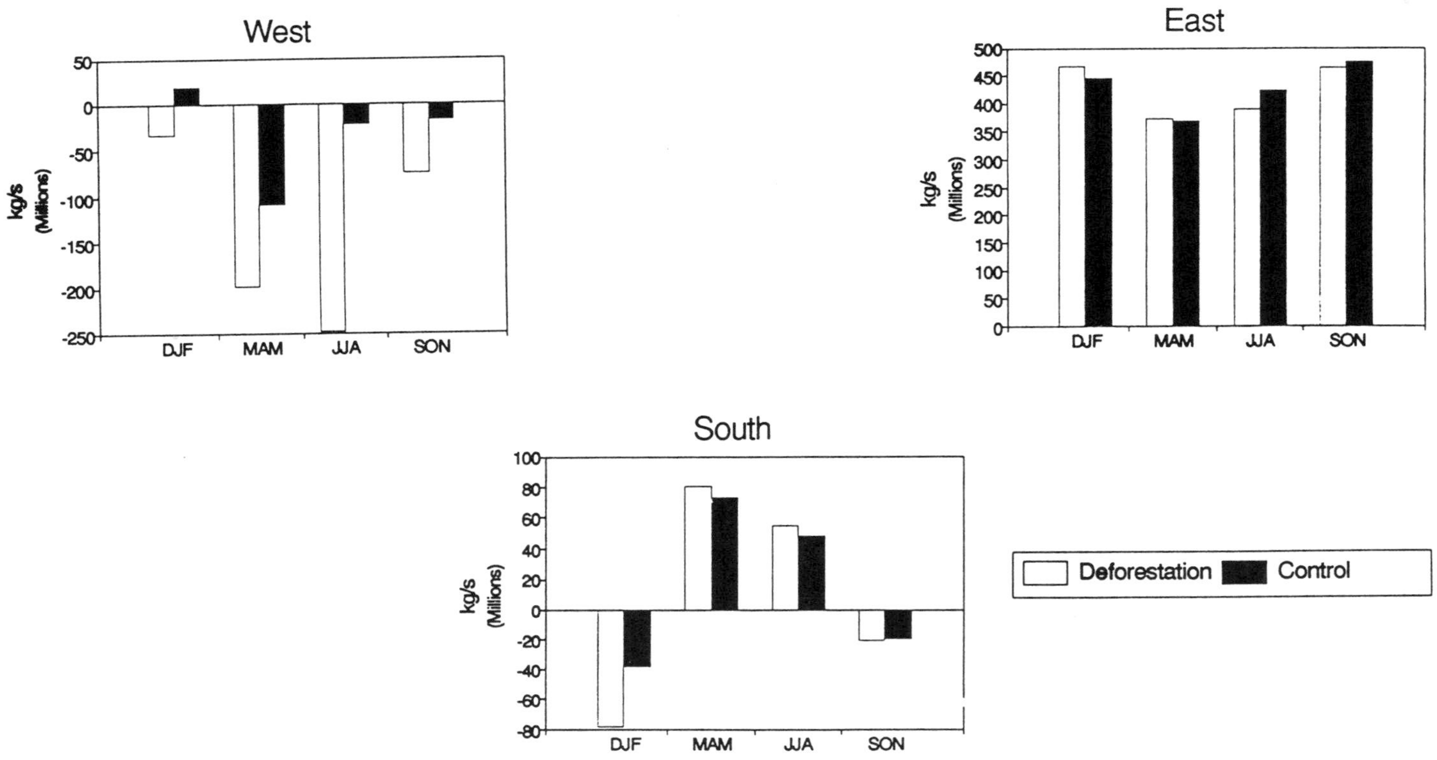

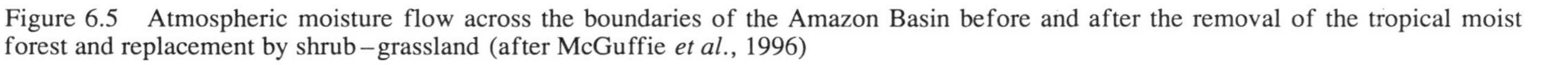

Figure 6.5 Atmospheric moisture flow across the boundaries of the Amazon Basin before and after the removal of the tropical moist forest and replacement by shrub–grassland (after McGuffie *et al.*, 1996)

simulated (reduction in the amount of moisture transported into the region and a reduction in precipitation) in the Amazon Basin, but the changes in SE Asia and tropical Africa were found to be smaller, especially the local precipitation changes. There is an interaction in the climate model between the changes in surface evapotranspiration and the changes in the total precipitation. Following deforestation, the reduction in precipitation appears to be initially due to the reduction in the surface evapotranspiration (because the net radiative energy in the atmosphere is increased over all three regions), this reduction in the total precipitation further reducing the evapotranspiration. At the same time, the reduction in the latent heat flux because of the weakened surface evapotranspiration leads to a net reduction in the atmospheric energy budget. Thus the regional atmospheric circulation is weakened and less water vapour is delivered into the deforested regions. Figure 6.5 shows how the amounts of moisture transported across the boundaries of the Amazon forest change as a result of the removal of the tropical forest. The scale of moisture convergence changes, and possibly also of cloud and convection changes, is such that there is a possibility that non-local climatic impacts may also occur.

Rather small changes in surface temperature were simulated with this particular model. This was concluded to be the result of the compensating effects of the reduction in the surface net radiation (due to the increased albedo) and the reduction in the surface evapotranspiration. However, the diurnal variation of surface temperature is enhanced following deforestation due to cooling during the night and warming during the day (Figure 6.6). These changes are due to a reduction in cloud cover and a reduction in evaporation. The changes in incoming solar radiation and outgoing longwave radiation from the surface are consistent with this interpretation.

The disturbance of the atmosphere induced by tropical deforestation may also interact with large scale circulation features and thereby have impacts in extra-tropical locations. A major topic of current study is whether tropical deforestation will have mid-latitude impacts similar to those of ENSO events. Future challenges for modellers include the simulation of the dynamic processes associated with vegetation destruction and (potential) regeneration. For example, after tropical deforestation, the climate of the regions may be so modified by the removal of the forest, that the forest will be unable to regenerate, even if the land-use disturbance was removed.

Mineral deposits

Palaeoclimate simulations with GCMs offer modellers the chance to test their models with parameters beyond the normal 'present-day' range (for which the model was probably constructed) and thereby increase confidence in their models. GCMs have been used in simulations of the last glacial maximum ~18 000 BP, and conditions ~6000 BP have been suggested as a useful complement to future climate simulations. Palaeoclimate modelling not only

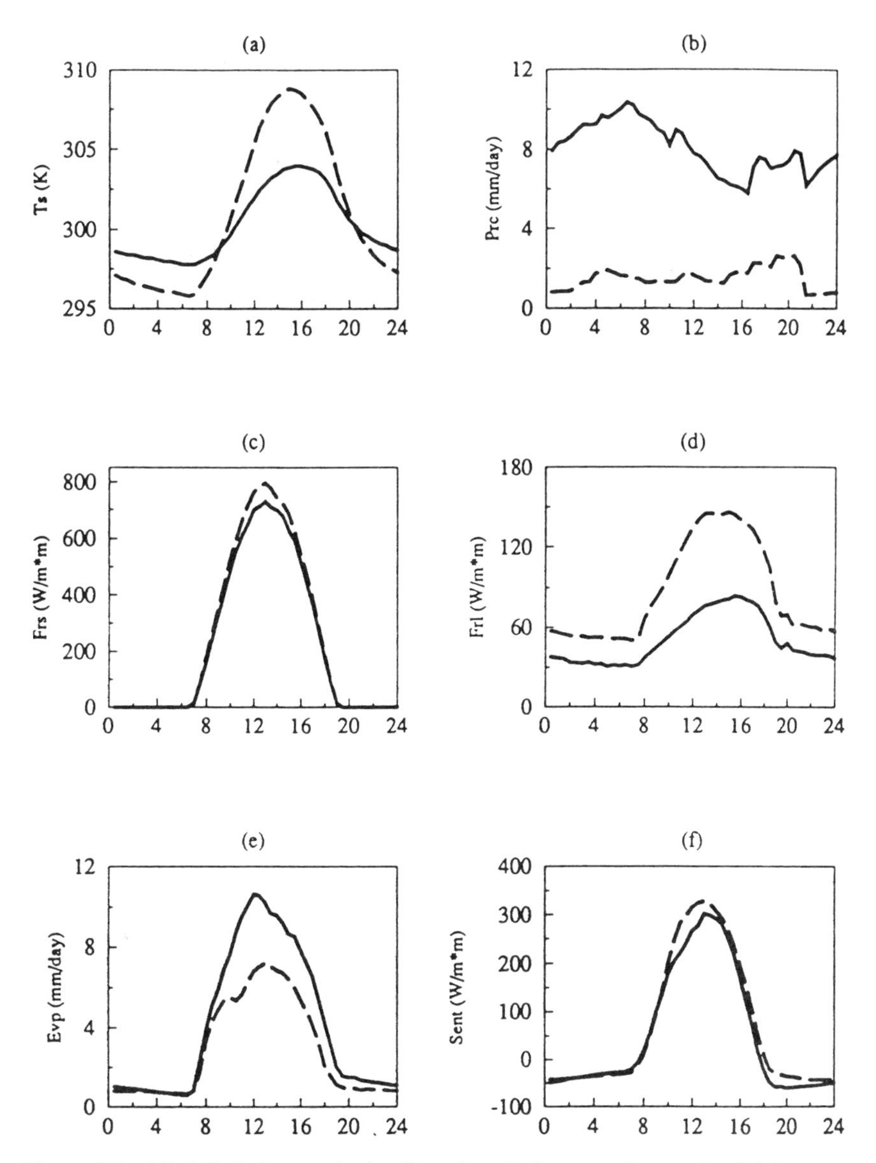

Figure 6.6 Modelled changes in the diurnal cycle (average January) of (a) canopy air temperature (K), (b) precipitation (mm d^{-1}), (c) net absorbed solar flux (W m^{-2}), (d) net infrared flux (W m^{-2}), (e) evaporation (mm d^{-1}) and (f) sensible heat flux (W m^{-2}) following deforestation of the Amazon Basin (after Zhang *et al.*, 1996a)

helps us understand the details of past climates, but it also encourages modelling based on physical processes rather than tuning to match present-day distributions. Palaeoclimatic data, which sometimes only offer classifications such as 'hot–dry' or 'humid' as descriptions of past climates, can be augmented by quantitative determinations of, for example, temperature and precipitation from model simulations. The model simulations can also offer indications of the past climate of areas where no proxy data are available. One such GCM experiment is described below as an example.

Evaporites (such as rock salt and gypsum) form near the boundaries of oceans and in shallow basins that are subject to frequent flooding and desiccation. The levels of salinity which are reached in the basins determine the nature of the evaporite deposits. Regions which are amenable to evaporite formation would be indicated in a GCM by regions where the total precipitation minus total evaporation ($P - E$) is negative. In one study, an evaporite basin model, consisting of a saline 'slab' of water with fixed depth and salinity, was run off-line (the model was run using output from the GCM as forcing data, rather than being coupled to the GCM). This offers no feedback to the GCM climate, and if the feedback is assumed to be small then this is acceptable. The evaporite basin model was used to determine whether evaporite potentially *could* form rather than to model the process of deposition. The evaporite model (which computed evaporation based on GCM forcing) was forced with GCM-simulated climate at all model grid points. Figure 6.7(a) shows the $P - E$ values derived from such a simulation of the Triassic (~225 Ma) when the basin salinity was 35‰. Areas with negative $P - E$ are potential locations where lakes could dry out. The dots show the locations of major known evaporite deposits in North America, South America (around 120°W), Arabia (10°S, 20°W) and the Western Tethys, Central Atlantic region (10°N, 50°–80°W). The basin model was run with two different levels of salinity: 35‰ and 300‰. The simulation with salinity at 300‰ (Figure 6.7(b), characteristic of the salinity required for the later stages of gypsum formation) therefore shows areas where sustained deposition might occur. Discrepancies between modelled and observed evaporite distribution can be attributed at least in part to small scale topography which is not resolved by the GCM, but are also an indication that the model-simulated climate is not adequate.

By using models such as these with different continental configurations and forcings, the modellers can make suggestions about the relative strength of other aspects of the climate (such as the salinity-driven deep ocean circulation) in different geological periods. Such assessments have been suggested as a useful component in determining the possible locations of oil-bearing rocks. The formation of such rocks is thought to be linked to sluggish thermohaline circulation, which results in reduced oxygen content of the deep ocean water and hence a reduction in the decay of organic matter. This matter can then accumulate as oil shale in the floors of ocean basins. The GCM is therefore able to provide tentative indications of oil-bearing strata where no direct palaeo-evidence exists.

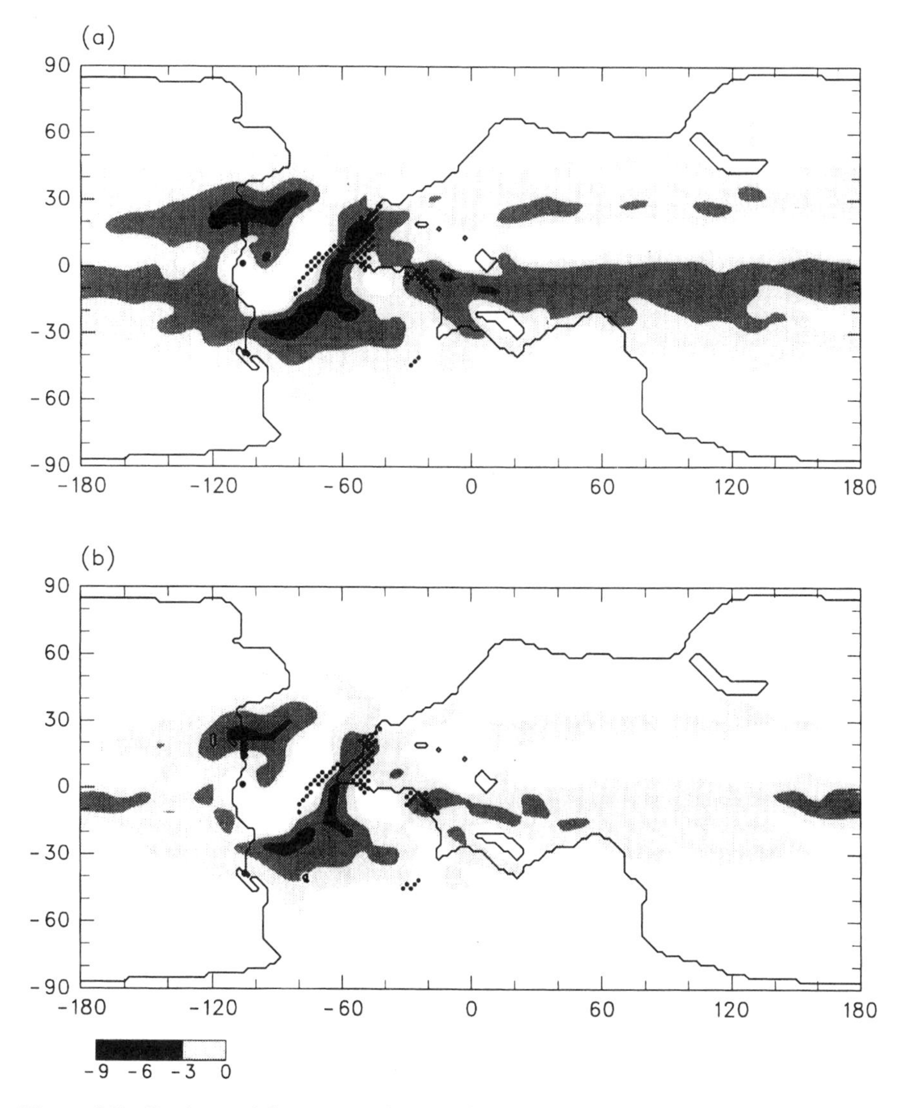

Figure 6.7 Basin model net annual $P-E$ for the Triassic (225 Ma). Only negative $P-E$ values are shown for clarity. (a) Basin salinity is 35‰. (b) Basin salinity is 300‰. At higher salinities evaporation is reduced, so the areas where high salinities are required for evaporite deposition are reduced (from Schutz and Pollard, 1995)

6.2.3 Regional climate prediction

It is clear from the two specific applications described in the previous section (tropical deforestation and mineral deposits) that climate model predictions need to be generated at finer scales than current models if their results are to be of use. The *need* is for regional climate scenarios, but the best models are the coupled GCMs run at coarse resolutions, as explained in Chapter 5. This is one of the major problems faced in trying to apply GCM projections to regional impact assessments.

'Downscaling', as the production of increased temporal and/or spatial resolution climates from GCM results has come to be termed, has three current forms: statistical, regional modelling and timeslice simulations (Figure 6.8). In model-based downscaling, a high resolution, limited-area model is run off-line using boundary forcing from the GCM, or a global atmospheric model is run for limited time 'slices' at high spatial resolution using sea surface temperatures and sea-ice distributions predicted by a lower resolution coupled OAGCM.

Statistical downscaling can take many forms, including the assignment of the nearest grid box estimate, statistical analysis of local climatic fields and the merging of several scenarios based on expert judgement. Although GCMs typically have high temporal resolutions, 'downscaling' has also been applied in the time domain because GCM results are found to have rather poor temporal characteristics when examined at time-scales less than about a month.

Unfortunately, all the currently available means of downscaling, and hence of producing climate change 'scenarios' at a 'useful' regional spatial resolution, are unsatisfactory. This was demonstrated most clearly by a European Union-sponsored comparison of regional model simulations from the Hadley Centre (RegCM), the Max Planck Institute (HIRHAM) and Météo-France's variable resolution GCM with high resolution over Europe which

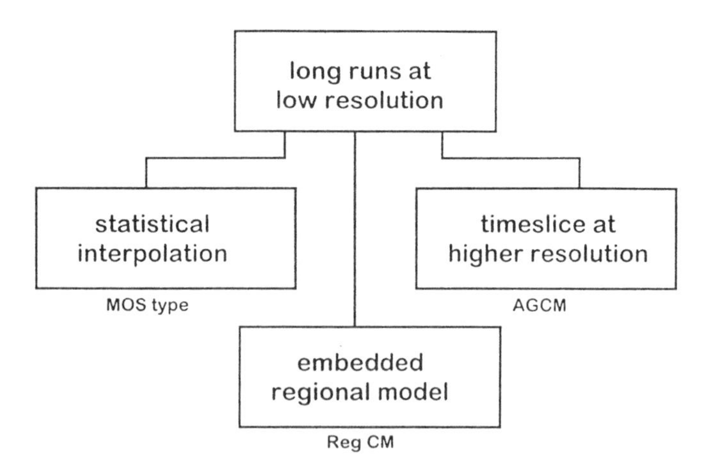

Figure 6.8 Three different techniques, statistical analysis, embedded high resolution models and timeslicing (high resolution global atmospheric model forced by the SST results from a lower resolution coupled model for a specific time period) are all employed to derive higher resolution results from low resolution model simulations for application to regional climate simulation

concluded that where there were poor regional climates (in the coarse resolution OAGCM) the dynamic embedding made things worse, and that neither downscaling nor embedding can solve problems inherent in the GCM simulation.

6.2.4 Policy development

The Framework Convention on Climate Change (FCCC), signed by 153 countries and the European Community at the United Nations Conference on Environment and Development (UNCED) in Brazil in June 1992, has as its central goal the stabilization of atmospheric greenhouse gas concentrations at a level that would prevent dangerous anthropogenic interference with the climate system. It also states that this goal should be realized soon enough that ecosystems could adapt naturally to climate change, that food production would not be threatened and that sustainable economic development could proceed. The Convention does not specify, however, the meaning of 'dangerous anthropogenic interference', how its occurrence or the risk of its occurrence should be detected or what measures, applied at what level of stringency, would be justified in avoiding it. The other central concepts in the objective, natural adaptation of ecosystems, threats to food production and sustainable economic development, are also not articulated precisely. These issues clearly transcend the capabilities of climate models which incorporate 'only' physics and biogeochemistry. The demands of the deliberations relating to the Framework Convention on Climate Change, together with other pressures, have prompted the application of climate predictions and, sometimes, the entire incorporation of climate models into a new class of numerical models called integrated assessment models.

Evaluating future climate in this context takes into account the ramifications of human health, food supply, population policies, national economies, international trade and relations, policy formulation and attendant political processes, national sovereignties, human rights, and international, inter-ethnic and intergenerational equity. In considering the estimated damages due to current emissions of greenhouse gases, the arguments for action are now extending beyond 'no regrets' measures, i.e. those whose benefits, such as reduced energy costs and reduced emissions of conventional pollutants, equal or exceed their cost. Decisions to be taken in the near future will necessarily have to be taken under great uncertainty. However, these decisions may be very sensitive to the level at which atmospheric concentrations are ultimately stabilized and to the environmental effects on ecosystems: the net productivity of the oceans, the response of trees and forests to carbon dioxide fertilization and climate change, and methane production by thawing tundra. It is clear that evaluation of this large suite of possible responses to the threat of future climatic projections must incorporate a myriad of issues outside the scope of conventional climate models but, perhaps, encompassed by integrated assessment models.

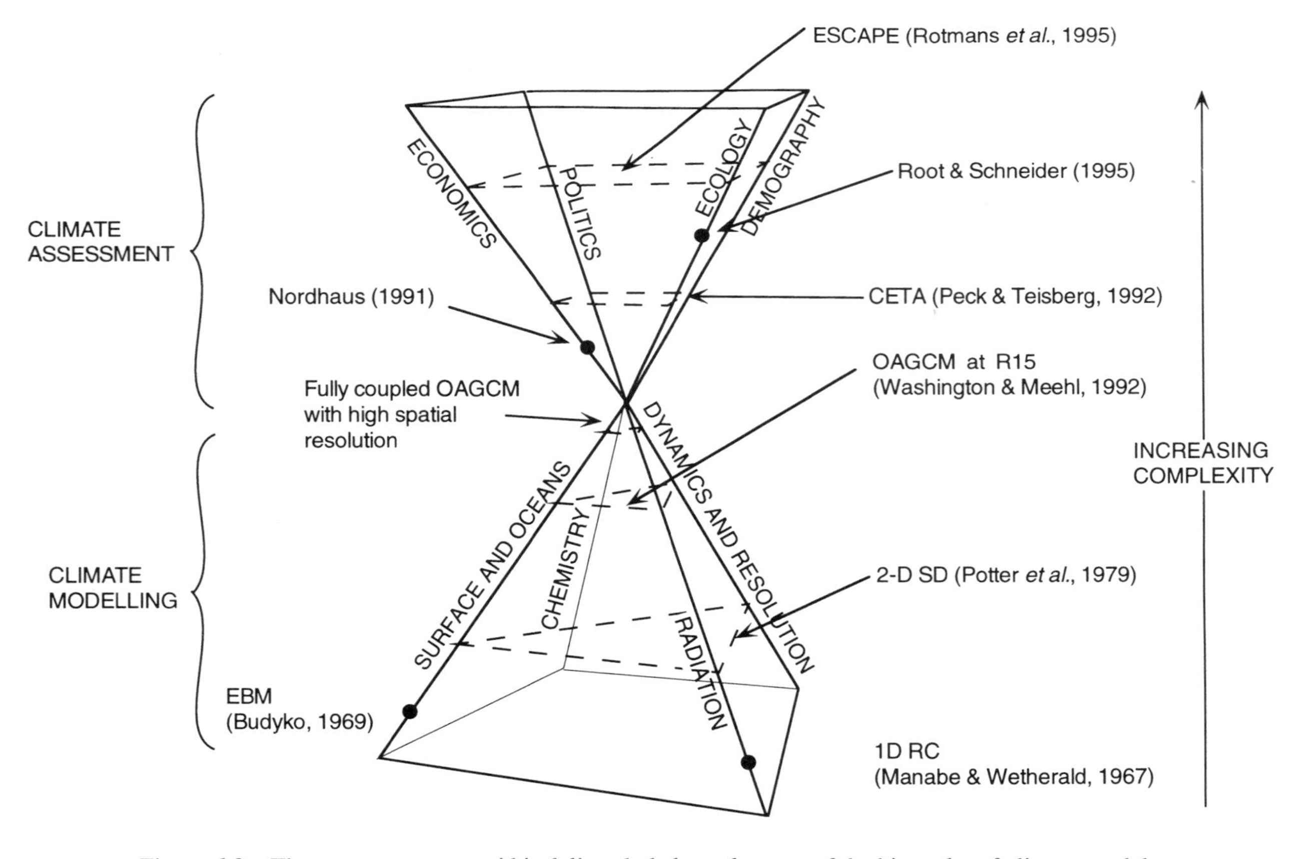

Figure 6.9 The assessment pyramid is delicately balanced on top of the hierarchy of climate models

It may be valuable to draw a distinction between 'integrated assessment' and 'integrated assessment modelling' analogous to the distinction between 'climate assessment' and 'climate modelling'. Most human activities that are affected by weather and climate (e.g. building or reservoir design; bathing suit and umbrella sales) are developed or designed in the context of a 'climate assessment'. However, few of these (at least as yet) draw on the results of climate model simulations. Similarly, all nations which are party to the Framework Convention on Climate Change are developing national assessments of the effects of, and possible responses to, climate change. These assessments are 'integrated' in the sense that they include societal, economic and ecological characteristics. Although most use results from climate models, relatively few, as yet, draw on the results of integrated assessment models. The developers of these models believe that this situation is changing and that integrated assessments will, in the future, draw much more fully on the results of integrated assessment models. Figure 6.9 depicts one uneasy representation of a future in which an inverted pyramid of integrated assessment is 'balanced' on the existing pyramid of climate modelling.

Among the criteria that governments are likely to consider when selecting future policies are distributional (including inter-generational) equity, flexibility in the face of technological change and new information, efficiency or cost-effectiveness, compatibility with the institutional structure and existing policies, and understandability to the general public. A mix of political instruments is likely to be needed in order to achieve the best results. Governments may apply different criteria with different weights to the selection of international and domestic policy instruments. Cost-effectiveness should always be a criterion for selecting policy instruments, but it becomes more important at both the international and domestic levels as the abatement effort becomes more stringent. The immense challenges for integrated assessment include identifying the political processes that will lead and guide change, recognizing the information needs of that process, conducting the physical and social-science research needed to fill those information needs, and presenting that information in a candid and understandable manner.

6.3 INTEGRATED ASSESSMENT MODELS

In some ways, the development of integrated assessment models is similar to the history of climate models, including different disciplinary perspectives and different views on the need for capturing all processes as compared with parameterizing behaviour in terms of a few better understood variables. There are also many differences between climate modelling and integrated assessment modelling, particularly the lack of 'laws' describing social, political, technological and economic changes, and the need, or preference, for multiple scenarios of the future as compared with the goal of predicting the climate.

The history of integrated assessment modelling is much shorter than that of climate modelling (less than 10 years as compared with more than 30 years) but, despite the relative youth of the subject, there are a substantial number of integrated assessment models (Table 6.3). There is, as yet, little documentation of the relative strengths and weaknesses of the available integrated assessment models, but some general characteristics can be recognized: (i) integrated assessment models should offer added value compared with single disciplinary (e.g. climate) orientated assessment; (ii) integrated assessment models should provide useful information to decision makers.

All integrated assessment models attempt to represent and predict the relationships between human society and the environment, with a primary focus on climatic change, its causes and effects (Figure 6.10). They have been roughly grouped in terms of top-down versus bottom-up schemes. 'Top-down' models are aggregate models, often based on macroeconomic models, that analyse how changes in one sector of the economy affect other sectors and regions. Early top-down models tended to contain little detail on energy consumption, especially at the technology-specific level, but included explicit treatment of behaviour and economic interactions. In contrast, 'bottom-up' models tended to describe energy consumption in detail, whilst consumer behaviour and interactions with other sectors of the economy originally tended to be dealt with much less completely. Recent integrated assessment models have tended to provide greater detail in areas that were previously less developed, so that differences in model results are increasingly driven by differences in input assumptions rather than in model structure. Nevertheless, differences in integrated assessment model structure remain important because, as is the case with climate models, different model types are better suited to answer different types of questions.

The similarities between climate models and models of integrated assessment also extend to the underlying philosophy of the model developers. These different philosophies can be characterized, in their extreme forms, as either (i) a small and highly generalized group of equations, or (ii) a very large number of equations. In climate modelling, these paradigms are represented by (i) EBMs or simple box models (Chapter 3), and (ii) coupled ocean–atmosphere global climate models (Chapter 5). In integrated assessment, they are represented by (i) simple economic forms (termed response surfaces or reduced form models), and (ii) very complex systems of equations designed to capture all the processes under consideration. Examples of these two extreme forms of models are DICE and TARGETS, which in Table 6.3 can be seen to capture very different numbers of processes explicitly.

There is a second characteristic of integrated assessment models which is shared with climate models: the issue keenly felt in the integrated assessment community of focus on precision or accuracy. Examples of the two extreme forms are IMAGE-2, which projects location-specific changes in $1/2°$ by $1/2°$ grid cells across the Earth, as compared with ICAM-2, which is designed to

Table 6.3 Summary characterization of integrated assessment models (after Dowlatabadi and Rotmans, 1997)

Model	Forcings	Geographic specificity	Socio-economic dynamics	Geophysical simulation	Impact assessment	Treatment of uncertainty	Treatment of decision making
	0. CO2 1. Other GHG 2. Aerosols 3. Land use 4. Other	0. Global 1. Continental 2. Countries 3. Grids/basins	0. Exogenous 1. Economics 2. Technological choice 3. Land use 4. Demographic 5. Cultural perspectives	0. ΔF 1. Global ΔT 2. 1D ΔT, ΔP 3. 2D ΔT, ΔP	0. ΔT indexed 1. Sea-level rise 2. Agriculture 3. Ecosystems 4. Health	0. None 1. Basic 2. Advanced	0. Optimization 1. Simulation 2. Simulation with adaptive decisions
AIM	0,1,2,3,4	2,3	1,2,3,4	2,3	0,1,2,3	0	1
CETA	0	0,1	1,2	1	0	1	0
Connecticut	0	0	1	1	0	1	0
CSERGE	0	0	1	1	0	1	0
DICE	0	0	1	1	0	1	0
CFUND	0,1	1	1	0	0,1,2,3,4	2	0
Grubb	0	0	1	1	0	1	0
ICAM-2	0,1,2,3	1,2	1,3,4	2,3	0,1,3	2	1,2
IMAGE 2.0	0,1,2,3	3	0,2,3	3	1,2,3	1	1
MERGE 2.0	0,1	1	1,2	1	0	1	0
MiniCAM	0,1,2,3	2,3	1,2,3	3	0	1	0
MIT	0,1,2,3	2,3	1	3	0,2,3	1	0,1
PAGE (EEC)	0,1	1,2	1	1	0,1,2,3,4	2	1
PEF	0,1	1	0	1	0	2	1
ProCAM	0,1,2,3	2,3	1,2,3,4	3	0,2	1	1
TARGETS	1,2,3,4,5	0	1,2,3,4,5	2	1,2,3,4,5	2	1,2

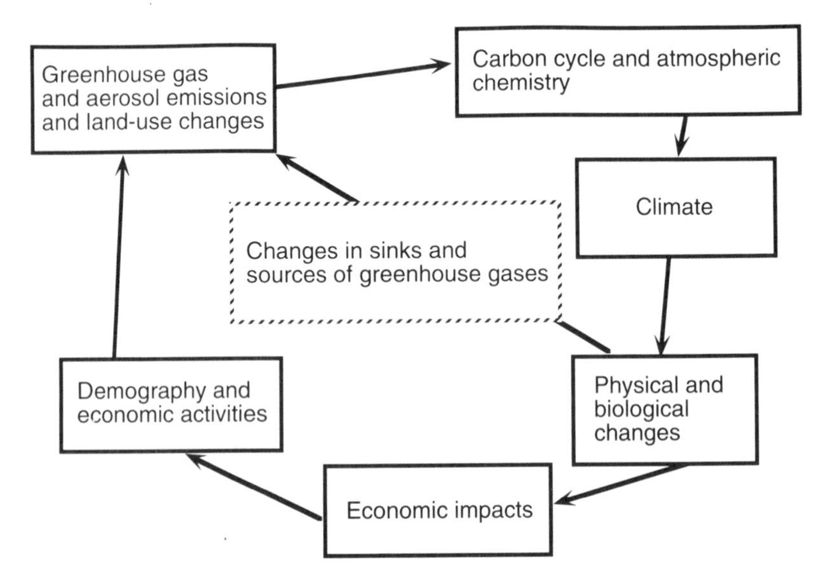

Figure 6.10 Elements of, and interactions in, the integrated assessment (IA) process

deliver probability estimates of likely futures but intentionally avoids location-specific projections, generating instead aggregated outcomes for the world. Both modelling paradigms have strengths and weaknesses, as is the case with climate modelling where the analogy might be with the 'accuracy' of the global responses predicted by a 1DRC (Chapter 4) as compared with the apparent precision of a high resolution OAGCM (Chapter 5). However, the contrast between model types appears, and probably is, very much more stark for integrated assessment models than for climate models because of the inherently 'unknowable' nature of the fully integrated human and environmental system.

Recently, simple integrated assessment models have been used to devise new stabilization profiles that explicitly (albeit qualitatively) incorporate economic considerations, estimate the corresponding anthropogenic emissions requirements and assess the significance of the profiles in terms of the global mean temperature and sea-level changes. These calculations are a response to Article 3 of the Framework Convention which states that 'policies and measures to deal with climate change should be cost-effective so as to ensure global benefits at the lowest possible cost' since they assume that if two greenhouse gas emission futures were indistinguishable in terms of their environmental implications, then the emission future with the lower mitigation (i.e. emissions reduction) costs would be preferred. Thus this integrated assessment model adds the additional constraint to the usual three: (i) prescribed initial (1990) greenhouse gas concentrations and rate of change of concentrations; (ii) a range of prescribed stabilization levels and attainment dates; (iii) the requirement that the implied emissions should not change too abruptly. To these is added (iv) that the resulting emissions trajectories initially track a business as usual path: an idealization of the assumption that the initial departure from the business as usual path would be slow, for economic and developmental reasons which include (a) the further into the future the economic burden lies, the smaller the present resource base required to undertake it, (b) time is therefore needed to re-optimize the capital stock, and (c) the availability of substitutes is likely to improve and their costs reduce over time.

The results of these global-mean calculations are pathway-related differentials up to about 0.2°C in global mean temperature and 4 cm in global mean sea-level change. In benefit–cost analyses of climate change policy options, the implications for market (e.g. agriculture, timber, fisheries) and non-market (e.g. bio-diversity, environmental quality, human health) impacts of these climatic differentials is unclear. The cost of a more economical transition away from fossil fuels depends on the regional details associated with the projected climate changes in these and other key climate variables. Great uncertainty surrounds these evaluations.

6.4 THE FUTURE OF CLIMATE MODELLING

This book has been written with the basic aim of building up a framework within which all types of climate models can be considered. The need to recruit scientists, social and economic researchers, and political and demographic planners and policymakers into climate modelling and assessment has never been greater. We have also drawn readers' attention to unexpected results of climate modelling and to the continuing need for a spectrum of models, from the simplest EBM to the most complex coupled ocean–atmosphere model, as a way of aiding understanding of the climate system.

Climate change imposes a variety of impacts on society. Future climate changes are likely to include effects on agriculture, forests, water resources, the costs of heating and cooling, the impact of a rising sea level on small island states and low-lying coastal areas, and this generation's choice of nuclear waste disposal sites.

Any climate change presents society as a whole with a set of formidable difficulties: large uncertainties, the potential for irreversible damages or costs, a very long planning horizon, long time lags between causes and effects, the potential for 'winners' as well as 'losers', an irreducibly global problem with very wide, but as yet unknowable, regional variations. Beyond these tangible impacts are a variety of intangible effects, including damages to existing ecosystems and the threat of species loss. Climate scientists agree that greenhouse gas emissions are rising and that both industrial- and biomass-derived aerosols will continue to be emitted. They agree on the mechanisms linking these changes to climate but do not yet agree on the speed of change, or the ultimate amount of change. In addition, social scientists do not agree on the size of the behavioural responses or economic effects that would follow, or on the effect of these changes on well-being. Climate change, by its nature, is a global challenge.

Despite the very considerable importance of, for example, human-induced warming and stratospheric ozone depletion and the relevance of climate model predictions to their understanding and ultimate solution, it needs to be recognized that a warmer planet and higher surface UV are not the only possible benefits or future hazards. Climate models also predict a nuclear winter and very similar climate catastrophes resulting from a meteorite impact on Earth. The Earth's climate has been shown to be susceptible to long-term changes in solar radiation and its distribution by models which also predict that human-induced land-use change can cause regional shifts in climate of similar magnitude to those likely to be caused by greenhouse warming and industrial aerosols.

These challenges continue to be the concern of climate modellers around the world. The range of application of climate models is great. Modellers cannot answer all the questions about the climate system, but the continuing search for a more complete understanding of the climate system is a most laudable and fascinating endeavour.

RECOMMENDED READING

Alcamo, J. (ed.) (1994) *Image 2.0: Integrated Modelling of Global Climate Change.* Kluwer Academic Publishers, Dordrecht, 328 pp.

Bruce, J.P., Lee, H. and J. Hailes, E.F. (1996) *Climate Change 1995. Economic and Social Dimensions of Climate Change. Contribution of working Group III to the Second Assessment Report of the Intergovernmental Panel on Climate Change,* Cambridge University Press, Cambridge, 608 pp.

Cess, R.D., Potter, G.L., Blanchet, J.P., Boer, G.J., Del Genio, A.D., Déqué, M., Dymnikov, V., Galin, V., Gates, W.L., Ghan, S., Kiehl, J.T., Lacis, A.A., Le Treut, H., Li, Z.-X., Liang, X.-Z., McAvaney, B.J., Meleshko, V.P., Mitchell, J.F.B., Morcrette, J.-J., Randall, D.A., Rikus, L., Roeckner, E., Royer, J.F., Schlese, U., Sheinin, D.A., Slingo, A., Sokolov, A.P., Taylor, K.E., Washington, W.M., Wetherald, R.T., Yagai, I. and Zhang, M.-H. (1990) Intercomparison and interpretation of climate feedback processes in 19 atmospheric general circulation models. *Journal of Geophysical Research* **95**, 16601–16615.

Dowladabati, H. and Morgan, M.G. (1993) A model framework for integrated assessment of the climate problem. *Energy Policy* **21**, 209–221.

Gates, W.L. (1992) AMIP: The atmospheric model intercomparison project. *Bulletin of the American Meteorological Society* **73**, 1962–1970.

Gates, W.L. (1995) *Proceedings of the First AMIP Scientific Conference (15–19 May 1995, Monterey, CA).* WMO Technical Document No 732, World Meteorological Organization, Geneva, 532 pp.

Henderson-Sellers, A. (1987) Effects of change in land-use on climate in the humid tropics. *In* R.E. Dickinson (ed.), *The Geophysiology of Amazonia.* John Wiley & Sons, New York, pp. 463–493.

Henderson-Sellers, A., Pitman, A.J., Love, P.K., Irannejad, P. and Chen, T.H. (1995) The project for intercomparison of land-surface parameterization schemes (PILPS): Phases 2 and 3. *Bulletin of the American Meteorological Society* **76**, 489–503.

Houghton, J.T., Mera Filho, L.G., Callander, B.A., Harris, N., Kattenberg, A. and Maskell, K. (eds) (1996) *Climate Change 1995. The Science of Climate Change. Contribution of Working Group I to the Second Assessment Report of the Intergovernmental Panel on Climate Change*. Cambridge University Press, Cambridge, 572 pp.

Howe, W. and Henderson-Sellers, A. (1996) The MECCA project – overview and evolution of the MECCA project. *In* W. Howe and A. Henderson-Sellers (eds) *Assessing Climate Change: The Story of the Model Evaluation Consortium for Climate Assessment*. Gordon and Breach, London, in press.

Luther, F.M., Ellingson, R.G., Fouqart, Y., Fels, S., Scott, N.A. and Wiscombe, W. (1988) Intercomparison of radiation codes in climate models (ICRCCM): Longwave clear-sky results – a workshop summary. *Bulletin of the American Meteorological Society* **69**, 40–48.

Nordhaus, W.D. (1991) To slow or not to slow: the economics of the greenhouse effect. *Economic Journal* **101**, 920–937.

Peck, S. and Teisberg, T.J. (1992) CETA: A model for carbon emissions trajectory assessment. *Energy Journal* **13**, No. 1.

Pollard, D. and Schulz, M. (1994) A model for the potential locations of Triassic evaporite basins driven by palaeoclimatic GCM simulations. *Global and Planetary Change* **9**, 233–249.

Root, T.L. and Schneider, S.H. (1995) Ecology and climate: research strategies and implications. *Science* **269**, 334–341.

Rotmans, J., Hulme, M. and Downing, T. (1995) Climate change implications for Europe. *Global Envronmental Change* **4,** 97–124.

Skiles, J.W. (1995) Modelling climate change in the absence of climate change data. *Climatic Change* **30**, 1–6.

Taplin, R. (1996) Climate science and politics: the road to Rio and beyond. *In* T. Giambelluca and A. Henderson-Sellers (eds), *Coupled Climate System Modelling: A Southern Hemisphere Perspective*. John Wiley & Sons, Chichester, pp. 377–395.

VEMAP (1995) Vegetation/Ecosystem Modelling and Analysis Project (VEMAP): Comparing Biogeography and Biogeochemistry Models in a Continental-Scale Study of Terrestrial Ecosystem Responses to Climate Change and CO_2 Doubling. *Global Biogeochemical Cycles* **9**, 407–437.

Watson, R.T., Zinyowera, M.C. and Moss, R.H. (1996) *Climate Change 1995. Impacts Adaption and Mitigation of Climate Change: Scientific-Technical Analysis. Contributions of Working Group II to the Second Assessment Report of the Intergovernmental Panel on Climate Change*, Cambridge University Press, Cambridge, 880 pp.

Wigley, T.M.L., Richels, R. and Edmonds, J.A. (1996) Stabilizing CO_2 concentrations: The choice of emissions pathway. *Nature*, **379**, 240–243.

Appendix A Glossary

Ablation. The collective description of the processes by which a cryosphere mass is diminished in size. The term is applied to the depletion of ice sheets, glaciers, sea-ice and to a snow-pack. Ablation occurs by melting, sublimation and by physical disruption (i.e. bits fall off, for example, icebergs breaking off from the Antarctic ice sheet).

Absorption bands. Molecules absorb radiation by being excited into vibration and rotation. In the case of water vapour, these absorption bands lie in the same region as the longwave radiation emitted by the Earth and are therefore of significance in climate studies. Carbon dioxide does not possess rotation bands but its main vibration band also lies in this region, as do the absorption bands of some trace gases such as chlorofluorocarbons and methane.

Absorptivity. The fraction of the radiation incident upon a body which the body absorbs is called its absorptivity; also known as absorptance, although this usually refers to a particular wavelength.

Advection. Horizontal transport, usually of energy, mass etc., in the atmosphere or ocean.

Aerosol. A suspension of small solid particles or liquid droplets in the atmosphere. Aerosols can be both natural (e.g. volcanic) and of human origin (e.g. industrial), and may have significant radiative effects on climate.

Albedo. From the Lain *albus*, meaning white. It is the reflected fraction of incident radiation. It is synonymous with hemispheric reflectance.

Aliasing. In the sampling of a continuous function to produce values at discrete points, the sampling frequency must be high enough to resolve the highest frequency present in the continuous function. If not, the high-frequency information will appear as a false enhancement of a related lower frequency. The high-frequency signal is then said to be aliased into a lower frequency.

AMIP. The Atmospheric Model Intercomparison Project is coordinated by the Program for Climate Model Diagnostics and Intercomparison (PCMDI) for the Working Group on Numerical Experimentation.

Attractor. The time evolution of a physical system can be characterized by system variables. If these variables converge to a single set of values, then this is the attractor of the system. Conservative dynamical systems never reach equilibrium and therefore do not exhibit attractors. The Poincaré section (or phase portrait) is commonly used to display attractors.

Attribution. The linking of cause and effect. For example, in terms of human-induced climate change it has to be determined, first, if change can be detected and then if the detected change can be unambiguously attributed to one or more causes.

Baroclinic. The atmosphere is baroclinic when isotherms are not parallel to isobars, i.e. there is a temperature gradient along the isobars.

Baroclinic waves. A disturbance in a smooth zonal flow which is equivalent barotropic will generate baroclinic waves. They are characteristic of the upper-level atmospheric flow in mid-latitudes.

Barotropic. The atmosphere is barotropic when there are horizontally uniform temperatures at all heights. Thus pressure gradients can exist but horizontal temperature gradients cannot. In an equivalent barotropoic atmosphere horizontal temperature gradients may exist but isotherms must be parallel to the isobars.

Bottom water. The cold dense water which lies at the bottom of the oceans in contact with the ocean floor. It is formed in high latitudes as a result of the formation of sea-ice and has distinctly different characteristics from the water immediately above it.

Boundary layer. The lowest 1 km or so of the atmosphere. It is the location of all interactions between the Earth's surface (land, ocean and ice) and the free atmosphere.

Bowen ratio. The ratio of sensible heat flux to latent heat flux from a surface.

Business as usual. A term coined by the IPCC (and others) to describe likely future human activities if steps are not taken to mitigate climate change. This prediction of social, technical and economic evolution is, of itself, open to considerable debate and dispute. In general terms, it always indicates increased emissions into the atmosphere.

Chaos, chaotic. The term has come to be applied to a system which is deterministic (i.e. its future state is determined by its present state), but which is governed by a range of processes which have non-linear behaviour. Small changes in the initial state of the system are amplified over time, so that the final states of two instances of a system with similar initial conditions are likely to be very different. The inability to specify accurately the initial conditions of a weather forecast model and the highly non-linear processes involved result in the forecast diverging from the real weather over a period of 10–15 days. Climate model simulations, because they are a forecast of an ensemble climate (e.g. mean January precipitation), rather than a single instance of weather (e.g. rainfall on 26 January), are not affected by chaos.

Clausius–Clapeyron (equation). The dependence of the saturation vapour content of the atmosphere upon its temperature.

Climate system. Variously defined. The atmosphere, hydrosphere, cryosphere, land surface and biosphere, i.e. all the components of the Earth's environment are, to a greater or lesser extent, components of the climate system.

Convection. When a parcel of air is warmer than its environment it will move upwards and carry energy with it. This is the process of convection.

Convergence. See **Divergence.**

Coriolis force. The apparent force experienced by an entity moving over a rotating body such as the Earth. On the surface of the Earth the direction of travel is deflected towards the right in the northern hemisphere and to the left in the southern hemisphere.

Cryosphere. A collective term describing the frozen water masses of the Earth. These are ice sheets, snow cover, glaciers, permafrost and sea-ice.

Deep water. A term used to describe water in the ocean which lies below the surface water layer and has different temperature and salinity characteristics. The formation and circulation of this dense, cold and saline, *deep water* are important components in the global ocean circulation. It may be possible to distinguish distinct layers in this water. Also: water that is 'taller than Eccles'. See also *bottom water*.

Degree day. Summer warmth can be expressed as the sum of degree days: the product excess of daily temperatures above a preset threshold, which is often the threshold for plant growth (about 5° C), and the number of days exceeding the threshold by that amount. Winter cold can also be summarized this way except that the temperature depression below a threshold is used.

Degrees of freedom. Each observation in a random sample of size n can be compared with $n-1$ other observations and hence there are $n-1$ degrees of freedom.

Dependent variables. If a quantity y is a function of a quantity x then y is the dependent variable. It is dependent upon x. It follows that x is the independent variable.

Detection. The determination of a climate change. Although this sounds straightforward it has proved to be an exceedingly fraught issue for IPCC. Detection of human-induced climate change is not a simple yes/no determination. It depends on a statistical evaluation of both model results and observations. Detection must occur before *attribution* (q.v.).

Diffuse radiation. When radiation passes through the atmosphere, it is scattered in interactions with molecules and particles in the air. This scattering alters the direction of travel of the radiation progressively and randomly. After a great deal of scattering the radiation becomes totally diffuse (i.e. the same quantity of radiation is travelling in all directions). Generally the radiation received at the Earth's surface is composed of a diffuse component and a direct (unscattered) component. The factor 1.66 is often introduced into calculations to allow for the larger distance (on average) travelled by diffuse radiation through an atmospheric layer. Thermal radiation is emitted from all directions within the atmosphere and therefore is totally diffuse to a good approximation.

Diffusion coefficient. This determines the rate of diffusion of a quantity along a gradient of that quantity.
Divergence. The 'spreading out' of a flow or the flux of a quantity away from a point in more than one direction. Convergence is its opposite.
Downscaling. The process by which coarse resolution GCM results are brought to a higher resolution. It generally implies a space-scale change but can also be used for temporal information. Different downscaling methods produce different results.
Eddy. A characteristic feature of turbulent fluid flow consisting of a coherent swirling motion.
Eddy fluxes. The flux of a quantity by means of turbulent disturbances propagating in a fluid. Such a flux is made possible by the process called eddy diffusion.
Eddy resolving model. A model (usually used with reference to the ocean) which explicitly includes the motions associated with eddies rather than simply parameterizing their effects.
Effective temperature. The temperature of the Earth, derived by simple energy balance considerations assuming the Earth to be a black body emitter.
Emissivity. The ratio of the emittance from a body to that of a black body (perfect) emitter at the same temperature is the emissivity of that body.
Ensemble average. An ensemble average can be thought of as an average of a group of instances of the same phenomenon. In climate modelling, an average of data for January from, for example, 20 years of a model simulation produces an ensemble average of January conditions for the model.
Enstrophy. The root-mean-square of the vorticity of a body (usually a fluid).
Entrainment. The process by which fluid external to a parcel is incorporated into the parcel.
Equilibration time. The time taken for a component of the climate system to reach equilibrium with one or more of the other components.
FANGIO. Feedback Analysis of GCMs and Intercomparison with Observations. The acronym honours Juan Manuel Fangio, who was world driving champion in 1951, 54, 55, 56 and 57. He had 24 Formula 1 wins in 51 starts.
FCCC. The UN Framework Convention on Climate Change. Signed at the UN Conference on Environment and Development in 1992 and ratified in 1994. The FCCC has defined climate change to be only the human-induced effects (i.e. not natural variability) for its negotiations.
Feedback. The phenomenon whereby the output of a system is fed into the input and the output is subsequently affected.
Fingerprint. A set of tests which together allow for the detection (q.v.) of climate change. The analogy is to a human fingerprint which has many characteristics which, together, permit unambiguous identification.
Flux. The flow of a quantity through a surface. The flux of, for example, energy is always *from* somewhere *to* somewhere else, i.e. its vector nature is important.
Forcing. A change in an internal or external factor which affects the climate.

Fourier transform. A continuous function can be represented mathematically either by its value at many grid points or as the sum of many waves of differing frequency, amplitude and phase. The two representations are formally equivalent, i.e. they contain the same information. (A good analogy is a number and its logarithm to any base.) The Fourier transform of a variable is the frequency domain equivalent of its space domain representation. A fast Fourier transform (FFT) is a computer algorithm for moving between the frequency and spatial domains.

Gain. The gain of a system is a measure of the amplification of the input to the system. In its simplest form it equals output divided by input.

Gaussian quadrature. A particular type of numerical computation of a one-dimensional integral from the knowledge of individual values of it.

General circulation of the atmosphere. Two factors control the general circulation of the atmosphere: the energy imbalance between the equator (net absorption of energy) and the poles (net emission of energy) and the rotation of the Earth. In low latitudes the direct Hadley cell circulations transfer energy polewards, but in the middle latitudes the rotation of the system causes a wave-like flow in the troposphere (Figure A.1). These Rossby waves travel in a westerly direction and energy transfer is via horizontal eddies which form the familiar depression systems and anticyclones of mid-latitude weather. In polar regions there are weak direct cellular flows but the seasonal variation in insolation from polar day to night dominates the pattern.

Gravity waves. Analogous to water waves. They occur in the atmosphere due to forcing by orography and manifest themselves as undulations in potential temperature surfaces.

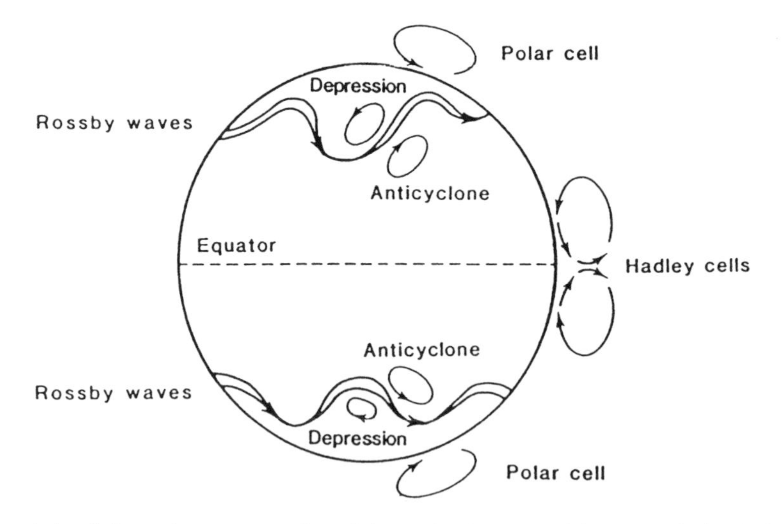

Figure A.1 Schematic representation of the major features of the general circulation of the atmosphere

Greenhouse effect. The effect of the atmosphere in re-radiating longwave radiation back to the surface of the Earth. It has nothing to do with glass-houses, which trap warm air at the surface.
Hadley cell. The thermally driven circulation comprising upward motion at the ITCZ and downward motion in the subtropics. The air moves poleward at high altitude and equatorward near the surface.
Halocline. A region of the ocean column where the salinity gradient is high. See also *pycnocline* and *thermocline*.
Heat capacity. The energy required to change the temperature of a body by 1 K.
Hydrostatic equation. The relationship between pressure (p) and height (z) in the atmosphere. $\Delta p = -\rho g \Delta z$.
ICRCCM. The International Comparison of Radiation Codes for Climate Models.
Ideal gas law. The statement that the ratio of the product of pressure and volume of a gas to its temperature is a constant.
Independent variable. See **Dependent variable**.
Infinite. Quite a lot – really an awful lot.
Integrated assessment. The incorporation of all aspects of climate change into the decision making process, for example, social, political, economic and technological as well as ecological, environmental and climatic.
Intertropical convergence zone (**ITCZ**). The zone of convergence between the two Hadley cells. Characterized by large amounts of vertical ascent.
IPCC. The Intergovernmental Panel on Climate Change. Established in 1988 and jointly sponsored by UNEP and WMO. Note that the IPCC is an assessment not a research organization.
Isothermal. Having the same temperature everywhere. Often used to describe the situation where no vertical temperature gradient exists in the atmosphere.
Isotropic. Having the same properties in all directions.
Lapse rate. The rate of decrease of temperature with height in the atmosphere.
Latent heat. The energy used in conversion between different phases of water. It is the energy used in evaporation of water (vaporization) or in the melting of ice (fusion).
Leaf area index (LAI). The area of foliage per unit area of ground. Usually only the upper side of the leaves is considered but an alternative definition includes both upper and lower leaf surfaces.
Little Ice Age. A period between the fifteenth and eighteenth centuries when temperatures were lower than today over many areas of the northern hemisphere. The timing differs from region to region and recently there has been some doubt cast upon whether a 'climatic epoch' really occurred.
Meridional circulation. The circulation of the atmosphere is dominated by meridional motions caused by the equator-to-pole temperature gradient.
Milankovitch mechanism. The orbital parameters of the Earth are constantly changing due to the influence of other planets. These changes in the orbital geometry of the Earth result in changes in the pattern of insolation at the Earth.

This may provide a forcing agent for climate variations, the Milankovitch mechanism.

Mixed layer. Because of wind and wave action and convection at the surface, the top part of the ocean is well mixed (very small gradients of temperature and salinity). This layer, the thickness of which varies geographically and seasonally, is called the mixed layer. Climate models have often assumed that the layer is 50–100 m thick globally.

Mixing ratio (of water vapour). The quantity of water vapour (in kg) per kilogram of air.

Model noise. In climate models, inaccuracies in calculations, rounding and truncation errors introduce variability into model simulations which are classed as model noise.

Momentum. The product of mass and velocity. In a closed system, momentum is conserved.

Montréal Protocol. This is an international agreement designed to protect stratospheric ozone. The treaty was originally signed in 1988 and substantially amended in 1990 and 1992. It stipulates that the production and consumption of compounds that deplete ozone in the stratosphere are to be phased out by 2000 (2005 for methyl chloroform).

Non-linear. An equation with dependent variable y containing terms in y^2 or higher is non-linear. Such equations exhibit a response to perturbations which is not related in a constant fashion to the forcing.

No regrets. If a community takes some course of action in response to a threat which will result in a benefit to the community even if the threat does not materialize, then that response can be considered a 'no regrets' measure.

Optical thickness. Radiation passing through a body is attenuated by a factor $e^{-\delta}$, where δ is the optical thickness of the body. It is a strong function of wavelength. Alternatively, the length of time required by any student of climate modelling to 'see the light'.

Oxygen isotope data. Because of their different masses, water molecules containing different oxygen isotopes are evaporated and precipitated at different rates. The ratio of ^{18}O to ^{16}O can be used to indicate the nature of palaeo-climates since it is related to ocean temperatures.

Ozone hole. A description of the appearance, over the poles (most pronounced over Antarctica), in ~1985–86 of an area of distinctly reduced ozone levels. The cause of these reduced levels is linked to increased levels of free chlorine and rapid reactions on solid particulates in the stratosphere.

Parameterization. The method of incorporating a process by representation as a simplified function of some other fully resolved variables without explicitly considering the details of the process.

PCMDI. The Program for Climate Model Diagnosis and Intercomparison. This was established in 1989 at the Lawrence Livermore National Laboratory in California. Its principal aim is to develop improved methods and tools for diagnosis, validation and intercomparison of global climate models.

Physical, physically based (**models**). Models which are constructed on the basis of physical relationships (and laws) and known processes rather than being based on correlations, where there is no clearly defined causal relationship. In climate models, this term has come to embrace aspects of chemistry and biology as well as physics.

PILPS. The Project for Intercomparison of Land-surface Parameterization Schemes.

Pinatubo. In 1992, Mount Pinatubo in The Philippines erupted and injected large amounts of dust and sulphate into the upper atmosphere. The eruption was important because it occurred at a time when satellite observing systems were well configured to provide the climate community with information on the dynamics and chemistry of the processes following a major eruption.

Planck function. The description of the amount of radiation emitted by a body at a given temperature as a function of the wavelength of the radiation.

Poincaré section. This term is used in chaotic (q.v.) dynamics. When a plot is made of two or more of the characterizing variables of a chaotic system, this is termed a Poincaré section. The two- or three-dimensional plot often has a distinct pattern, which cannot be seen if only one variable is examined. The term phase portrait is often used.

Potential evaporation. Potential evaporation can be defined as the evaporation which would occur given a free and plentiful supply of water. Clearly such a situation would occur only over water bodies and wet vegetation.

Potential temperature. In the atmosphere the potential temperature of a parcel of air is the temperature it would have if moved under adiabatic conditions (without gaining or losing energy) to a reference level (usually 1000 hPa). Potential temperature is more useful than actual temperature because the compressibility of gases means that temperature increases as pressure increases and vice versa.

Prognistic variable. A variable in a model which is utilized in prediction of itself at an advanced time point.

Pycnocline. A boundary in the ocean which is characterized by a large change in density. The density difference between two bodies of water may be due to differences in temperature or salinity or both.

Radius of deformation. A description of the characteristic size of eddies in the atmosphere or ocean. It is a function of the density of the fluid and the Coriolis parameter.

Richardson number. A measure of the stability of a fluid layer: a ratio of buoyancy to inertial forces.

Scattering. In the atmosphere the redirection of light waves due to reflection and diffraction by atmospheric molecules (Rayleigh) and cloud droplets (Mie).

Sensitivity. The sensitivity of a model to a perturbation. Usually described as a unit of response per unit change.

Solar constant. The amount of radiation from the Sun incident on a surface at the top of the atmosphere perpendicular to the direction of the Sun. Currently taken to be 1370 W m^{-2} and known to be variable! Note that S can denote both 1370 W m^{-2}, one quarter of this and the instantaneous top-of-the-atmosphere solar flux at a particular location. Context usually indicates which is meant, but beware confusing algebra.

Solar elevation. The angular distance above the horizon of the Sun at any time.

Solar zenith angle. The angular distance between the solar position and the true zenith (the point directly overhead). It is the complement of the elevation angle.

Spherical harmonic. For a variable defined on the surface of a sphere the natural method of representation in the frequency domain is by means of functions called spherical harmonics. These can be thought of as the extension to spherical coordinates of the concept of Fourier transformation (q.v.).

Stability. A measure of the capacity of a system to resist perturbation. The ability to recover the original position after displacement.

Stability (of the atmosphere). Consider the situation in Figure A.2(a) where the dashed line shows the critical lapse rate, γ_c. The dotted lines show both stable and unstable temperature profiles. If a parcel of air is displaced from the surface it will rise such that its temperature decreases at a rate given by γ_c. If the environmental lapse rate is stable, then the displaced parcel will be colder (and thus denser) than the surrounding air (Point A) and will sink back. In the unstable case the parcel will be warmer (less dense) than its surroundings (Point B) and will continue to rise.

Statistical significance. A means of trying to determine the reality of either an observed change or a model result. In the most general terms, statistical significance is based on a ratio of the 'change' compared with the normal

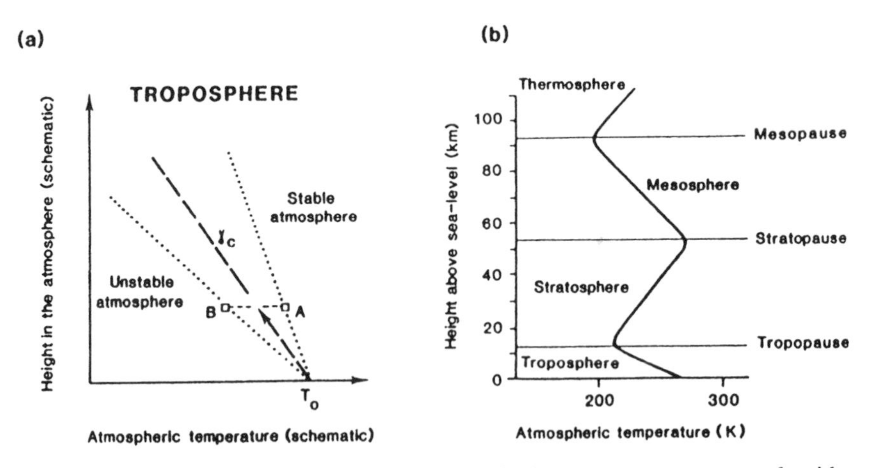

Figure A.2 (a) Stable and unstable tropospheric lapse rates as compared with a predetermined critical lapse rate (γ_c). (b) The four 'regions' of the atmosphere

variability, or 'noise', in the modelled or real climate. Statistical tests are usually considered as a prelude to an interpretation of results based on physical processes.

Stefan–Boltzmann constant. σ, having a value of 5.67×10^{-8} W m^{-2} K^{-4}, the constant of proportionality in Stefan's law.

Stefan's law. This is the relationship between the amount of energy radiated by a body and its absolute temperature and is given by $E = \sigma T^4$ where E is in W m^{-2} and σ is the Stefan–Boltzmann constant.

Stratosphere. The zone of the atmosphere above the troposphere (Figure A.2(b)). Atmospheric temperature increases with height due to absorption of radiation by ozone.

Stream function. A description of the amount of volume transport taking place at a particular point (in space) in an ocean or atmospheric model per unit time. It has units of m^3 s^{-1}. In the ocean the term Sverdrup is often used where 1 Sverdrup = 10^6 m^3 s^{-1}.

Student's *t* test. One type of statistical test used by climate modellers. Although not strictly valid for testing statistical significance, it provides a general identification of the location of important changes which have occurred and thus the basis for cause and effect evaluations.

Sulphate aerosol. When sulphur dioxide is further oxidized, the result is solid sulphate particles. These act to scatter solar radiation back to space and also act as cloud condensation nuclei.

SVAT. A Soil–Vegetation–Atmosphere Transfer scheme. One type of land surface parameterization scheme, which describes the transfer of energy, moisture and momentum from the land surface to the lower layer of the atmosphere. These models are focused primarily on short-period physics, rather than on chemistry or ecology.

Thermocline. A region of the ocean column where the vertical temperature gradient is high. It indicates a layer of warmer water (which is less dense) on top of colder water.

Thermodynamics. The science of the movement of heat. Usually concerned with calculations involving fluids.

Thermohaline. The term is usually applied to the deep ocean circulation which is driven by the heat and salt budgets of the ocean. An example of this is the creation of deep water in the North Atlantic where cold saline water, created when sea-ice forms, sinks to the bottom. This water flows around the bottom of the ocean and surfaces again in regions of oceanic upwelling.

Timestep. The base unit of temporal resolution in a numerical model.

Transitivity. The phenomenon whereby a system (in this case the climate) evolves from an initial state to another, different, state and stays there.

Transmissivity. The portion of the radiation, incident upon a body, which passes through it.

Troposphere. Lowest region of the atmosphere (Figure A.2(b)). The atmosphere is well mixed and temperature decreases with height. Clouds and weather systems are confined below its upper boundary, the tropopause, at a global average height of about 10 km.

Truncation. When a function, which is currently represented as a summation of many terms, is reduced in length and complexity by removing a number of small terms.

UNCED. The United Nations Conference on Environment and Development.

UNEP. The United Nations Environment Programme.

Variance. A measure of the spread of a set of results (about a mean). Used in climate modelling as a measure of the variability of the model climate.

Vorticity. The vorticity of a body is twice its angular velocity about the vertical. The Coriolis parameter $f = 2\Omega \sin \phi$ is the vorticity of the Earth. The absolute vorticity is the vorticity of the body plus the vorticity of the Earth at latitude ϕ. The potential vorticity of an air column is the absolute vorticity divided by the height of the column. Potential vorticity is always conserved.

WCRP. The World Climate Research Programme.

Weak Sun–warm early Earth paradox. It is known that the solar constant was 20–30% lower in the Earth's early history. Despite this, the surface temperature has never been as low as would be expected from simple calculations. It is possible that the larger amounts of CO_2 in the early atmosphere and different cloud and surface characteristics may have compensated for the lower solar luminosity.

WGNE. The World Climate Research Programme's Working Group on Numerical Experimentation.

Working Groups I, II and III. The three parts of the IPCC assessments in 1990 and 1995. The role of these working groups changed somewhat between the first and second assessments but the general responsibilities were: I, science; II, impacts and responses; III, social, economic and technological issues.

Appendix B About the CD

The CD which accompanies this book was prepared in ISO9660 format and should be readable on most CDROM drives. Not all of the resources are usable on a single computer platform.

Hardware requirements
Computer with ISO9660 capable CDROM drive. For best results utilize a 486 machine, Macintosh with 68030 and System 7.1 or better, or Unix™ workstation. Some hypertext links are external and require an Internet connection.

Software requirements
The on-line guide requires an HTML viewer such as Netscape™ or Mosaic™. Additional commonly available items are discussed in the documentation. These are usually available on local computer systems or can be downloaded free (some involve a shareware fee). The following programs are also required:

- GrADS (for displaying and manipulating gridded data)
- a FORTRAN compiler
- Microsoft Excel
- a BASIC interpreter
- an MPEG player (for the animations)

A WWW Browser (like Netscape™ or Mosaic™) is essential for the basic operation of the CD.

How to navigate the CDROM
Insert the CDROM into the drive. Start your Web Browser (e.g. Mosaic or Netscape) and choose 'open file' from the FILE menu. Select the file guide.htm in the top level directory of the CDROM. This file contains links to the rest of the disk.

If you have difficulties with the CDROM, consult the documentation for your computer and CDROM or ask your local computer administrator for assistance. We are sorry, but because of the range of possible configurations of hardware, software and network connections, we cannot help you with getting the CD running.

The CD contains the following items:

Models

(a) A suite of BASIC models, which were written for the first edition of the book.
- EBM
- 1D BOX DIFFUSION MODEL
- LAND USE CHANGE MODEL
- DAISYWORLD MODEL

(b) The full source for Version 1.02 of the GENESIS GCM.
(c) Source and executable for a 1D radiative convective model.
(d) Some information on where to get other GCMs and associated information.
(e) A spreadsheet EBM and instructions on how to build your own.

Data
Results from a GCM deforestation experiment suitable for analysis with GrADS on a Unix™ platform.

Images and movies
MPEG animations and images showing the results of a range of climate models.

If you have any comments about the CD or if you come across any errors, please let us know by e-mail to 100233.1554@compuserve.com. If you would like to be included on a list for updates or corrections to the CD, please let us know.

General Bibliography

Abbot, E.A. (1963) *Flatland: A Romance of Many Dimensions*, 5th edn. Barnes and Noble, New York, 108 pp.
Abramopoulos, F., Rosenzweig, C. and Choudhury, B. (1988) Improved ground hydrology calculations for global climate models (GCMs): Soil water movement and evapotranspiration. *Journal of Climate* **1**, 921–941.
Adem, J. (1965) Experiments aiming at monthly and seasonal numerical weather prediction. *Monthly Weather Review* **93**, 495–503.
Adem, J. (1979) Low resolution thermodynamic grid models. *Dynamics of Atmosphere and Oceans* **3**, 433–451.
Alcamo, J. (ed.) (1994) *Image 2.0: Integrated Modelling of Global Climate Change.* Kluwer Academic Publishers, Dordrecht, 328 pp.
Alexander, R.C. and Mobley, R.L. (1976) Monthly average sea-surface temperatures and ice-pack limits on a 1 degrees global grid. *Monthly Weather Review* **104**, 143–148.
Allaby, M. and Lovelock, J.E. (1984) *The Greening of Mars.* André Deutsch, London, 165 pp.
Allan, M.R. (1990) Modelling climate change. *Energy Policy* **18**, 681–682.
Alexander, R.C. and Mobley, R. L. (1974) Monthly average sea surface temperature and ice pack limbs over a $1° \times 1°$ global grid. *Research Report 4-1310-APPA*, Rand Corporation, Santa Monica, CA, 22 pp.
Anthes, R. (1977) A cumulus parameterization scheme utilizing a 1-D cloud model. *Monthly Weather Review* **105**, 270–286.
Anthes, G.H. (1994) Lifting the cloud from weather predictions. *Computerworld* **28**, 65.
Arakawa, A. and Schubert, W.H. (1974) Interaction of a cumulus cloud ensemble with the large-scale environment. Part I. *Journal of the Atmospheric Sciences* **31**, 674–701.
Arakawa, A. and Suarez , M.J. (1983) Vertical differencing of the primitive equations in sigma coordinates. *Monthly Weather Review* **111**, 34–45.
Baran, A.J., Foot, J.S. and Dibben, P.C. (1993) Satellite detection of volcanic sulphuric acid aerosol. *Geophysical Research Letters* **20**, 1799–1801.
Barron, E.J. and Washington, W.M. (1984) The role of geographic variables in explaining paleoclimates: Results from Cretaceous climate model sensitivity studies. *Journal of Geophysical Research* **69**, 1267–1279.
Barry, R.G., Henderson-Sellers, A. and Shine, K.P. (1984) Climate sensitivity and the marginal cryosphere. In J.E. Hansen and T. Takahashi (eds), *Climate Processes and Climate Sensitivity.* Maurice Ewing Series 5, American Geophysical Union, Washington, DC, pp. 221–237.
Bath, L.M., Dias, M.A., Williamson, D.L., Williamson, G.S. and Wolski, R.J. (1987)

Users' Guide to NCAR CCM1. NCAR Technical Note NCAR/TN-286+IA, National Center for Atmospheric Research, Boulder, CO, 173 pp.

Bath, L.M., Rosinski, J. and Olson, J. (1992) *User's Guide to NCAR CCM2*, NCAR Technical Note NCAR/TN-379+IA, National Center for Atmospheric Research, Boulder, CO, 156 pp.

Bengtsson, L. (1994) Climate of the 21st century. *Agriculture and Forest Meteorology* **72**, 3–29.

Berger, A. (1979) Insolation signature of quaternary climate changes. *Il Nuovo Cimento* **2C(1)**, 63–87.

Berger, A., Tricot, C., Gallee, H. and Loutre, M.F. (1993) Water vapour, CO_2 and insolation over the last glacial interglacial cycles. *Philosophical Transactions of the Royal Society of London, Series B* **341**, 253–261.

Berner, R.A., Lasaga, A.C. and Garrels, R.M. (1983) The carbonate–silicate geochemical cycle and its effect on atmospheric carbon dioxide over the last 100 million years. *American Journal of Science* **283**, 641–683.

Betts, A.K. and Miller, M.J. (1993) The Betts–Miller Scheme. *In* K.A. Emanuel and D.J. Raymond (eds), *The Representation of Cumulus Convection in Numerical Models of the Atmosphere*. American Meteorological Society, Boston, MA, pp. 107–121.

Boer, G.J., McFarlane, N.A., Laprise, R., Henderson, J.D. and Blanchet, J.-P. (1984) The Canadian Climate Centre spectral atmospheric general circulation model. *Atmosphere–Ocean* **22**, 397–429.

Boer, G.J., McFarlane, N.A. and Lazare, M. (1992) Greenhouse gas-induced climate change simulated with the CCC second-generation general circulation model. *Journal of Climate* **5**, 1045–1077.

Bolin, B., Döös, B., Jäger, J. and Warrick, R.A. (1986) *SCOPE 29. The Greenhouse Effect, Climatic Change, and Ecosystems*. John Wiley and Sons, Chichester, 541 pp.

Bonan, G.B. (1994) Comparison of land surface climatology of the NCAR Community Climate Model 2 at R15 and T42 resolutions. *Journal of Geophysical Research* **99**, 10357–10364.

Bonan, G.B., Pollard, D. and Thompson, S.L. (1993) Influence of subgrid-scale heterogeneity in leaf area index, stomatal resistance and soil moisture on grid-scale land–atmosphere interactions, *Journal of Climate* **6**, 1882–1897.

Borisenkov, Ye. P., Tsvetkov, A.V. and Agaponov, S.V. (1983) On some characteristics of insolation changes in the past and the future. *Climatic Change* **5**, 237–244.

Bourke, W., McAvaney, B., Puri, K. and Thurling, R. (1977) Global modelling of atmospheric flow by spectral methods. *In* J. Chang (ed.), *Methods in Computational Physics*, Vol. 17. Academic Press, New York, pp. 267–324.

Bower, K.N., Choularton, T.W., Latham, J., Nelson, J., Baker, M.B. and Jensen, J. (1994) A parameterization of warm clouds for use in atmospheric general circulation models. *Journal of the Atmospheric Sciences* **51**, 2722–2732.

Bowman, K. P. (1982) Sensitivity of an annual mean diffusive energy balance model with an ice sheet. *Journal of Geophysical Research* **87**, 9667–9674.

Brady, P.V. and Carroll, S.A. (1994) Direct effects of CO_2 and temperature on silicate weathering – possible implications for climate control. *Geochimica et Cosmochimica Acta* **58**, 1853–1856.

Bretherton, F.P. (1982) Ocean climate modeling. *Progress in Oceanography* **II**, 93–129.

Briegleb, B.P. (1992) Delta-Eddington approximation for solar radiation in the NCAR community climate model. *Journal of Geophysical Research* **97**, 7603–7612.

Briegleb, B.P. and Ramanathan, V. (1982) Spectral and diurnal variation in clear-sky planetary albedo. *Journal of Applied Meteorology* **21**, 1160–1171.

Broccoli, A.J. and Manabe, S. (1992) The effects of orography on midlatitude northern hemisphere dry climates. *Journal of Climate* **5**, 1181–1201.

Broecker, W.S. and Takahashi, T. (1984) Is there a tie between atmospheric CO_2 content and ocean circulation? *In* J.E. Hansen and T. Takahashi (eds), *Climate Processes and Climate Sensitivity*. Maurice Ewing Series 5. American Geophysical Union, Washington, DC, pp. 314–326.

Broecker, W.S. and Van Donk, J. (1970) Insolation changes, ice volumes and the O^{18} record in deep sea cores. *Reviews of Geophysics* **8**, 169–196.

Broecker, W.S., Takahashi, T., Simpson, H.S. and Peng, T.-H. (1979) Fate of fossil fuel carbon dioxide and the global carbon budget. *Science* **206**, 409–422.

Bruce, J.P., Lee, H. and Hailes, E.F. (1996) *Climate Change 1995. Economic and Social Dimensions of Climate Change. Contribution of Working Group III to the Second Assessment Report of the Intergovernmental Panel on Climate Change*, Cambridge University Press, Cambridge, 608 pp.

Bryan, K. (1969) A numerical method for the study of the world ocean. *Journal of Computational Physics* **4**, 347–376.

Bryan, K., Komro, F.G., Manabe, S. and Spelman, M.J. (1982) Transient climate response to increasing atmospheric carbon dioxide. *Science* **215**, 56–58.

Bryson, R.A. and Goodman, B.M. (1980) Volcanic activity and climatic change. *Science* **207**, 1041–1044.

Budyko, M.I. (1969) The effect of solar radiation variations on the climate of the Earth. *Tellus* **21**, 611–619.

Carbon Dioxide Assessment Committee (1983) *Changing Climate*. National Academy Press, Washington, DC, 490 pp.

Cess, R.D. (1976) Climatic change: An appraisal of atmospheric feedback mechanisms employing zonal climatology. *Journal of the Atmospheric Sciences* **33**, 1831–1843.

Cess, R.D. (1983) Arctic aerosols: Model estimates of interactive influences upon the surface–atmosphere clear-sky budget. *Atmospheric Environment* **17**, 2555–2564.

Cess, R.D. (1985) Nuclear war: Illustrative effects of atmospheric smoke and dust upon solar radiation. *Climatic Change* **7**, 237–251.

Cess, R.D. and Goldenberg, S.D. (1981) The effect of ocean heat capacity upon global warming due to increasing atmospheric carbon dioxide. *Journal of Geophysical Research* **86**, 498–509.

Cess, R.D., Briegleb, B.P. and Lian, M.S. (1982) Low-latitude cloudiness and climate feedback: Comparative estimates from satellite data. *Journal of the Atmospheric Sciences* **39**, 53–59.

Cess, R.D., Potter, G.L., Blanchet, J.P., Boer, G.J., Del Genio, A.D., Déqué, M., Dymnikov, V., Galin, V., Gates, W.L., Ghan, S., Kiehl, J.T., Lacis, A.A., Le Treut, H., Li, Z.-X., Liang, X.-Z., McAvaney, B.J., Meleshko, V.P., Mitchell, J.F.B., Morcrette, J.-J., Randall, D.A., Rikus, L., Roeckner, E., Royer, J.F., Schlese, U., Sheinin, D.A., Slingo, A., Sokolov, A.P., Taylor, K.E., Washington, W.M., Wetherald, R.T., Yagai, I. and Zhang, M.-H. (1990) Intercomparison and interpretation of climate feedback processes in 19 atmospheric general circulation models. *Journal of Geophysical Research* **95**, 16601–16615.

Cess, R.D., Zhang, M.H., Potter, G.L., Barker, H.W., Colman, R.A., Dazlich, D.A., Del Genio, A.D., Esch, M., Fraser, J.R., Galin, V., Gates, W.L., Hack, J.J., Ingram, W.J., Kiehl, J.T., Lacis, A.A., LeTreut, H., Li, Z.X., Mahfouf, J.F., McAvaney, B.J., Meleshko, V.P., Morcrette, J.J., Randall, D.A., Roeckner, E. and Royer, J.F. (1993) Uncertainties in carbon dioxide radiative forcing in atmospheric general circulation models. *Science* **262**, 1252–1255.

Cess, R.D., Zhang, M.H., Minnis, P., Corsetti, L., Dutton, E.G., Forgan, B.W., Garber, D.P., Gates, W.L., Hack, J.J., Harrison, E.F., Jing, X., Kiehl, J.T., Long, C.N., Morcrette, J.-J., Potter, G.L., Ramanathan, V., Subasilar, B., Whitlock, C.H., Young, D.F. and Zhou, Y. (1995) Absorption of solar radiation by clouds: Observations versus models. *Science* **267**, 496–499.

Chalita, S. and Le Treut, H. (1994) The albedo of temperate and boreal forest and the northern hemisphere climate: A sensitivity experiment using the LMD GCM. *Climate Dynamics* **10**, 231–240.
Chan, D., Higuchi, K. and Lin, C.A. (1995) The sensitivity of the simulated normal and enhanced CO_2 climates to different heat transport parameterisations in a two dimensional multilevel energy balance model. *Journal of Climate* **8**, 844–852.
Chang, J. (ed.) (1977) General circulation models of the atmosphere. *Methods in Computational Physics*, Vol. 17, Academic Press, New York, 337 pp.
Charlson, R.J. and Heintzenberg, J. (1995) *Aerosol Forcing of Climate*. John Wiley & Sons, Chichester, 399 pp.
Charlson, R. J., Lovelock, J. E., Andreae, M. O. and Warren, S. G. (1987) Ocean phytoplankton, atmospheric sulphur, cloud albedo and climate. *Nature* **326**, 655–661.
Charney, J. G. (1975) Dynamics of deserts and drought in the Sahel. *Quarterly Journal of the Royal Meterological Society* **101**, 193–202.
Charnock, H. (1955) Wind stress on a water surface. *Quarterly Journal of the Royal Meterological Society* **81**, 639–640.
Chen, T.H., Henderson-Sellers, A., Milly, P.C.D., Pitman, A.J., Beljaars, A.C.M, Abramopolous, F., Boone, A., Chang, S., Chen, F., Dai, Y., Desborough, C.E., Dickinson, R.E., Dumenil, L., Ek, M., Garratt, J.R., Gedney, N., Gusev, Y.M., Kim, J., Koster, R., Kowalczyk, E.A., Laval, K., Lean, J., Lettenmaier, D., Liang, X., Mahfouf, J.-F., Mengelkamp, H.-T., Nasonova, O.N., Noilhan, J., Polcher, J., Robock, A., Rosenzweig, C., Schaake, J., Schlosser, C.A., Schulz, J-P., Shao, Y., Shmakin, A.B., Verseghy, D., Wetzel, P., Wood, E.F., Xue, Y., Yang, Z-L. and Zeng, Q. (1996) Cabauw experiment results from the project for intercomparison of land-surface parameterization schemes (PILPS). *Journal of Climate* **9**, in press.
Chervin, R. M. (1980) On the simulation of climate and climate change with general circulation models. *Journal of the Atmospheric Sciences* **37**, 1903–1913.
Chervin, R. M. (1981) On the comparison of observed and GCM simulated climate ensembles. *Journal of the Atmospheric Sciences* **38**, 885–901.
Chervin, R.M. (1986) Interannual variability and seasonal climate predictability. *Journal of the Atmospheric Sciences* **43**, 233–251.
Chervin, R.M. and Schneider, S.H. (1976) On determining the statistical significance of climate experiments with general circulation models. *Journal of the Atmospheric Sciences* **33**, 405–412.
Chou, M-D., Peng, L. and Arking, A. (1982) Climate studies with a multi-layer energy balance model. Part II. The role of feedback mechanisms in the CO_2 problem. *Journal of the Atmospheric Sciences* **39**, 2657–2666.
Chylek, P. and Kiehl, J.T. (1981) Sensitivities of radiative–convective models. *Journal of the Atmospheric Sciences* **38**, 1105–1110.
Clapp, R.B. and Hornberger, G.M. (1978) Empirical equations for some soil hydraulic properties. *Water Resources Research* **14**, 601–604.
Climate Research Board (1979) *Carbon Dioxide and Climate: Scientific Assessment*. National Academy of Sciences, Washington, DC. 22 pp.
Cogley, J.G. and Henderson-Sellers, A. (1984) The origin and earliest state of the Earth's hydrosphere. *Reviews of Geophysics and Space Physics* **22**, 131–175.
Colman, R.A., McAvaney, B.J., Fraser, J.R., Rikus, L.J. and Dahni, R.R. (1994) Snow and cloud feedbacks modelled by an atmospheric general circulation model. *Climate Dynamics* **9**, 253–265.
Cox, C. and Munk, W. (1956) Slopes of the sea surface deduced from photographs of the sun glitter. *Bulletin of the Scripps Institute of Oceanography* **6**, 401–488.
Cressman, G.P. (1959) An operative objective analysis scheme. *Monthly Weather Review* **86**, 293–297.

Cressman, G.P. (1960) Improved terrain effects in barotropic forecasts. *Monthly Weather Review* **88**, 327–342.
Crowley, T.J. (1993) Geological assessment of the greenhouse effect. *Bulletin of the American Meteorological Society* **74**, 2363–2373.
Crowley, T.J. and Baum, S.K. (1995) Reconciling late Ordovician (440Ma) glaciation with very high (14×) CO_2 levels. *Journal of Geophysical Research* **100**, 1093–1101.
Crowley, T.J. and Kim, K.Y. (1993) Towards development of a strategy for determining the origin of decadal–centennial scale climate variability. *Quarterly Science Review* **12**, 375–385.
Crowley, T.J. and Kim, K.Y. (1994) Milankovitch forcing of the last interglacial sea level. *Science* **265**, 1566–1568.
Crowley, T.J. and Kim, K.Y. (1995) Comparison of longterm greenhouse projections with the geologic record. *Geophysical Research Letters* **22**, 933–936.
Crowley, T.J., Yip, K.J.J. and Baum, S.K. (1994) Snowline instability in a general circulation model – application to carboniferous glaciation. *Climate Dynamics* **10**, 363–376.
Cubasch, U., Santer, B.D., Hellbach, A., Hegerl, G., Hock, H., Maierreimer, E., Mikolajewicz, U., Stossel, A. and Voss, R. (1994) Monte-Carlo climate change forecasts with a global coupled ocean–atmosphere model. *Climate Dynamics* **10**, 1–19.
Cullen, M.J.P. (1991) Positive definite advection scheme. Unified Model Documentation Paper No. 11, United Kingdom Meteorological Office, Bracknell, Berkshire RG12 2SZ, UK.
Cullen, M.J.P. (1993) The unified forecast/climate model. *Meteorological Magazine* **122**, 81–94.
Deardorff, J. (1978) Efficient prediction of ground temperature and moisture with inclusion of a layer of vegetation. *Journal of Geophysical Research* **83**, 1889–1903.
DeBruin, H.A.R. (1983) A model for the Priestley–Taylor parameter α. *Journal of Climatology and Applied Meterology* **22**, 572–578.
De Haan, B.J., Jonas, M., Klepper, O., Krabec, J., Krol, M.S. and Olendrzynski, K. (1994) An atmosphere–ocean model for integrated assessment of global change. *Water, Air and Soil Pollution* **76**, 283–318.
Del Genio, A.D. and Yao, M.S. (1988) Sensitivity of a global climate model to the specification of convective updraft and downdraft mass fluxes. *Journal of the Atmospheric Sciences* **45**, 2641–2668.
Del Genio, A.D., Yao, M.-S. and Wendell, C.E. (1993) GCM feedback sensitivity to interactive cloud water budget parameterizations. *Preprints of the Fourth Symposium on Global Change Studies*. American Meteorological Society, Anaheim, CA, pp. 176–181.
Dewey, K.F. (1987) Satellite-derived maps of snow cover frequency for the northern hemisphere. *Journal of Climate and Applied Meterology* **26**, 1210–1229.
Dickinson, R.E. (1981) Convergence rate and stability of ocean–atmosphere coupling schemes with a zero-dimensional climate model. *Journal of the Atmospheric Sciences* **38**, 2112–2120.
Dickinson, R.E. (1982) Modeling climate changes due to carbon dioxide increases. *In* W.C. Clark (ed.), *Carbon Dioxide Review 1982*. Oxford University Press, pp. 101–133.
Dickinson, R.E. (1983) Land surface processes and climate–surface albedos and energy balance *Advances in Geophysics* **25**, 305–353.
Dickinson, R.E. (1984) Modeling evapotranspiration for three dimensional global climate models. *In* J.E. Hansen and T. Takahashi (eds), *Climate Processes and*

Climate Sensitivity. Maurice Ewing Series 5, American Geophysical Union, Washington, DC, pp. 58–72.

Dickinson, R.E. (1985) Climate sensitivity. *In* S. Manabe (ed.), *Issues in Atmospheric and Oceanic Modelling. Part A. Climate Dynamics*. Advances in Geophysics, Vol. 28. Academic Press, New York, pp. 99–129.

Dickinson, R.E. (ed.) (1987) *The Geophysiology of Amazonia*. John Wiley and Sons, New York, 526 pp.

Dickinson, R.E. (1995) Land processes in climate models. *Remote Sensing of Environment* **51**, 27–38.

Dickinson, R.E., Henderson-Sellers, A., Kennedy, P.J. and Wilson, M.F. (1986) *Biosphere–Atmosphere Transfer Scheme (BATS) for the NCAR Community Climate Model*. NCAR Technical Note NCAR/TN–275+STR, National Center for Atmospheric Research, Boulder, CO, 69 pp.

Dolman, A.J. and Gregory, D. (1992) The parameterization of rainfall interception in GCMs. *Quarterly Journal of the Royal Meteorological Society* **118**, 455–467.

Dorman, J.L. and Sellers, P.J. (1989) A global climatology of albedo, roughness length and stomatal resistance for atmospheric general circulation models as represented by the simple biosphere model (SiB). *Journal of Applied Meteorology* **28**, 833–855.

Douglas, R.G. and Woodruff, F. (1981) Deep sea foraminifera. *In* C. Emilian (ed.), *The Sea*. Vol. 7. Wiley Interscience, New York, pp. 1233–1327.

Dowlatabadi, H. (1995) Integrated assessment models of climate change: An incomplete overview. *Energy Policy* **23(4)**, 1–8.

Dowlatabadi, H. and Morgan, M.G. (1993) A model framework for integrated assessment of the climate problem. *Energy Policy* **21**, 209–221.

Dowlatabadi, H. and Rotmans, J. (1974) *in Human Choice and Climate Change* (eds) Benedick, S., Edmonds, J.A.E. and Rayner, B.) in preparation.

Dumenil, L. and Todini, E. (1992) A rainfall–runoff scheme for use in the Hamburg climate model. *In* J.P. O'Kane (ed.), *Advances in Theoretical Hydrology: A Tribute to James Dooge*. European Geophysical Society Series on Hydrological Sciences. Vol. 1. Elsevier, Amsterdam, 129–157.

Eagleson, P.S. (1978) Climate, soil and vegetation. *Water Resources Research* **14**, 705–776.

Eagleson, P.S. (ed.) (1982) *Land Surface Processes in Atmospheric General Circulation Models*. Cambridge University Press, Cambridge, 560 pp.

Ebert, E.E. and Curry, J.A. (1993) An intermediate one-dimensional thermodynamic sea ice model for investigating ice atmosphere interactions. *Journal of Geophysical Research* **98**, 10085–10109.

Eddy, J.A., Gililand, R.L. and Hoyt, D.V. (1982) Changes in the solar constant and climatic effects. *Nature* **300**, 689–693.

England, M.H. and Garcon, V. (1994) South Atlantic circulation in a world ocean model. *Annals of Geophysics–Atmospheric Hydrology and Space Science* **12**, 812–825.

England, M.H., Garcon, V. and Minster, J.F. (1994) Chlorofluorocarbon uptake in a world ocean model. I. Sensitivity to the surface gas forcing. *Journal of Geophysical Research* **99**, 25215–25233.

Fels, S.B., Mahlman, J.D., Schwarzkopf, M.D. and Sinclair, R.W. (1980) Stratospheric sensitivity to perturbations in ozone and carbon dioxide: Radiative and dynamical response. *Journal of the Atmospheric Sciences* **37**, 2265–2297.

Flato, G.M. and Hibler, W.D. III (1992) Modelling sea-ice as a cavitating fluid. *Journal of Physical Oceanography* **22**, 626–651.

Foukal, P.V. (1980) Solar luminosity variation in directly observable time scales: Observational evidence and basic mechanisms. *In Sun and Climate*. CNES/CNRS/DGRST Conference Proceedings, pp. 275–284.

Fouquart, Y. (1988) Radiative transfer in climate modeling. *In* M.E. Schlesinger (ed.), *Physically Based Modelling and Simulation of Climate and Climate Change. Part 1.* Kluwer Academic Publishers, Dordrecht, pp. 223–283.

Gallee, H., van Ypersele, J.P., Fichefet, T., Tricot, C. and Berger, A. (1992) Simulation of the last glacial cycle by a coupled, sectorially averaged climate–ice sheet model 2. Response to insolation and CO_2 changes, *Journal of Geophysical Research* **97**, 15713–15740.

Garcia, R.R., Stordal, F., Solomon, S. and Kiehl, J.T. (1992) A new numerical model of the middle atmosphere. 1. Dynamics and transport of tropospheric source gases. *Journal of Geophysical Research* **97**, 12967–12991.

GARP (1975) The physical basis of climate and climate modelling. *GARP Publication Series No. 16.* WMO/ICSU, Geneva.

Gates, W.L. (1979) The effect of the ocean on the atmospheric general circulation. *Dynamics of Atmospheres and Oceans* **3**, 95–109.

Gates, W.L. (1985) Modelling as a means of studying the climate system. In M.C. MacCracken and F.M. Luther (eds), *Projecting the Climatic Effects of Increasing Carbon Dioxide.* DOE/ER-0237, US Department of Energy, Washington, DC, pp. 57–79.

Gates, W.L. (1992) AMIP: The Atmospheric Model Intercomparison Project. *Bulletin of the American Meteorological Society* **73**, 1962–1970.

Gates, W.L. (1995) *Proceedings of the First AMIP Scientific Conference (15–19 May 1995, Monterey, CA).* Technical Document No 732 World Meteorological Organization, Geneva, 532 pp.

Genthon, C. (1994) Antarctic climate modelling with general circulation models of the atmosphere. *Journal of Geophysical Research* **99**, 12953–12961.

Genthon, C. and Armengaud, A. (1995) GCM simulations of atmospheric tracers in the polar latitudes – south pole (Antarctica) and summit (Greenland) cases. *Science of the Total Environment* **161**, 101–116.

Gentilli, J. (ed.) (1971) Climates of Australia and New Zealand. *World Survey of Climatology.* Elsevier, Amsterdam, 405 pp.

Gilchrist, A. (1983) Increased carbon dioxide concentrations and climate: The equilibrium response. *In* W. Bach *et al.* (eds), *Carbon Dioxide: Current Views and Developments in Energy/Climate Research.* D. Reidel, Dordrecht.

Gilliland, R.L. (1980) Solar luminosity variations. *Nature* **286**, 838.

Gilliland, R.L. (1982) Modeling solar variability. *Astrophysics Journal* **253**, 399–405.

Giorgi, F., Marinucci, M.R. and Bates, G.T. (1993) Development of a second generation regional climate model (REGCM2) I: Boundary layer and radiative transfer processes. *Monthly Weather Review* **121**, 2794–2813.

Giorgi, F., Brodeur, C.S. and Bates, G.T. (1994) Regional climate change scenarios over the United States produced with a nested regional climate model. *Journal of Climate* **7**, 375–399.

Glecker, P.J., Randall, D.A., Boer, G., Colman, R., Dix, M., Galin, V., Helfand, M., Kiehl, J., Kitoh, A., Lau, W., Liang, X.Y., Lykossov, V., McAvaney, B., Miyakoda, K., Planton, S. and Stern, W. (1995) Cloud radiative effects on implied oceanic energy transports as simulated by atmospheric general circulation models. *Geophysical Research Letters* **22**, 791–794.

Gleick, J. (1987) *Chaos: The Making of a New Science.* Viking Penguin, New York, 352 pp.

Golitsyn, G.S. and Mokhov, I.I. (1978) Sensitivity estimates and the role of clouds in simple models of climate. *Izvestiya, Atmospheric and Ocean Physics* **14**, 569–576.

Gordon, C.T. (1992) Comparison of 30 day integrations with and without cloud–radiation interaction. *Monthly Weather Review* **120**, 1244–1277.

Gordon, C.T. and Stern, W.F. (1982) A description of the GFDL global spectral model. *Monthly Weather Review* **110**, 625–644.

Greatbach, R.J. and Zhang, S. (1995) An interdecadal oscillation in an idealized ocean basin forced by constant heat flux. *Journal of Climate* **8**, 81–91.

Green, J.S.A. (1970) Transfer properties of the large-scale eddies and the general circulation of the atmosphere. *Quarterly Journal of the Royal Meteorological Society* **96**, 157–185.

Gregory, D. and Rowntree, P.R. (1990) A mass flux convection scheme with representation of cloud ensemble characteristics and stability dependent closure. *Monthly Weather Review* **118**, 1483–1506.

Griffith, K.T., Cox, S.K. and Knollenberg, R.C. (1980) Infrared radiative properties of tropical cirrus clouds inferred from aircraft measurements. *Journal of the Atmospheric Sciences* **37**, 1073–1083.

Hack, J.J. (1994) Parameterization of moist convection in the NCAR Community Climate Model (CCM2). *Journal of Geophysical Research* **99**, 5551–5568.

Hack, J.J., Boville, B.A., Briegleb, B.P., Kiehl, J.T., Rasch, P.J. and Williamson D.L. (1993) *Description of the NCAR Community Climate Model (CCM2)*. NCAR Technical Note, NCAR/TN–382+STR, National Center for Atmospheric Research, Boulder, CO, 108 pp.

Hammer, C.U. (1977) Past volcanism revealed by Greenland Ice Sheet impurities. *Nature* **270**, 482–486.

Hammer, C.U., Clausen, H.B. and Dansgaard, W. (1980) Greenland ice sheet evidence of post-glacial volcanism and its climatic impact. *Nature* **288**, 230–235.

Hansen, J.E., Wang, W.-C. and Lacis, A.A. (1978) Mount Agung eruption provides test of a global climate perturbation. *Science* **199**, 1065–1068.

Hansen, J.E., Lacis, A., Lee, P. and Wang, W.-C. (1980) Climatic effect of atmospheric aerosols. *Annals of the New York Academy of Sciences* **338**, 575–586.

Hansen, J., Johnson, D., Lacis, A., Lebedeff, S., Lee, P., Rind, D. and Russell, G. (1981) Climate impact of increasing carbon dioxide. *Science* **213**, 957–966.

Hansen, J., Russell, G., Rind, D., Stone, P., Lacis, A., Lebedeff, S., Ruedy, R. and Travis, L. (1983) Efficient three-dimensional global models for climate studies: Models I and II. *Monthly Weather Review* **111**, 609–622.

Hansen, J., Lacis, A., Rind, D., Russell, G., Stone, P., Fung, I., Ruedy, R. and Lerner, J. (1984) Climate sensitivity: Analysis of feedback mechanisms. *In* J.E. Hansen and T. Takahashi (eds), *Climate Processes and Climate Sensitivity*. Maurice Ewing Series 5. American Geophysical Union, Washington, DC, 368 pp.

Hansen, J., Russell, G., Lacis, A., Fung, I., Rind, D. and Stone, P. (1985) Climate response times: Dependence on climate sensitivity and ocean mixing. *Science* **229**, 857–859.

Hansen, J.E., Lacis, A., Ruedy, R. and Sato, M. (1992) Potential climatic impact of Mount Pinatubo eruption. *Geophysics Research Letters* **19**, 215-218.

Hansen, J.E., Lacis, A., Ruedy, R, Sato, M. and Wilson, H. (1993) How sensitive is the world's climate? *National Geographic Research and Exploration* **9**, 142–158.

Hare, K.F. (1983) Climate and desertification: A revised analysis. *World Climate Programme*, Vol. 44, 149 pp.

Harshvardhan, Davies, R., Randall, D.A. and Corsetti, T.G. (1987) A fast radiation parameterization for general circulation models. *Journal of Geophysical Research* **92**, 1009–1016.

Harshvardhan, Randall, D.A., Corsetti, T.G. and Dazlich, D.A. (1989) Earth radiation budget and cloudiness simulations with a general circulation model. *Journal of the Atmospheric Sciences* **40**, 1922–1942.

Hart, T.L., Gay, M.J. and Bourke, W. (1988) Sensitivity studies with the physical parameterizations in the BMRC global atmospheric spectral model. *Australian Meteorological Magazine* **36**, 47–60.

Hart, T.L., Bourke, W., McAvaney, B.J. and Forgan, B.W. (1990) Atmospheric general circulation simulations with the BMRC global spectral model. The impact of revised physical parameterizations. *Journal of Climate* **3**, 436–459.

Hartmann, D.L. (1984) On the role of global-scale waves in ice-albedo and vegetation-albedo feedback. *In* J.E. Hansen and T. Takahashi (eds), *Climate Processes and Climate Sensitivity*. Maurice Ewing Series 5. American Geophysical Union, Washington, DC, pp. 18–28.

Hartmann, D.L. and Short, D.A. (1979) On the role of zonal asymmetries in climate change. *Journal of the Atmospheric Sciences* **36**, 519–528.

Harvey, L.D.D. (1994) Transient temperature and sea-level response of a two dimensional ocean–climate model to greenhouse gas increases. *Journal of Geophysical Research* **99**, 18447–18466.

Harvey, L.D.D. and Schneider, S.H. (1985) Transient climate response to external forcing on 10^0–10^4 year time scales. 2. Sensitivity experiments with a seasonal, hemispherically averaged, coupled atmosphere, land, and ocean energy balance model. *Journal of Geophysical Research* **90**, 2207–2222.

Hasselmann, K. (1976) Stochasric climate models, Part I. Theory. *Tellus* **28**, 473–485

Hasselmann, K. (1979) On the signal-to-noise problem in atmospheric response studies. *In Meteorology over Tropical Oceans*. Royal Meteorological Society, Bracknell, pp. 251–259.

Hayashi, Y. (1982) Confidence intervals of a climatic signal. *Journal of the Atmospheric Sciences* **39**, 1895–1905.

Hecht, M., Holland, W., Artale, V. and Pindardi, N. (1996) North Atlantic model sensitivity to Mediterranean waters. *In* W. Howe and A. Henderson-Sellers (eds), *Assessing Climate Change: The Story of the Model Evaluation Consortium for Climate Assessment*, Gordon and Breach, London.

Held, I.M. (1982) Climate models and the astronomical theory of the ice ages. *Icarus* **50**, 449–461.

Held, I.M. and Suarez, M.J. (1974) Simple albedo feedback models of the icecaps. *Tellus* **26**, 613–629.

Held, I.M. and Suarez, M.J. (1978) A two-level primitive equation model designed for climate sensitivity experiments. *Journal of the Atmospheric Sciences* **35**, 206–229.

Held, I.M. and Suarez, M.J. (1994) A proposal for the intercomparison of the dynamical cores of atmospheric general circulation models. *Bulletin of the American Meteorological Society* **75**, 1825–1830.

Held, I.M., Hemler, R.S. and Ramaswamy, V. (1993) Radiative convective equilibrium with explicit 2-dimensional moist convection. *Journal of the Atmospheric Sciences* **50**, 3909–3927.

Henderson-Sellers, A. (1987) Effects of change in land-use on climate in the humid tropics. In R.E. Dickinson (ed.), *The Geophysiology of Amazonia*. John Wiley and Sons, New York. pp. 463–493.

Henderson-Sellers, A. (ed) (1995) *Future Climates of the World*. World Survey of Climatology. Vol. 16. Elsevier, Amsterdam, 568 pp.

Henderson-Sellers, A. and Gornitz, V. (1984) Possible climatic impacts of land cover transformations, with particular emphasis on tropical deforestation. *Climatic Change* **6**, 231–256.

Henderson-Sellers, A. and McGuffie, K. (1994) Land surface characterisation in greenhouse climate simulations. *International Journal of Climatology* **14**, 1065–1094.

Henderson-Sellers, A. and McGuffie, K. (1995), Global climate models and 'dynamic' vegetation changes. *Global Change Biology* **1**, 63–75.

Henderson-Sellers, A. and Robinson, P.J. (1986) *Contemporary Climatology*. Longman, Harlow, 439 pp.
Henderson-Sellers, A. and Wilson, M.F. (1983) Surface albedo for climate modelling. *Reviews of Geophysics and Space Physics* **21**, 1743–1778.
Henderson-Sellers, A., Pitman, A.J., Love, P.K., Irannejad, P. and Chen, T.H. (1995) The project for intercomparison of land surface parameterization schemes (PILPS): Phases 2 & 3. *Bulletin of the American Meteorological Society* **76**, 489–503.
Hering, W.S. and Borden, T.R. Jr. (1965) *Mean Distributions of Ozone Density Over North America 1963–1964.*, Environmental Research Paper 162, US Air Force Cambridge Research Laboratory, Hanscom Field, Bedford, MA, 19 pp.
Hewitson, B. (1994) Regional climates in the GISS general circulation model – surface air temperature. *Journal of Climate* **7**, 283–303.
Hibler, W.D., III (1984) The role of sea ice dynamics in modeling CO_2 increases. *In* J.E. Hansen and T. Takahashi (eds), *Climate Processes and Climate Sensitivity*. Maurice Ewing Series 5. American Geophysical Union, Washington, DC, pp. 238–253.
Hibler, W.D. and Bryan, K. (1984) Ocean circulation: Its effects on seasonal sea-ice simulations. *Science* **224**, 489–491.
Hirschboeck, K.K. (1980) A new worldwide chronology of volcanic eruptions. *Palaeogeography, Palaeoclimatology, Palaeoecology* **29**, 223–241.
Hoffert, M.I., Flannery, B.P., Callegari, A.J., Hsieh, C.T. and Wiscombe, W. (1983) Evaporation-limited tropical temperatures as a constraint on climate sensivity. *Journal of the Atmospheric Sciences* **40**, 1659–1668.
Hollingsworth, A. (1994) Validation and diagnosis of atmospheric models. *Dynamics of Atmosphere and Oceans* **20**, 227–246.
Hoskins, B.J. and Simmons, A.J. (1975) A multi-layer spectral model and the semi-implicit method. *Quarterly Journal of the Royal Meteorological Society* **101**, 637–655.
Houghton, J.T. (ed.) (1984) *The Global Climate*. Cambridge University Press, Cambridge, 233 pp.
Houghton, J.T., Jenkins, G.J. and Ephraums, J J. (1990) *Climate Change: The IPCC Scientific Assessment*. Cambridge University Press, Cambridge, 403 pp.
Houghton, J.T., Callander, B.A. and Varney, S.K. (1992) *Climate Change 1992: The Supplementary Report to the IPCC Scientific Assessment*. Cambridge University Press, Cambridge.
Houghton, J.T., Mera Filho, L.G., Callander, B.A., Harris, N., Kattenberg, A. and Maskell, K. (eds) (1996) *Climate Change 1995. The Science of Climate Change Contribution of Working Group I to the Second Assessment Report of the Intergovernmental Panel on Climate Change*. Cambridge University Press, Cambridge.
Hovine, S. and Fichefet, T. (1994) A zonally averaged, 3-basin ocean circulation model for climate studies. *Climate Dynamics* **10**, 313–331.
Hovis, W.A. and Callahan, W.R. (1966) Infrared reflectance spectra of igneous rocks, tuffs, and red sandstone from 0.5 to 22 microns. *Journal of the Optical Society of America* **56**, 639–643.
Howe, W. and Henderson-Sellers, A. (1996) The MECCA project – overview and evolution of the MECCA project. *In* W. Howe and A. Henderson-Sellers (eds), *Assessing Climate Change: The Story of the Model Evaluation Consortium for Climate Assessment*. Gordon and Breach, London.
Hughes, T.M.C. and Weaver, A.J. (1994) Multiple equilibria of an asymmetric two-basin ocean model. *Journal of Physical Oceanography* **24**, 619–637.
Hummel, J.R. and Kuhn, W.B. (1981) An atmospheric radiative–convective model with interactive water vapour transport and cloud development. *Tellus* **33**, 372–381.
Hunt, B.G. (1973) Zonally symmetric global general circulation models with and without the hydrologic cycle. *Tellus* **25**, 337–354.

Hyde, W.T. and Peltier, W.R. (1993) Effect of altered boundary conditions on GCM studies of the last glacial maximum. *Geophysics Research Letters* **20**, 939–942.

Imbrie, J. and Imbrie, K.P. (1979) *Ice Ages: Solving the Mystery*. Macmillan, London, 224 pp.

Iwasaki, T., Yamada, S. and Tada, K. (1989) A parameterization scheme of orographic gravity wave drag with the different vertical partitionings. Part II. Zonally averaged budget analyses based on transformed Eulerian-mean method. *Journal of the Meteorological Society of Japan* **67**, 29–41.

Jaeger, L. (1976) Monatskarten des Niederschlags für die ganze Erde. *Bericht Deutsche Wetterdienst*, **18**, no. 139.

Jäger, J. (1983) *Climate and Energy Systems. A Review of their Interactions*. Wiley, Chichester, 231 pp.

Jakobchien, R., Hack, J.J. and Williamson, D.L. (1995) Spectral transform solutions to the shallow water test set. *Journal of Computational Physics* **119**, 164–187.

Jenkins, G.S. (1993) A general circulation model study of the effects of faster rotation rate, enhanced CO_2 concentration, and reduced solar forcing – implications for the faint young sun paradox. *Journal of Geophysical Research* **98**, 20803–20811.

Ji, M. and Smith, T.M. (1995) Ocean model response to temperature data assimilation and varying surface wind stress – intercomparisons and implications for climate forecast. *Monthly Weather Review* **123**, 1811–1821.

Joseph, J.H., Wiscombe, W.J. and Weinman, J.A. (1976) The delta-Eddington approximation for radiative flux transfer. *Journal of the Atmospheric Sciences* **33**, 2452–2459.

Kanamitsu, M., Tada, K., Kudo, T., Sata, N. and Isa, S. (1983) Description of the JMA operational spectral model. *Journal of the Meteorological Society of Japan* **61**, 812–827.

Katayama, A. (1978) *Parameterization of the Planetary Boundary Layer in Atmospheric General Circulation Models*. Kisyo Kenkyu Note No. 134, Meteorological Society of Japan, pp. 153–20 (in Japanese).

Katz, R.W. (1982) Statistical evaluation of climate experiments with general circulation models: A parametric time series modeling approach. *Journal of the Atmospheric Sciences* **39**, 1446–1455.

Katz, R.W. (1983) Statistical procedures for making inferences about precipitation changes simulated by an atmospheric general circulation model. *Journal of the Atmospheric Sciences* **40**, 2193–2201.

Kazakov, A.L. and Lykossov, V.N. (1980) Parameterization of heat and moisture exchange during storms with application to problems of atmosphere–ocean interaction. *Soviet Meteorology and Hydrology* **8**, 45–50.

Kazakov, A.L. and Lykossov, V.N. (1982) On parameterization of the interaction between the atmosphere and the underlying surface for numerical modelling of the atmospheric processes. *Trudy Zapsinnii* **55**, Gidrometeoizdat, Moscow, 3–20.

Keepin, W., Mintzer, I. and Kristofersan, L. (1986) Emission of CO_2 into the atmosphere. The rate of release of CO_2 as a function of future energy developments. *In* B. Bolin, B. Döös, J. Jäger and R.A. Warrick (eds), *SCOPE 29. The Greenhouse Effect, Climatic Change, and Ecosystems*. John Wiley and Sons, Chichester, pp. 35–92.

Kelly, P.M. and Sear, C.B. (1984) The climatic impact of explosive volcanic eruptions. *Nature* **311**, 740–743.

Kiehl, J.T. and Briegleb, B.P. (1991) A new parameterization of the absorptance due to the 15 micron band system of carbon dioxide. *Journal of Geophysical Research* **96**, 9013–9019.

Kiehl, J.T. and Ramanathan, V. (1983) CO_2 radiative parameterization used in climate

models: Comparison with narrow band models and with laboratory data. *Journal of Geophysical Research* **88**, 5191–5202.
Kiehl, J.T., Hack, J.J. and Briegleb, B.P. (1994) The simulated earth radiation budget of the NCAR CCM2 and comparison with the Earth Radiation Budget Experiment. *Journal of Geophysical Research* **99**, 20815–20827.
Kim, K.Y. and Crowley, T.J. (1994) Modelling the climate effect of unrestricted greenhouse emissions over the next 10 000 years. *Geophysical Research Letters* **21**, 681–684.
Kita, K. and Sumi, A. (1986) Reference ozone models for middle atmosphere. Meteorological Research Report 86–2, Division of Meteorology, Geophysical Institute, University of Tokyo, 26 pp.
Kitoh, A., Yamazaki, K. and Tokioka, T. (1988) Influence of soil moisture and surface albedo changes over the African tropical rain forest on summer climate investigated with the MRI GCM-I. *Journal of the Meteorological Society of Japan* **66**, 65–86.
Kumar, A. and Hoerling, M.P. (1995) Prospects and limitations of seasonal atmospheric GCM predictions. *Bulletin of the American Meteorological Society* **76**, 335–345.
Kutzbach, J.E. and Otto-Bliesner, B.L. (1982) The sensitivity of the African–Asian monsoonal climate to orbital parameter changes for 9000 years BP in a low-resolution general circulation model. *Journal of the Atmospheric Sciences* **39**, 1177–1188.
Kutzbach, J.E. and Ziegler, A.M. (1993) Simulation of late Permian climate and biomes with an atmosphere ocean model – comparison with observations. *Philosophical Transactions of the Royal Society Series B*. **341**, 327–340.
Labitzke, K., Naujokiat, B. and McCormack, M.P. (1983) Temperature effects on the stratosphere of the April 4 1982 eruption of El Chichon, Mexico. *Geophysical Research Letters* **10**, 24–27.
Lamb, H.H. (1970) Volcanic dust in the atmosphere: With a chronology and assessment of its meteorological significance. *Philosophical Transactions of the Royal Society London, Series A* **266**, 425–533 (see also the updates given by Lamb in *Climate Monitoring* **6**, 57–67, and *Climate Monitoring* **12**, 76–90, 1983).
Lau, K.M., Sui, C.H. and Tao, W.K. (1993) A preliminary study of the tropical water cycle and its sensitivity to surface warming. *Bulletin of the American Meteorological Society* **74**, 1313–1321.
Lee, W.H. and North, G.R. (1995), Small ice cap instability in the presence of fluctuations. *Climate Dynamics* **11**, 242–246.
Legates, B.R. (1987) A climatology of global precipitation. *Publications in Climatology* **40**, 85 pp.
Leith, C.E. (1973) The standard error of time-average estimates of climatic means. *Journal of Applied Meteorology* **12**, 1066–1069.
Lempert, R.J., Schlesinger, M.E. and Hammitt, J.K. (1994) The impact of potential abrupt climate changes on near-term policy choices. *Climatic Change* **26**, 351–376.
Lian, M.S. and Cess, R.D. (1977) Energy balance climate models: A reappraisal of ice-albedo feedback. *Journal of the Atmospheric Sciences* **34**, 1058–1062.
Lindzen, R.S. (1994a) On the scientific basis for global warming scenarios. *Environmental Pollution* **83**, 125–134.
Lindzen, R.S. (1994b) Climate dynamics and global change. *Annual Review of Fluid Mechanics* **26**, 353–378.
Lindzen, R.S., Hou, A.Y. and Farrell, B.F. (1982) The role of convective model choice in calculating the climate impact of doubling CO_2. *Journal of the Atmospheric Sciences* **39**, 1189–1205.
Liou, K.-N. (1980) *An Introduction to Atmospheric Radiation*. International Geophysics Series, No. 25. Academic Press, New York, 392 pp.
Livezey, R.E. and Chen, W.Y. (1983) Statistical field significance and its determination by Monte Carlo techniques. *Monthly Weather Review* **111**, 46–59.

Lovelock, J.E. (1979) *Gaia. A New Look at Life on Earth.* Oxford University Press, Oxford, 157 pp.
Lovelock, J.E. (1991) Gaia: *The Practical Science of Planetary Medicine.* GAIA Books, London.
Lovelock, J.E. and Watson, A.J. (1982) The regulation of carbon dioxide and climate. Gaia or geochemistry? *Planetary and Space Science* **30**, 793–802.
Luther, F.M. and Cess, R.D. (1985) Review of the recent carbon dioxide–climate controversy. *In The Potential Climatic Effects of Increasing Carbon Dioxide.* United States Department of Energy, Washington, DC, pp. 321–335.
Luther, F.M., Ellingson, R.G., Fouquart, Y., Fels, S., Scott, N.A. and Wiscombe, W. (1988) Intercomparison of radiation codes in climate models (ICRCCM): Longwave clear-sky results – a workshop summary, *Bulletin of the American Meteorological Society* **69**, 40–48.
Lynch, P. (1987) Techniques of initialization. *Weather* **42**, 66–70.
Lynn, B.H., Rind, D. and Avissar, R. (1995) The importance of mesoscale circulations generated by subgrid-scale landscape heterogeneities in general circulation models. *Journal of Climate* **8**, 191–205.
MacCracken, M.C. and Ghan, S. (1987) *Design and Use of Zonally-Averaged Climate Models.* UCRL-94338, University of California, Livermore, CA, 44 pp.
MacCracken, M.C., Ellis, J.S., Ellsaesser, H.W., Luther, F.M. and Potter, G.L. (1981) *The Livermore Statistical Dynamical Climate Model.* Lawrence Livermore National Laboratory (UCID-19060).
MacKay, R.M. and Khalil, M.A.K. (1991) Theory and development of a one dimensional time dependent radiative convective model. *Chemosphere* **22**, 383–417.
MacKay, R.M. and Khalil, M.A.K. (1994) Climate simulations using the GCRC 2-D zonally averaged statistical dynamical climate model. *Chemosphere* **29**, 2651–2683.
Madden, R.A. and Shea, D.J. (1978) Estimates of the natural variability of time-averaged temperatures over the United States. *Monthly Weather Review* **106**, 1695–1703.
Manabe, S. (ed.) (1985) *Issues in Atmospheric and Oceanic Modelling. Part A. Climate Dynamics.* Advances in Geophysics, Vol. 28. Academic Press, New York, 591 pp.
Manabe, S. and Bryan, K. (1969) Climate calculations with a combined ocean atmosphere model. *Journal of the Atmospheric Sciences* **26**, 786–789.
Manabe, S. and Bryan, K. (1985) CO_2 induced changes in a coupled ocean–atmosphere model and its palaeoclimatic implications. *Journal of Geophysical Research* **90**, 1689–1707.
Manabe, S. and Möller, F. (1961) On the radiative equilibrium and heat balance of the atmosphere. *Monthly Weather Review* **89**, 503–532.
Manabe, S. and Stouffer, R.J. (1980) Sensitivity of a global clmate model to an increase of CO_2 concentration in the atmosphere. *Journal of Geophysical Research* **85**, 5529–5554.
Manabe, S. and Strickler, R.F. (1964) Thermal equilibrium of the atmosphere with a convective adjustment. *Journal of the Atmospheric Sciences* **21**, 361–385.
Manabe, S. and Wetherald, R.T. (1967) Thermal equilibrium of the atmosphere with a given distribution of relative humidy. *Journal of the Atmospheric Sciences* **24**, 241–259.
Manabe, S. and Wetherald R.T. (1975) The effect of doubling the CO_2 concentration on the climate of a general circulation model. *Journal of the Atmospheric Sciences* **32**, 3–15.
Manabe, S. and Wetherald R.T. (1980) On the distribution of climate change resulting from an increase in CO_2-content of the atmosphere. *Journal of the Atmospheric Sciences* **37**, 99–118.
Manabe, S. and Wetherald, R.T. (1982) Simulation of global cloud cover. In *Cloud/*

Radiation Interaction. ICSU/WMO Meeting in Dublin, 1982, WCP-34, WMO, Geneva, pp. 1–18.
Manabe, S., Bryan, K. and Spelman, M.J. (1979) A global ocean–atmosphere climate model with seasonal variation for future studies of climate sensitivity. *Dynamics of Atmosphere and Oceans* **3**, 393–426.
Manabe, S., Wetherald, R.T. and Stouffer, R.J. (1981) Summer dryness due to an increase of atmospheric CO_2 concentration. *Climatic Change* **3**, 347–386.
Marengo, J.A. and Druyan, L.M. (1994) Validation of model improvements for the GISS GCM. *Climate Dynamics* **10**, 163–179.
Matteucci, G. (1993) Multiple equilibria in a zonal energy balance climate model – the thin ice cap instability. *Journal of Geophysical Research* **98**, 18515–18526.
Matthews, E. (1983) Global vegetation and land use: New high-resolution data bases for climate studies. *Journal of Climatology and Applied Meteorology* **22**, 474–487.
Maykut, G.A. and Untersteiner, N. (1971) Some results from a time-dependent thermodynamic model of sea ice. *Journal of Geophysical Research* **76**, 1550–1575.
McAvaney, B.J., Bourke, W. and Puri, K. (1978) A global spectral model for simulation of the general circulation. *Journal of the Atmospheric Sciences* **35**, 1557–1583.
McClatchey, R.A., Fenn, R.W., Selby, J.E.A., Voltz, F.E. and Garing, J.S. (1972) *Optical Properties of the Atmosphere*. 3rd edn. Environmental Research Papers No. 411, Air Force Cambridge Research Laboratories.
McFarlane, N.A. (1987) The effect of orographically excited gravity-wave drag on the circulation of the lower stratosphere and troposphere. *Journal of the Atmospheric Sciences* **44**, 1775–1800.
McFarlane, N.A., Boer, G.J., Blanchet, J.-P. and Lazare, M. (1992) The Canadian Climate Centre second-generation general circulation model and its equilibrium climate. *Journal of Climate* **5**, 1013–1044.
McGuffie, K., Henderson-Sellers, A. Zhang, H., Durbidge, T.B. and Pitman A.J. (1995) Global climate sensitivity to tropical deforestation. *Global amd Planetary Change* **10**, 97–128
McGuffie, K., Henderson-Sellers, A. and Zhang, H. (1996) Modelling climatic impacts of tropical deforestation. *In* B. Maloney (ed.), *Development and Destruction of the Tropical Rainforest*. Kluwer Scientific, Dordrecht.
McNaughton, K.G. and Jarvis, P.G. (1983) Predicting effects of vegetation changes on transpiration and evaporation. *In Water Defects and Plant Growth*. Vol VII. Academic Press, New York, pp. 1–47.
McPeters, R.D., Heath, D.F. and Bhartia, P.K. (1984) Averaged ozone profiles for 1979 from the NIMBUS 7 SBUV instrument. *Journal of Geophysical Research* **89**, 5199–5214.
Meehl, G.A. (1990) Development of global coupled ocean–atmosphere general circulation models. *Climate Dynamics* **5**, 19–33.
Meehl, G.A. (1995) Global coupled general circulation models. *Bulletin of the American Meteorological Society* **76**, 951–957.
Meehl, G.A. and Washington, W.M. (1995) Cloud albedo feedback and the super greenhouse effect in a global coupled GCM. *Climate Dynamics* **11**, 399–411.
Meehl, G.A., Wheeler, M. and Washington, W.M. (1994) Low frequency variability and CO_2 transient climate change. 3. Intermonthly and interannual variability. *Climate Dynamics* **10**, 277–303.
Mellor, G.L. and Yamada, T. (1982) Development of a turbulent closure model for geophysical fluid problems. *Reviews of Geophysics Space and Physics* **20**, 851–875.
Michael, P., Hoffert, M., Tobias, M. and Tichler, J. (1981) Transient climate response to changing carbon dioxide concentration. *Climatic Change* **3**, 137–153.
Miller, M.J., Palmer, T.N. and Swinbank, R. (1989) Parameterization and influence of

subgridscale orography in general circulation and numerical weather prediction models. *Meteorology and Atmospheric Physics* **40**, 84–109.
Miller, R.L. and Del Genio, A.D. (1994) Tropical cloud feedbacks and natural variability of climate. *Journal of Climate* **7**, 1388–1402.
Milne, A.A. (1928) *The House at Pooh Corner*. Methuen, London, 176 pp.
Mintz, Y. (1984) The sensitivity of numerically simulated climates to land-surface boundary conditions. *In* J.T. Houghton (ed.), *The Global Climate*. Cambridge University Press, Cambridge, pp. 79–106.
Mitchell, J.F.B. (1983) The seasonal response of a general circulation model to changes in CO_2 and sea temperatures. *Quarterly Journal of the Royal Meteorological Society* **109**, 113–152.
Mitchell, J.F.B. (1993) Modelling of palaeoclimates – examples from the recent past. *Philosophial Transactions of the Royal Society, Series B* **341**, 267–275.
Mitchell, J.F.B. and Lupton, G. (1984) A $4 \times CO_2$ integration with prescribed changes in sea surface temperatures. *Progress in Biometeorology* **3**, 353–374.
Mitchell, J.F.B., Davis, R.A., Ingram, W.J. and Senior, C.A. (1995) On surface temperature, greenhouse gases and aerosols: Models and observations. *Journal of Climate* **10**, 2364–2386.
Michell, J.M., Jr. (1983) Empirical modeling of effects of solar variability, volcanic events and carbon dioxide on global-scale average temperature since AD 1880. *In* B.M. McCormac (ed.), *Weather and Climate Responses to Solar Variations*. Colorado Associated University Press, pp. 265–273.
Mokhov, I.I. (1981) Effect of CO_2 on the thermal regime of the earth's climatic system. *Meteorologiya i Gidrologiya* **4**, 24–34.
Monahan, E.C. (1968) Sea spray as a function of low elevation wind speed. *Journal of Geophysical Research* **73**, 1127–1137.
Moorthi, S. and Suarez, M.J. (1992) Relaxed Arakawa–Schubert: A parameterization of moist convection for general circulation models. *Monthly Weather Review* **120**, 978–1002.
Morantine, M.C. and Rao, K.A. (1994) Timescales in energy balance climate models. II. The intermediate time solutions. *Journal of Geophysical Research* **99**, 3643–3653.
Morcrette, J.-J. (1990) Impact of changes to the radiation transfer parameterizations plus cloud optical properties in the ECMWF model. *Monthly Weather Review* **118**, 847–873.
Morcrette, J.-J. (1991) Radiation and cloud radiative properties in the ECMWF operational weather forecast model. *Journal of Geophysical Research* **96**, 9121–9132.
Murphy, J.M. (1995a) Transient response of the Hadley Centre coupled ocean–atmosphere model to increasing carbon dioxide. I. Control climate and flux adjustment. *Journal of Climate* **8**, 36–56.
Murphy, J.M. (1995b) Transient response of the Hadley Centre coupled ocean–atmosphere model to increasing carbon dioxide. Part III. Analysis of global mean response using simple models. *Journal of Climate* **8** 496–514.
Nagai, T., Kitamura, Y., Endoh, M. and Tokioka, T. (1995) Coupled atmosphere ocean model simulations of El Nino southern oscillation with and without an active Indian Ocean. *Journal of Climate* **8**, 3–14.
National Research Council (1982) *Carbon Dioxide and Climate: A Second Assessment, Report of the CO_2/Climate Review Panel*. National Academy Press, Washington, DC, 72 pp.
Neelin, J.D. and Marotzke, J. (1994) Representing ocean eddies in climate models. *Science* **264**, 1099–1100.
Newell, R.E. and Deepak, A. (eds) (1982) Mount St. Helens Eruptions of 1980: Atmospheric effects and potential climatic impact. NASA SP-458, NASA Scientific and Technical Information Branch, Washington, DC, 119 pp.

Newhall, C.G. and Self, S. (1982) The volcanic explosivity index (VEI): An estimate of explosive magnitude for historical volcanism. *Journal of Geophysical Research* **87**, 1231–1238.
Newkirk, G., Jr. (1983) Variations in solar luminosity. *Annual Review of Astronomy and Astrophysics* **21**, 429–467.
Noda, A. and Tokioka, T. (1989) The effect of doubling the CO_2 concentration on convective and non–convective precipitation in a general circulation model coupled with a simple mixed layer ocean model. *Journal of the Meteorological Society of Japan* **67**, 1057–1069.
Nordhaus, W.D. (1991) To slow or not to slow: The economics of the greenhouse effect. *Economic Journal* **101**, 920–937.
North, G.R. (1975) Theory of energy balance climate models. *Journal of the Atmospheric Sciences* **32**, 3–15.
North, G.R. Cahalan, R.F. and Coakley, J.A. (1981) Energy-balance climate models. *Reviews of Geophysics and Space Physics* **19**, 91–122.
North, G.R., Mengel, J.G. and Short, D.A. (1983) Simple energy balance model resolving the seasons and the continents: Application to the astronomical theory of the ice ages. *Journal of Geophysical Research* **88**, 6576–6586.
North, G. R., Mengel, J.G. and Short, D.A. (1984) On the transient response patterns of climate to time-dependent concentrations of atmospheric CO_2. *In* J.E. Hansen and T. Takahashi (eds), *Climate Processes and Climate Sensitivity*. Maurice Ewing Series 5. American Geophysical Union, Washington, DC, pp. 164–170.
Oerlemans, J. (1982) Glacial cycles and ice-sheet modeling. *Climatic Change* **4**, 353–374.
Oerlemans, J. and van der Veen, C.J. (1984) *Ice Sheets and Climate*. Reidel, Dordrecht, 217 pp.
Oerlemans, J. and Vernekar, A.D. (1981) A model study of the relation between northern hemispheric glaciation and precipitation. *Contributions to Atmospheric Physics* **54**, 352–361.
Oeschger, H., Siegenthaler, U., Schotterer, U. and Gugelmann, A. (1975) A box diffusion model to study the carbon dioxide exchange in nature. *Tellus* **27**, 168–192.
Ohring, G. and Adler, S. (1978) Some experiments with a zonally averaged climate model. *Journal of the Atmospheric Sciences* **35**, 186–205.
Oliver, R.C. (1976) On the response of hemispheric mean temperature to stratospheric dust: An empirical approach. *Journal of Applied Meteorology* **15**, 933–950.
Oort, A.H. (1983) *Global Atmospheric Circulation Statistics, 1958–1973*. NOAA Professional Paper 14, US Department of Commerce, Washington, DC, 180 pp + 24 fiche.
Oort, A.H. and Peixóto, J.P. (1983) Global angular momentum and energy balance requirements from observations. *Advances in Geophysics* **25**, 355–490.
Owen, T., Cess, R.D. and Ramanathan, V. (1979) Enhanced CO_2 greenhouse to compensate for reduced solar luminosity on the early earth. *Nature* **277**, 640–642.
Palmén, E. (1949) Meridional circulations and the transfer of angular momentum in the atmosphere. *Journal of Meteorology* **6**, 429–430.
Parker, D.E. (1985) The influence of the southern oscillation and volcanic eruptions on temperature in the tropical troposphere. *Journal of Climate* **5**, 273–282.
Parker, D.E. and Brownscombe, J.L. (1983) Stratospheric warming following the El Chichon volcanic eruption. *Nature* **301**, 406–408.
Parkinson, J.H., Morrison, L.V. and Stephenson, F.R. (1980) The constancy of the solar diameter over the past 250 years. *Nature* **28**, 548–551.
Parry, M.L. (1978) *Climatic Change, Agriculture and Settlement*. Dawson, Folkestone, 214 pp.
Peck, S. and Teisberg, T.J. (1992) CETA: A model for carbon emissions trajectory assessment. *Energy Journal* **13**, No.1.

Peixóto, J.P. and Oort, A.H. (1984) Physics of climate. *Reviews of Modern Physics* **56**, 365–430.

Peixoto, J.P. and Oort, A.H. (1991) *Physics of Climate*. American Institute of Physics, 520 pp.

Penner, J.E., Charlson, R.J., Hales, J.M., Laulainen, N.S., Leifer, R., Novakov, T., Ogren, J., Radke, L.F., Schwartz, S.E. and Travis, L. (1994) Quantifying and minimizing uncertainty of climate forcing by anthropogenic aerosols. *Bulletin of the American Meteorological Society* **75**, 375–400.

Pitman, A.J., Durbidge, T.B., McGuffie, K. and Henderson-Sellers, A. (1993) Assessing climate model sensitivity to prescribed deforested landscapes. *International Journal of Climatology* **13**, 879–898.

Pittock, A.B. (1978) A critical look at long-term sun–weather relationships. *Reviews of Geophysics and Space Physics* **16**, 400–420.

Pittock, A.B. (1983) Solar variability, weather and climate: An update. *Quarterly Journal of the Royal Meteorological Society* **109**, 23–57.

Pivovarova, Z.I. (1977) *Radiation Characteristics of Climate in the USSR*. Gidrometeoizdat, Leningrad, 355 pp. (in Russian).

Platt, C.M.R., Reynolds, D.W. and Abshire, N.L. (1980) Satellite and lidar observations of the albedo, emittance and optical depth of cirrus compared to model calculations. *Journal of the Atmospheric Sciences* **108**, 195–204.

Pollack, J.B., Rind, D., Lacis, A., Hansen, J.E., Sato, M. and Ruedy, R. (1993) GCM simulations of volcanic aerosol forcing. I. Climate changes induced by steady state perturbations. *Journal of Climate* **6**, 1719–1742.

Pollard, D. and Shulz, M. (1994) A model for the potential locations of Triassic evaporite basins driven by palaeoclimatic GCM simulations. *Global and Planetory Change* **9**, 233–249.

Pollard, D. and Thompson, S.L. (1995) Use of the land-surface transfer scheme (LSX) in a global climate model (GENESIS): The response to doubling stomatal resistance. *Global and Planetory Change* **10**, 129–161.

Potter, G.L. and Cess, R.D. (1984) Background tropospheric aerosols: Incorporation within a statistical dynamical climate model. *Journal of Geophysical Research* **89**, 9521–9526

Potter, G.L. and Gates, W.L. (1984) A preliminary intercomparison of the seasonal response of two atmospheric climate models. *Monthly Weather Review* **112**, 909–917.

Potter, G.L., Ellsaesser, H.W., MacCracken, M.C. and Luther, F.M. (1979) Performance of the Lawrence Livermore Laboratory zonal atmospheric model. *In Proceedings of the GARP JOC Study Conference on Climate Models, GARP Publication Series No. 22 (Vol. 2)*, WMO, Geneva, pp. 995–1001.

Potter, G.L., Ellsaesser, H.W., MacCraken, M.C. and Mitchell, C.S. (1981) Climate change and cloud feedback: The possible radiative effects of latitudinal redistribution. *Journal of the Atmospheric Sciences* **38**, 489–493.

Power, S.B. and Kleeman, R. (1993) Multiple equilibria in a global general circulation model. *Journal of Physical Oceanography* **23**, 1670–1681.

Quiroz, R.S. (1983) The climate of the 'El Niño' winter of 1982–1983 – a season of extraordinary climatic anomalies. *Monthly Weather Review* **111**, 1685–1706.

Rahmstorf, S. (1994) Rapid climate transitions in a coupled ocean atmosphere model. *Nature* **372**, 82–85.

Ramanathan, V. (1977) Interactions between ice-albedo, lapse-rate, and cloud-top feedbacks: An analysis of the nonlinear response of a GCM climate model. *Journal of the Atmospheric Sciences* **34**, 1885–1897.

Ramanathan, V. (1981) The role of ocean–atmosphere interactions in the CO_2 climate problem. *Journal of the Atmospheric Sciences* **38**, 918–930.

Ramanathan, V. and Coakley, J.A., Jr. (1978) Climate modeling through radiative-convective models. *Reviews of Geophysics and Space Physics* **16**, 465–489.
Ramanathan, V., Singh, H.B., Cicerone, R.J. and Kiehl, J.T. (1985) Trace gas trends and their potential role on climate change. *Journal of Geophysical Research* **90** (**D3**), 5547–5566.
Ramaswamy, V. and Ramanathan,V. (1989) Solar absorption by cirrus clouds and the maintenance of the tropical upper troposphere thermal structure. *Journal of the Atmospheric Sciences* **46**, 2293–2310.
Rampino, M.R. and Self, S. (1984) Sulphur-rich volcanic eruptions and stratospheric aerosols. *Nature* **310**, 677–679.
Ramsden, D. and Fleming, G. (1995) Use of a coupled model to investigate the sensitivity of the arctic ice cover to doubling carbon dioxide. *Journal of Geophysical Research* **100**, 6817–6828.
Randall, D.A., Abeles, J.A. and Corsetti, T.G. (1985) Seasonal simulations of the planetary boundary layer and boundary-layer stratocumulus clouds with a general circulation model. *Journal of the Atmospheric Sciences* **42**, 641–676.
Randall, D.A., Harshvardhan, Corsetti, T.G. and Dazlich, D.A. (1989) Interactions among clouds, radiation, and convection in a general circulation model. *Journal of the Atmospheric Sciences* **46**, 1943–1970.
Randall, D.A., Harshvardhan, and Dazlich, D.A. (1990) Diurnal variability of the hydrological cycle in a general circulation model. *Journal of Atmospheric Science* **48**, 40–62.
Randall, D.A., Cess, R.D., Blanchet, J.P., Chalita, S., Colman, R., Dazlich, D.A., Del Genio, A.D., Keup, E., Lacis, A., LeTreut, H., Liang, X.Z., McAvaney, B., Mahfouf, J.F., Meleshko, V.P., Morcrette, J.J., Norris, P.M., Potter, G.L., Rikus, L., Roeckner, E., Royer, J.F., Schlese, U., Sheinin, D.A., Sokolov, A.P., Taylor, K.E., Wetherald, R.T. *et al.* (1994) Analysis of snow feedback in 14 general circulation models. *Journal of Geophysical Research* **99**, 20757–20771.
Rasch, P.J. and Williamson, D.L. (1990) Computational aspects of moisture transport in global models of the atmosphere. *Quarterly Journal of the Royal Meteorological Society* **116**, 1071–1090.
Renno, N.O., Emanuel, K.A. and Stone, P.H. (1994a) Radiative convective model with an explicit hydrological cycle. I. Formulation and sensitivity to model parameters. *Journal of Geophysical Research* **99**, 14429–14441.
Renno, N.O., Stone, P.H. and Emanuel, K.A. (1994b) Radiative convective model with an explicit hydrological cycle. II. Sensitivity to large changes in solar forcing. *Journal of Geophysical Research* **99**, 17001–17020.
Rind, D. and Lebedeff, S. (1984) *Potential Impacts of Increasing Atmospheric CO_2 with Emphasis on Water Availability and Hydrology in the United States*. Report Prepared for the Environmental Protection Agency, NASA Goddard Space Flight Center Institute for Space Studies, New York, 96 pp.
Rind, D., Healy, R., Parkinson, C. and Martinson, D. (1995) The role of sea ice in $2 \times CO_2$ climate model sensitivity. I. The total influence of sea ice thickness and extent. *Journal of Climate* **8**, 449–463.
Ritchie, H. (1985) Application of a semi-Lagrangian integration scheme to the moisture equation in a regional forecast model. *Monthly Weather Review* **113**, 424–435.
Robock, A. (1978) Internally and externally caused climate change. *Journal of the Atmospheric Sciences* **35**, 1111–1122.
Robock, A. (1983a) Ice and snow feedbacks and the latitudinal and seasonal distribution of climate sensitivity. *Journal of the Atmospheric Sciences* **40**, 986–997.
Robock, A. (1983b) The dust cloud of the century. *Nature* **301**, 373–374.
Rockel, B., Raschke, E. and Weyres, B. (1991) A parameterization of broad band clouds. *Beiträge zur Physik der Atmosphäre*. **64**, 1–12.

Rodgers, C.D. (1967) The use of emissivity in atmospheric radiation calculations. *Quarterly Journal of the Royal Meteorological Society* **93**, 43–54.

Rodgers, C.D. (1968) Some extension and applications of the new random model for molecular band transmission. *Quarterly Journal of the Royal Meteorological Society* **94**, 99–102.

Root, T.L. and Schneider, S.H. (1995) Ecology and climate: Research strategies and implications. *Science* **269**, 334–341.

Rossow, W.B., Henderson-Sellers, A. and Weinreich, S.K. (1982) Cloud-feedback – a stabilizing effect for the early Earth. *Science* **217**, 1245–1247.

Rossow, W.B., Mosher, F., Kinsella, E., Arking, A., Desbois, M., Harrison, E., Minnis, P., Ruprecht, E., Sèze, G. and Summer, C. (1985) ISCCP cloud algorithm intercomparison. *Journal of Climate and Applied Meteorology* **24**, 877–903.

Rotmans, J., Hulme, M. and Downing, T. (1995) Climate change implications for Europe. *Global and Environmental Change* **4,** 97–124.

Rowntree, P. and Walker, J. (1978) The effects of doubling the CO_2 concentration radiative–convective equilibrium. *In* J. Williams (ed.), *Carbon Dioxide, Climate and Society*. Pergamon Press, New York, pp. 181–192.

Sagan, C., Toon, O.B. and Pollack, J.B. (1979) Anthropogenic albedo changes and the Earth's climate. *Science* **206**, 1363–1368.

Saltzman, B. (1978) A survey of statistical dynamical models of terrestrial climate. *Advances in Geophysics* **20**, 183–304.

Saltzman, B. (ed.) (1983) *Theory of Climate*. Advances in Geophysics Vol. 25. Academic Press, New York, 505 pp.

Sassen, K. and Dodd, G.C. (1989) Haze particle nucleation simulations in cirrus cloud and applications for numerical and lidar studies. *Journal of the Atmospheric Sciences* **46**, 3005–3014.

Sato, N., Sellers, P.J., Randall, D.A., Schneider, E.K., Shukla, J., Kinter, III, J.L., Hou, Y.-T. and Albertazzi, E. (1989) Implementing the simple biosphere model in a general circulation model. *Journal of the Atmospheric Sciences* **46**, 2757–2782.

Satoh, M. (1994) Hadley circulations in radiative convective equilibrium in an axially symmetric atmosphere. *Journal of the Atmospheric Sciences* **51**, 1947–1968.

Schlesinger, M.E. (1983) A review of climate models and their simulation of CO_2-induced warming. *International Journal of Environmental Studies* **20**, 103–114.

Schlesinger, M.E. (ed.) (1987) *Physically Based Climate Models and Climate Modelling*. Proceedings of a NATO ASI, Reidel, Dordrecht.

Schlesinger, M.E. (1993) Greenhouse policy. *National Geographic Research and Exploration* **9**, 159–172.

Schlesinger, M.E. and Mitchell, J.F.B. (1985) Model projections of the equilibrium climatic response to increased carbon dioxide. In *Projecting the Climatic Effects of Increasing Carbon Dioxide*. United States Department of Energy, DOE/ER 0237, Washington, DC, pp. 81–148.

Schneider, S.H. (1972) Cloudiness as a global climatic feedback mechanism: The effects on the radiation balance and surface temperature of variations in cloudiness. *Journal of the Atmospheric Sciences* **29**, 1413–1422.

Schneider, S.H. (1986) A goddess of the Earth?: The debate on the Gaia hypothesis – an editorial. *Climatic Change* **8**, 1–4.

Schneider, S.H. (1994) Detecting climate change signals: Are there any fingerprints? *Science* **263**, 341–347.

Schneider, S.H. and Dickinson, R.E. (1974) Climate modeling. *Reviews of Geophysics and Space Physics* **12**, 447–493.

Schneider, S.H. and Thompson, S.L. (1981) Atmospheric CO_2 and climate: Importance of the transient response. *Journal of Geophysical Research* **86,** 3135–3147.

Schneider, S.H., Washington, W.M. and Chervin, R.M. (1978) Cloudiness as a climatic feedback mechanism: Effects on cloud amounts of prescribed global and regional surface temperature changes in the NCAR GCM. *Journal of Atmospheric Science* **35**, 2207–2221.

Schutz, C and Gates, W. L. (1971) Global climatic data for surface, 800 mb, 400 mb: January. Advanced Research Projects Agency, Report R-915-ARPA, Rand Corp, Santa Monica, CA, 173 pp.

Schutz, C. and Gates, W.L. (1972) Global climatic data for surface, 800 mb, 400 mb: July. Advanced Research Projects Agency, Report R-1029-ARPA, Rand Corp., Santa Monica, CA, 180 pp.

Sedlacek, W.A., Mroz, E.J., Lazrus, A.L. and Gandrud, B.W. (1983) A decade of stratospheric sulphate measurements compared with observations of volcanic eruptions. *Journal of Geophysical Research* **88**, 3741–3776.

Seidel, S. and Keyes, D. (1983) Can we delay a greenhouse warming? *In The Effectiveness and Feasibility of Options to Slow a Build-up of Carbon Dioxide in the Atmosphere.* Strategic Studies Staff, Office of Policy Analysis, Office of Policy and Resources Management, EPA, Washington, DC 20460.

Sela, J. (1980) Spectral modeling at the National Meteorological Center. *Monthly Weather Review* **108**, 1279–1292.

Self, S., Rampino, M.R. and Barbera, J.J. (1981) The possible effects of large 19th century volcanic eruptions on zonal and hemispheric surface temperatures. *Journal of Volcanology and Geothermal Research* **11**, 41–60.

Sellers, P.J., Mintz, Y., Sud, Y.C. and Dalcher, A. (1986) A simple biosphere model (SiB) for use within general circulation models. *Journal of the Atmospheric Sciences* **43**, 505–531.

Sellers, W.D. (1969) A global climatic model based on the energy balance of the Earth–atmosphere system. *Journal of Applied Meteorology* **8**, 392–400.

Sellers, W.D. (1973) A new global climate model. *Journal of Applied Meteorology* **12**, 241–254.

Sellers, W.D. (1976) A two-dimensional global climate model. *Monthly Weather Review* **194**, 233–248.

Semtner, A.J. (1976) A model for the thermodynamic growth of sea ice in numerical investigations of climate. *Journal of Physical Oceanography* **6**, 379–389.

Semtner, A.J., Jr. (1984a) On modelling the seasonal thermodynamic cycle of sea ice in studies of climate change. *Climatic Change* **6**, 27–38.

Semter, A.J., Jr. (1984b) Development of efficient, dynamical ocean–atmosphere models for climate studies. *Journal of Climate and Applied Meteorology* **23**, 353–374.

Semtner, A.J. (1995) Modelling ocean circulation. *Science* **269**, 1379–1385.

Semtner, A.J. and Chervin, R.M. (1988) A simulation of the global ocean with resolved eddies. *Journal of Geophysical Research* **93**, 15502–15522, 15767–15775.

Semtner, A.J. and Chervin, R.M. (1992) Ocean general circulation from a global eddy-resolving model. *Journal of Geophysical Research* **97**, 5493–5550.

Shen, S.S.P., North, G.R. and Kim, K.Y. (1994) Spectral approach to optimal estimation of the global average temperature. *Journal of Climate* **7**, 1999–2007.

Shine, K.P. and Henderson-Sellers, A. (1983) Modelling climate and the nature of climate models: A review. *Journal of Climatology* **3**, 81–94.

Shukla, J. and Mintz, Y. (1982) Influence of land-surface evapotranspiration on the earth's climate. *Science* **215**, 1498–1501.

Shuttleworth, W.J. (1988) Macrohydrology: The new challenge for process hydrology. *Journal of Hydrology* **100**, 31–56.

Shuttleworth, W.J. and Calder, I.R. (1979) Has the Priestley–Taylor equation any relevance to forest evaporation? *Journal of Applied Meteorology* **18**, 639–646.

Shutts, G. J. (1983) Parameterization of travelling weather systems in a simple model of large-scale atmospheric flow. *In* B. Saltzman (ed.), *Theory of Climate*. Advances in Geophysics, Vol. 25. Academic Press, New York, pp. 117–172.

Simkin, T., Sibert, L., McClelland, L., Bridge, D., Newhall, C. and Latter, J.H. (1981) *Volcanoes of the World*. Hutchinson Ross Publishing Company, Stroudsburg, 233 pp.

Simmons, A.J. and Bengtsson, L. (1987) Atmospheric general circulation models: Their design and use for climate studies. *In* M.E. Schlesinger (ed.), *Physically Based Climate Models and Climate Modelling*. Proceedings of a NATO ASI, Reidel, Dordrecht.

Simmons, A.J. and Burridge, D.M. (1981) An energy and angular-momentum conserving vertical finite difference scheme and hybrid vertical coordinates. *Monthly Weather Review* **109**, 758–766.

Simmons, A.J., Hoskins, B.J. and Burridge, D.M. (1978) Stability of the semi-implicit method of time integration. *Monthly Weather Review* **106**, 405–412.

Simmons, A.J., Burridge, D.M., Jarraud, M., Girard, C. and Wergen, W. (1989) The ECMWF medium-range prediction models: Development of the numerical formulations and the impact of increased resolution. *Meteorology and Atmospheric Physics* **40**, 28–60.

Simpson, J. and Wiggard, V. (1969) Models of precipitating cumulus towers. *Monthly Weather Review* **97**, 471–489.

Sinha, A. and Shine, K.P. (1994) A one dimensional study of possible cirrus cloud feedbacks. *Journal of Climate* **7**, 158–173.

Skiles, J.W. (1995) Modelling climate change in the absence of climate change data. *Climatic Change* **30**, 1–6.

Slingo, A. (ed.) (1985) *Handbook of the Meteorological Office 11-Layer Atmospheric General Circulation Model. Vol. 1. Model Description*. Dynamical Climatology, Technical Note 29, UK Meteorological Office (Met O20), 155 pp.

Slingo, A. (1989) A GCM parameterization for the shortwave radiative properties of water clouds. *Journal of Geophysical Research* **46**, 1419–1427.

Slingo, A. and Slingo, J.M. (1991) Response of the National Center for Atmospheric Research Community Climate Model to improvements in the representation of clouds. *Journal of Geophysical Research* **96**, 15341–15357.

Slingo, J.M. (1987) The development and verification of a cloud prediction model for the ECMWF model. *Quarterly Journal of Royal Meteorological Society* **113**, 899–927.

Smagorinsky, J. (1963) General circulation experiments with the primitive equations. I. The basic experiment. *Monthly Weather Review* **91**, 99–164.

Smagorinsky, J. (1983) The beginnings of numerical weather prediction and general circulation modeling: Early recollections. *In* B. Saltzman (ed.), *Theory of Climate*. Academic Press, New York, pp. 3–38.

Smith, E.A., von der Haar, T.H., Hicket, J.R. and Maschhoff, R. (1983) The nature of short-period fluctuations in solar irradiance received by the Earth. *Climatic Change* **5**, 211–235.

Smith, N.R. (1993) Ocean modelling in a global observing system. *Review of Geophysics* **31**, 281–317.

Smith, R.N.B. (1990) A scheme for predicting layer clouds and their water content in a general circulation model. *Quarterly Journal of the Royal Meteorological Society* **116**, 435–460.

Smith, S.D. (1980) Wind stress and heat flux over the ocean in gale force winds. *Journal of Physical Oceanography* **10**, 709–726.

Somerville, R.C.J. and Remer, L.A. (1984) Cloud optical thickness feedbacks in the CO_2 climate problem. *Journal of Geophysical Research* **89**, 9668–9672.

Sonett, C.P. (1984) Very long solar periods and the radiocarbon record. *Reviews of Geophysics and Space Physics* **22**, 239–254.

Spelman, M.J. and Manabe, S. (1984) Influence of oceanic heat transport upon the sensitivity of a model climate. *Journal of Geophysical Research* **89**, 571–586.

Spivakovsky, C.M., Yevich, R., Logan, J.A., Wofsy, S.C. and McElroy, M.B. (1990) Tropospheric OH in a three-dimensional chemical tracer model: An assessment based on observations of CH_3CCl_3. *Journal of Geophysical Research* **95**, 18441–18471.

Stern, W.F. and Miyakoda, K. (1995) Feasibility of seasonal forecasts inferred from multiple GCM simulations. *Journal of Climate* **8**, 1071–1085.

Stone, P.H. (1973) The effects of large-scale eddies on climatic change. *Journal of the Atmospheric Sciences* **30**, 521–529.

Stone, P.H. and Miller, D. A. (1980) Empirical relations between seasonal changes in meridional temperature gradients and meridional fluxes of heat. *Journal of the Atmospheric Sciences* **37**, 1708–1721.

Stuiver, M. (1980) Solar variability and climatic change during the current millennium. *Nature* **286**, 868–871.

Stuiver, M. and Quay, P.D. (1980) Changes in atmospheric carbon-14 attributed to a variable sun. *Science* **207**, 11–19.

Sud, Y.C. and Molod, A. (1988) The roles of dry convection, cloud-radiation feedback processes and the influence of recent improvements in the parameterization of convection in the GLA GCM. *Monthly Weather Review* **116**, 2366–2387.

Sundqvist, H. (1978) A parameterization scheme for non-convective condensation including prediction of cloud water content. *Quarterly Journal of the Royal Meteorological Society* **104**, 677–690.

Sundqvist, H. (1981) Prediction of stratiform clouds: Results from a 5-day forecast with a global model. *Tellus* **33**, 242–253.

Sundqvist, H. (1988) Parameterization of condensation and associated clouds in models for weather prediction and general circulation simulation. *In* M.E. Schlesinger (ed.), *Physically-Based Modelling and Simulation of Climate and Climatic Change. Part 1.* Kluwer Academic Publishers, Dordrecht, pp. 433–462.

Tang, B.Y. and Weaver, A.J. (1995) Climate stability as deduced from an idealized coupled atmosphere ocean model. *Climate Dynamics* **11**, 141–150.

Taplin, R. (1996) Climate science and politics: the road to Rio and beyond. *In* T. Giambelluca and A. Henderson-Sellers (eds), *Coupled Climate System Modelling: A Southern Hemisphere Perspective*. John Wiley & Sons, Chichester, pp. 377–395.

Taylor, B.L., Gal-Chen, T. and Schneider, S.H. (1980) Volcanic eruptions and long-term temperature records: An empirical search for cause and effect. *Quarterly Journal of the Royal Meteorological Society* **106**, 175–199.

Taylor, K.E. (1980) The roles of mean meridional motions and large-scale eddies in zonally averaged circulations. *Journal of the Atmospheric Sciences* **37**, 1–19.

Thomas, G. and Henderson-Sellers, A. (1991) An evaluation of proposed representations of subgrid hydrologic processes in climate models. *Journal of Climate* **4**, 898–910.

Thompson, S.L. and Pollard, D. (1995) A global climate model (GENESIS) with a land-surface-transfer scheme (LSX). Part 1. Present climate simulation. *Journal of Climate* **8**, 732–761.

Thompson S.L. and Schneider, S.H. (1979) A seasonal zonal energy balance climate model with an interactive lower layer. *Journal of Geophysical Research* **84**, 2401–2414.

Thompson, S.L. and Schneider, S.H. (1982) Carbon dioxide and climate: The importance of realistic geography in estimating the transient temperature response. *Science* **217**, 1031–1033.

Tiedtke, M. (1984) The effect of penetrative cumulus convection on the large-scale flow in a general circulation model. *Beiträge zur Physik der Atmosphäre* **57**, 216–239.

Tiedtke, M. (1988) Parameterization of cumulus convection in large-scale models. *In* M.E. Schlesinger (ed.), *Physically Based Modelling and Simulation of Climate and Climatic Change. Part 1*. Kluwer Academic Publishers, Dordrecht, pp. 375–431.

Tiedtke, M. (1989) A comprehensive mass flux scheme for cumulus parameterization in large-scale models. *Monthly Weather Review* **117**, 1779–1800.

Trenberth, K.E. (1983) Interactions between orographically and thermally forced planetary waves. *Journal of the Atmospheric Sciences* **40**, 1126–1153.

Trenberth, K.E. (1992) *Coupled Climate System Modelling*. Cambridge University Press, Cambridge, 600 pp.

Troen, I. and Mahrt, L. (1986) A simple model of the atmospheric boundary layer: Sensitivity to surface evaporation. *Boundary-Layer Meteorology* **37**, 129–148.

Tselioudis, G., Lacis, A.A., Rind, D. and Rossow, W.B. (1993) Potential effects of cloud optical thickness on climate warming. *Nature* **366**, 670–672.

Tsonis, A.A. (1992) *Chaos: From Theory to Applications*. Plenum Press, New York, 274 pp.

Turco, R.P., Toon, O.B., Ackerman, T., Pollack, J.B. and Sagan, C. (1983) Nuclear winter: Global consequences of multiple nuclear explosions. *Science* **222**, 1283–1292.

US National Academy of Sciences (1975) *Understanding Climatic Change: A Program for Action*. Washington, DC, 239 pp.

Valdes, P. (1993) Atmospheric general circulation models of the Jurassic. *Philosophical Transactions of the Royal Society Series B*. **341**, 317–326.

VEMAP (1995) Vegetation/ecosystem modelling and analysis project (VEMAP): Comparing biogeography and biogeochemistry models in a continental-scale study of terrestrial ecosystem responses to climate change and CO_2 doubling. *Global Biogeochemical Cycles* **9**, 407–437.

Viner, D., Hulme, M. and Raper, S.C.B. (1995) Climate change scenarios for the assessments of the climate change on regional ecosystems. *Journal of Thermal Biology* **20**, 175–190.

Walker, G.K., Sud, Y.C. and Atlas, R. (1995) Impact of the ongoing Amazonian deforestation on local precipitation – a GCM simulation study. *Bulletin of the American Meteorological Society* **76**, 346–361.

Walker, J.C.G., Hays, P. and Kasting, J.F. (1981) A negative feedback mechanism for the long-term stabilization of earth's surface temperature. *Journal of Geophysical Research* **26**, 9776–9782.

Wang, W.-C. and Isaksen, I.S. (1995) *Atmospheric Ozone as a Climate Gas: General Circulation Model Simulations*. NATO ASI Series I; Vol. 32. Springer, New York, 459 pp.

Wang, W.-C. and Stone, P.H. (1980) Effects of ice-albedo feedback on global sensitivity in a one-dimensional radiative–convective model. *Journal of the Atmospheric Sciences* **37**, 545–552.

Wang, W.-C., Dudek, M.P., Liang, X.-Z. and Kiehl, J.T. (1991a) Inadequacy of effective CO_2 as a proxy in simulating the greenhouse effect of other radiatively active gases, *Nature* **350**, 573–577.

Wang, W.-C., Shi, G.-Y. and Kiehl, J.T. (1991b) Incorporation of the thermal radiative effect of CH_4, N_2O, CF_2Cl_2, and $CFCl_3$ into the NCAR Community Climate Model. *Journal of Geophysical Research* **96**, 9097–9103.

Warren, S.G. and Schneider, S.H. (1979) Seasonal simulation as a test for uncertainties in the parameterization of a Budyko–Sellers zonal climate model. *Journal of the Atmospheric Sciences* **36**, 1377–1391.

Washington, W.M. and Meehl, G.A. (1983) General circulation model experiments on the climatic effects due to a doubling and quadrupling of carbon dioxide concentration. *Journal of Geophysical Research* **88**, 6600–6610.

Washington, W.M. and Meehl, G.A. (1984) Seasonal cycle experiment on the climate sensitivity due to a doubling of CO_2 with an atmospheric general circulation model coupled to a simple mixed layer ocean model. *Journal of Geophysical Research* **89**, 9475–9503.

Washington, W.M. and Parkinson, C.L. (1986) *An Introduction to Three-Dimensional Climate Modelling*. University Science Books, Mill Valley, CA, 422 pp.

Washington, W.M., Semtner, A.J., Jr., Meehl, G.A., Knight, D.J. and Meyer, T. A. (1980) A general circulation experiment with a coupled atmosphere, ocean and sea ice model. *Journal of Physical Oceanography* **10**, 1887–1908.

Washington, W.M., Meehl, G.A., Verplank, L. and Bettge, T.W. (1994) A world ocean model for greenhouse sensitivity studies – resolution intercomparison and the role of diagnostic forcing. *Climate Dynamics* **9**, 321–344.

Watson, A.J. and Lovelock, J.E. (1983) Biological homeostasis of the global environment: The parable of Daisyworld. *Tellus* **35**, 284–288.

Watson, R.T., Zinyowera, M.C. and Moss, R.H. (1996) *Climate Change 1995. Impacts Adaptation and Mitigation of Climate Change: Scientific–Technical Analysis. Contributions of Working Group II to the Second Assessment Report of the Intergovernmental Panel on Climate Change*, Cambridge University Press, Cambridge 880 pp.

Watts, R.G., Morantine, M.C. and Rao, K.A. (1994) Timescales in energy balance climate models. I. The limiting case solutions. *Journal of Geophysical Research* **99**, 3631–3641.

Weaver, A.J. and Hughes, T.M.C. (1994) Rapid interglacial climate fluctuations driven by North Atlantic ocean circulation. *Nature* **367**, 447–450.

Webb, R.S., Rosenzweig, C.E. and Levine, E.R. (1993) Specifying land surface characteristics in general circulation models: Soil profile data set and derived water-holding capacities. *Global Biogeochemical Cycles* **7**, 97–108.

Weisse, R., Mikolajewicz, U. and Maier-Reimer, E. (1994) Decadal variability of the North Atlantic in an ocean general circulation model. *Journal of Geophysical Research* **99**, 12411–12421.

Wells, N.C. (1979) A coupled ocean–atmosphere experiment: The ocean response. *Quarterly Journal of the Royal Meteorological Society* **105**, 355–370.

Wetherald, R.T. and Manabe, S. (1975) The effects of changing the solar constant on the climate of a general circulation model. *Journal of the Atmospheric Sciences* **32**, 2044–2059.

Wetherald, R.T. and Manabe, S. (1980) Cloud cover and climate sensitivy. *Journal of the Atmospheric Sciences* **37**, 1485–1510.

Wetherald, R.T. and Manabe, S. (1981) Influence of seasonal variation upon the sensitivity of a model climate. *Journal of Geophysical Research* **86**, 1194–1204.

Wigley, T.M.L. (1983) The pre-industrial carbon dioxide level. *Climatic Change* **5**, 315–320.

Wigley, T.M.L. and Raper, S. (1992) Implications for climate and sea level of revised IPCC emissions scenarios. *Nature* **357**, 293–300.

Wigley, T.M.L. and Schlesinger, M.E. (1985) Analytical solution for the effect of increasing CO_2 on global mean temperature. *Nature* **315**, 649–652.

Wigley, T.M.L., Richels, R. and Edmonds, J.A. (1996) Stabilizing CO_2 concentrations: The choice of emissions pathway. *Nature* in press.

Wiin-Nielsen, A.C. and Sela, J. (1971) On the transport of quasi-geostrophic potential vorticity. *Monthly Weather Review* **99**, 447–459.

Williams, L.D., Wigley, T.M.L. and Kelly, P.M. (1980) *Climatic Trends at High Northern Latitudes During the Last 4000 Years Compared with ^{14}C Fluctuations, in Sun and Climate*. CNES/CNRS/DGRST Conference Proceedings, pp. 11–20.

Williamson, D.L. (1983) Description of NCAR Community Climate Model (CCMOB). *NCAR Technical Note TN-210 + STR*, 88 pp.

Williamson, D.L. (1988) The effect of vertical finite difference approximations on simulations with the NCAR Community Climate Model. *Journal of Climate* **1**, 40–58.

Williamson, D.L. and Rasch, P.J. (1989) Two-dimensional semi-Lagrangian transport with shape-preserving interpolation. *Monthly Weather Review* **117**, 102–129.

Willson, R.C., Gulkis, S., Janssen, M., Hudson, H.S. and Chapman, G.C. (1981) Observations of solar irradiance variability. *Science* **21**, 200–202.

Wilson, M.F. and Henderson-Sellers, A. (1985) A global archive of land cover and soils data sets for use in general circulation models. *International Journal of Climatology* **5**, 119–143.

Wilson, M.F., Henderson-Sellers, A., Dickinson, R.E. and Kennedy, P.J. (1987) Sensitivity of the biosphere–atmosphere transfer scheme (BATS) to the inclusion of variable soil characteristics. *Journal of Climate and Applied Meteorology* **26**, 341–362.

Woods, J.D. (1984) The upper ocean and air–sea interactions. *In* J.T. Houghton (ed.), *The Global Climate*. Cambridge University Press, Cambridge, pp. 79–106.

Woods, J. and Barkmann, W. (1993) The plankton multiplier – positive feedback in the greenhouse. *Journal of Plankton Research.* **15**, 1053–1074.

World Climate Research Programme (1983) Report of the experts meeting on erosols and their climatic effects. *World Climate Programme*, Vol. 55, 101 pp.

World Meteorological Organization (1982) WMO Global Ozone Research and Monitoring Project, Report No. 14 of the *Meeting of Experts on Potential Climatic Effects of Ozone and other Minor Trace Gases.* Boulder, CO, 13–17 September 1982. Geneva, WMO.

World Meteorological Organization (WMO) (1986) *Report of the International Conference on the Assessment of the Role of Carbon Dioxide and of other Greenhouse Gases in Climate Variations and Associated Impacts.* Villach, Austria, 9–15 October 1985. WMO No. 661, Geneva, 78 pp.

Xue, Y.K. and Shukla, J. (1993) The influence of land surface properties on Sahel climate. I. Desertification. *Journal of Climate* **6**, 2232–2245.

Yang, Z.-L., Dickinson, R.E., Henderson-Sellers, A. and Pitman, A.J. (1995) Preliminary study of spin-up processes in Land Surface models with the first stage data of the Project for the Intercomparison of Land Surface Parameterization schemes Phase 1(a). *Journal of Geophysical Research* **100**, 16553–16578.

Yao, M.S. and Del Genio, A.D. (1989) Effects of cumulus entrainment and multiple cloud types on a January global climate model simulation. *Journal of Climate* **2**, 850–863.

Yeh, T.-C., Wetherald, R.T. and Manabe, S. (1983) A model study of the short-term climatic and hydrologic effects of sudden snow-cover removal. *Monthly Weather Review* **111**, 1013–1024.

Zhang, H., Henderson-Sellers, A. and McGuffie, K. (1996a) Impacts of tropical deforestation I. Process analysis of local climatic change. *Journal of Climate* **9**, 1497–1517.

Zhang, H., McGuffie, K. and Henderson-Sellers, A. (1996b) Impacts of tropical deforestation. II. The role of large-scale dynamics. *Journal of Climate* **9**, 2498–2521.

Index

Page numbers in italics refer to tables or figures